THIRD-GENERATION EGFR INHIBITORS

THIRD-GENERATION EGFR INHIBITORS

Overcoming EGFR Resistance and Toxicity Problems

HARUN M. PATEL
R. C. Patel Institute of Pharmaceutical Education and Research, Shirpur, Maharashtra, India

RAHUL PAWARA
R. C. Patel Institute of Pharmaceutical Education and Research, Shirpur, Maharashtra, India

SANJAY J. SURANA
R. C. Patel Institute of Pharmaceutical Education and Research, Shirpur, Maharashtra, India

ELSEVIER

Elsevier
Radarweg 29, PO Box 211, 1000 AE Amsterdam, Netherlands
The Boulevard, Langford Lane, Kidlington, Oxford OX5 1GB, United Kingdom
50 Hampshire Street, 5th Floor, Cambridge, MA 02139, United States

Library of Congress Cataloging-in-Publication Data
A catalog record for this book is available from the Library of Congress

British Library Cataloguing-in-Publication Data
A catalogue record for this book is available from the British Library

ISBN: 978-0-08-102661-8

Publisher: Susan Dennis
Acquisition Editor: Emily M. McCloskey
Editorial Project Manager: Carly Demetre
Production Project Manager: Maria Bernard
Cover Designer: Matthew Limbert

Typeset by SPi Global, India

DEDICATION

This book is dedicated to my late grandmother, Amirsahab Patel, who left this world because of cancer and to all those cancer patients, who have lost the fight and to those who won't quit the fight yet.

CONTENTS

PREFACE

The present textbook entitled *Third-Generation EGFR Inhibitors: Overcoming EGFR Resistance and Toxicity Problems* mainly focuses on third-generation drugs and their development for the treatment of the nonsmall cell lung cancer (NSCLC). The purpose of this book is to update the researcher on the current drug progress in the treatment of nonsmall cell lung cancer by targeting the epidermal growth factor receptor (EGFR)-tyrosine kinase. Lung cancer has become the leading cause of cancer-related deaths worldwide. More than 80% of all lung cancers are identified as nonsmall cell lung cancer (NSCLC). The epidermal growth factor receptor (EGFR) has been established as one of the most important therapeutic targets for NSCLC. Gefitinib and erlotinib, the first-generation, ATP-competitive, and reversible EGFR inhibitors, were effective for NSCLC patients harboring somatic EGFR mutations L858R and delE746_A750, which account for 90% of all EGFR mutations in NSCLC. Unfortunately, due to acquired secondary mutations such as T790M in EGFR leading to drug resistance in 50% NSCLC patients. Second-generation EGFR inhibitors, such as canertinib, afatinib, neratinib, and dacomitinib, have been developed with the intent to overcome the T790M mutation–related resistance. However, their ATP affinity of EGFR T790M is similar to that of WT EGFR, leading to the dose-limiting toxicities such as diarrhea and skin rash.

Today, third-generation EGFR TKIs consisting of WZ4002, EGF816, Rociletinib (CO-1686), and Osimertinib (AZD9291) are evolved to target mutant EGFR-harboring T790M. This elegance of inhibitors ties covalently to Cys797 and to a great extent spares WT EGFR, which accordingly diminishes toxicity and allows for the utilization of doses that completely smother T790M.

Third-generation inhibitors AZD9291 (Osimertinib) and CO1686 (Rociletinib) both form a covalent bond with Cys797 in the active site and have activity against the resistance mutations (T790M/L858R, TMLR; T790M/del (746–750), TMdel) as well as the two initial activating mutations. Despite the clinical efficacy of the pyrimidine-based *EGFR* inhibitors rociletinib and osimertinib, it is fully anticipated that patients will ultimately develop acquired resistance to these agents. Unfortunately the *C797S* mutation, in the presence of Del19 or *L858R* and *T790M*, causes resistance to all current *EGFR* inhibitors. Additionally, studies with third-generation resistant cell lines have shown that the allelic context of the activating gatekeeper and *C797S* mutations affects the sensitivity to the three generations of inhibitors with no *EGFR* TKIs alone or in combination able to suppress activity when the mutations are in cis. This data suggests that there is a need for drugs which do not rely on covalent reaction with Cys797 for potency or selectivity.

Harun M. Patel, Rahul Pawara, Sanjay Surana

ACKNOWLEDGMENTS

It is our pleasure to acknowledge the enthusiastic support and assistance given to us by many people at various stages during the writing of this book. We express our gratitude to Dr. A.A. Shirkhedkar, Dr. C.R. Patil, and Dr. M.N. Noolvi for their quick and helpful reviews of this book. We especially wish to thank our friends Mr. Azim Ansari and Mr. Iqrar Ahmad for their suggestions during the writing this book.

We owe our sincere thanks to the Department of Science and Technology, New Delhi, India (SERB File No. YSS/201S/002017), for financial support. We are especially indebted to Ms. Naziya Patel for her unwavering support and enthusiasm that helped make this book a reality.

We would like to give special thanks to the Elsevier team for actively participating in the feedback and contributions for this book via the mailing lists. Without their insights and contributions, portions of this book may not have been possible.

Harun M. Patel
Rahul Pawara
Sanjay J. Surana

ABBREVIATIONS

$(Boc)_2O$	di-*tert*-butyl dicarbonate
$(COCl)_2$	oxalyl chloride
$(EtO)_3CMe$	1,1,1-triethoxy ethane
$(MeO)_3CH$	*tert*-methoxy methane
$(PhCO)_2O$	benzoic anhydride
^1H NMR	proton nuclear magnetic resonance
7-AAD	7-aminoactinomycin D
A	alanine
ABL1	Abelson murine leukemia viral oncogene homologue 1
AcOEt	ethyl acetate
AcOH	acetic acid
ADME	administration, distribution, metabolization, and excretion
AIBN	azobisisobutyronitrile
AKT	protein kinase B
ALPB	analytical linearized Poisson-Boltzmann
AM	activating mutant
AM1-BCC	semiempirical (AM1) with bond charge correction (BCC)
AO	aldehyde oxidase
APCI	atmospheric pressure chemical ionization
ArCHO	benzaldehyde
ATP	adenosine triphosphate
AUC	area under the blood concentration time curve
AUC_{last}	area under the curve from 0 h to last time
B3LYP	Becke three parameter, Lee-Yang-Parr
BA	bioavailability
BBr_3	boron tribromide
BF_4	tetrafluoroborate
BLK	blymphocyte kinase
Blk	β-lymphocyte kinase
BMX	bone marrow kinase in chromosome X
BnCl	benzyl chloride
Boc	*tert*-butyloxycarbonyl protecting group
Boc_2O	*tert*-butyloxycarbonyl protecting group
BrettPhos	2-(dicyclohexylphosphino)3,6-dimethoxy-2′,4′,6′-triisopropyl-1,1′-biphenyl

BTK	Bruton's tyrosine kinase
BuXPhos	[(2-di-*tert*-butylphosphino-2′,4′,6′-triisopropyl-1, 1′-biphenyl)-2-(2′-amino-1,1′-biphenyl)] palladacycle
C max	maximum concentration
C797S	cysteine797serine
$CaCl_2$	calcium chloride
CBr_4	tetrabromomethane
$CBrCl_3$	bromotrichloromethane
CbzCl	benzyl chloroformate
CCl_4	carbon tetrachloride
CDC7	cell division cycle 7-related protein kinase
CDI	1,1′-carbonyldiimidazole
CDK2	cyclin-dependent kinase 2
CF_3 Ph	trifluorotoluene
CF_3CH_2	1,1,1-trifluoroethane
CH_2Cl_2	dichloromethane
CH_3CN	acetonitrile
CH_3I	methyl iodide
$CHCl_3$	chloroform
Cl hep	hepatic clearance
CL	clearance
Clhep	predicted hepatic clearance
clint	intrinsic clearance
$ClOCOCH_3$	1-chlorosylethan-1-one
Clp	plasma clearance
Clu	free clearance
cm/s	centimeter/second
Cmax	maximum serum concentration
CNBr	cyanogen bromide
CRIs	covalent reversible inhibitors
CS_2	carbon disulfide
Cs_2CO_3	cesium carbonate
CSD	Cambridge structural database
cSrc-DM cSrc	(T338M/S345C)
CuCl	copper chloride
CuI	copper iodide
Da	dalton
DABCO	1,4-diaza[2.2.2]bicyclooctane
DAST	diethylaminosulfur trifluoride

DBU	1,8-diazabicyclo[5.4.0]undec-7-ene
DCC	dicyclohexylcarbodiimide
DCC	*N,N'*-dicyclohexylcarbodiimide
DDR1	discoidin domain receptor 1
Del	deletion
DFG	Asp-Phe-Gly
DHP	dihydropyran
DIEA	*N,N*-diisopropylethylamine
DM	double mutant
DMA	*N,N*-dimethylacetamide
DMAP	4-dimethylaminopyridine
DMF	*N,N*-dimethylformamide
DMSO	dimethyl sulfoxide
DNA	deoxyribonucleic acid
E	glutamic acid
EC-RISM	embedded cluster reference interaction site model
EDC	1-ethyl-3-(3-dimethylaminopropyl) carbodiimide hydrochloride
EGFR	epidermal growth factor receptor
ELSD	evaporative light-scattering detector
EPHA3	erythropoietin-producing human hepato cellular carcinoma receptor A3
Equiv	equivalent
ERK	extracellular signal-regulated kinase
Et_3N	triethylamine
EtOAc	ethyl acetate
EtOH	ethanol
F	oral bioavailability
FDA	Food and Drug Administration
Fe	iron
FLT3	FMS-related tyrosine kinase 3
FRK	Fyn-related kinase
GAFF	general AMBER force field
GAK	cyclin G-associated kinase
GI	growth inhibitory
GSH	glutathione
h	hour
H_2SO_4	sulfuric acid
HATU	*N,N,N',N'*-tetramethyl-*O*-(7-azabenzotriazol-1-yl) uroniumhexafluorophosphate

HATU 1-[bis(dimethylamino)methylene]-1*H*-1,2,3-triazolo[4,5-*b*] pyridinium3-oxid hexafluorophosphate
Hb hemoglobin
HBr hydrobromic acid
HBTU 2-(1*H*-benzotriazol-1-yl)-1,1,3,3-tetramethyluronium hexafluorophosphate
HCl hydrochloric acid
HEPES 4-(2-hydroxyethyl)-1-piperazineethanesulfonic acid
HER human epidermal growth factor receptor
HH human hepatocyte
HIF-1α hypoxia-inducible factor-1α
HLM human liver microsome
HNO_3 nitric acid
HOBt 1-hydroxybenzotriazole monohydrate
HPLC high-performance liquid chromatography
HTRF homogeneous time-resolved fluorescence
HT-sol high throughput solubility
I isoleucine
IC_{50} half-maximal inhibitory concentration
IGF1R insulin-like growth factor 1 (IGF-1) receptor
ILD interstitial lung disease
IOD integrated optical density
IP intraperitoneal
***i*Pr** isopropyl
***i*Pr2Net** *N*,*N*-diisopropylethylamine
***i*PrNH2** isopropylamine
***iso*Pr2EtN** *N*,*N*-diisopropylethylamine
iv intravenous
IVIVC in vitro-in vivo correlation
K_2CO_3 potassium carbonate
Kd binding constant
KDR kinase insert domain receptor
KF potassium fluoride
kg kilogram
KHMDS Potassium hexamethyldisilazide solution
Kit mast/stem cell growth factor receptor
L leucine
L858R leucine858arginine
LAH lithium aluminum hydride

LCK	lymphocyte-specific protein tyrosine kinase
LCMS	liquid chromatography-mass spectrometry
Li_2CO_3	lithium carbonate
LiAlH4	lithium aluminum hydride
LiBr	lithium bromide
LiHMDS	lithium bis(trimethylsilyl)amide
LiOH	lithium hydroxide
LipE	lipophilic efficiency
LLE	lipophilic ligand efficiency
Lys745	lysine745
M	methionine
MAPK	mitogen-activated protein kinase signaling pathway
mCPBA	3-chloroperbenzoic acid
MDCK	Madin-Darby canine kidney
Me	methyl
$MeCO_2CH_2COCl$	methyl 3-chloro-3-oxopropanoate
MEK5	mitogen-activated protein kinase 5
MeO	methoxy
MeOH	methanol
Met790	methionine790
Met793	methionine793
MFDS	Korean Ministry of Food and Drug Safety
mg/kg	milligram/kilogram
mg	milligram
$MgSO_4$	magnesium sulfate
MKNK1	MAP kinase-interacting serine/threonine-protein kinase 1
mL/min/kg	milliliter/minute/kilogram
MnO_2	magnesium oxide
MoA	mode of action
MS	mass spectroscopy
MS	metabolic stability
MsOH	methanesulfonic acid
MTD	maximum tolerated dose
MTT	3-(4,5-dimethylthiazol-2-yl)-2,5-diphenyltetrazolium bromide
$N(CH_3)_2$	*N*,*N*-dimethylamine
Na_2CO_3	sodium carbonate
$NaB(OAc)_3H$	sodium triacetoxyborohydride

$NaBH_4$	sodium borohydride
$NaClO_2$	sodium chlorite
NaH	sodium hydride
NaH_2PO_4	sodium dihydrogen phosphate
$NaHCO_3$	sodium bicarbonate
NaHMDS	sodium hexamethyldisilazane
NaI	sodium iodide
NaN_3	sodium azide
$NaNO_2$	sodium nitrite
$NaO\ (CH_2)_2OCH_3$	sodium 2-methoxyethan-1-olate
NaOAc	sodium acetate
$NaOCH_3$	sodium methoxide
NaOEt	sodium ethoxide
NaOH	sodium hydroxide
NaOMe	sodium methoxide
NaOtBu	sodium tertbutoxide
NBS	*N*-bromosuccinimide
NCI	National Cancer Institute
ND	no data
NF-κB	nuclear factor-κB
NH_3	ammonia
NH_4Cl	ammonium chloride
nM	nanomole
NMP	1-methyl-2-pyrrolidone
NO	nitric oxide
NSCLC	nonsmall cell lung cancer
OCH3	methoxy
OS	overall survival
PBS	phosphate-buffered saline
PCM	polarizable continuum model
$Pd(dppf)Cl_2$	[1,1′-bis(diphenylphosphino)ferrocene]-dichloropalladium (II)
$Pd(OH)_2$	palladium hydroxide
$Pd(PPh_3)_2Cl_2$	bis(triphenylphosphine)palladium(II) dichloride
$Pd(PPh_3)_4$	tetrakis(triphenylphosphine)palladium
Pd/C	palladium on carbon
Pd	palladium
Pd2(dba)3	tris(dibenzylideneacetone)dipalladium(0)
PDA	photodiode array detector

PDB	protein data bank
Ph	phenyl
$PhMe_3NBr_3$	phenyltrimethylammonium tribromide
PhOH	phenol
PI	pseudoirreversibility
PK	pharmacokinetic
pKa	acid dissociation constant
$POCl_3$	phosphorus oxychloride or phosphoryl chloride
PPh_3	triphenylphosphine
ppm	parts per million
PR	preconfiguration ratio
PrOH	propanol
PSE	partial series expansion
Pt/C	platinum/carbon
PtO_2/H_2	platinum dioxide
pTsOH	*p*-toluenesulfonic acid
Q	glutamine
QPPCaco2	cell permeability method
R	arginine
RH	rat hepatocyte
RIPK2	receptor-interacting serine/threonine-protein kinase 2
RLM	rat liver microsome
RMSD	root mean square deviation
RP-HPLC	reverse phase high-performance liquid chromatography
rt.	room temperature
RTK	receptor tyrosine kinase
SAR	structure-activity relationships
SBDD	structure-based drug design
SD	Sprague-Dawley
SD	standard deviation
SEM	2-(trimethylsilyl)ethoxymethyl
SFC	supercritical fluid chromatograph
SNAr	nucleophilic aromatic substitution
$SOCl_2$	thionyl chloride
SPhos	2-dicyclohexylphosphino-2′,6′-dimethoxybiphenyl
STK10	serine/threonine kinase 10
T/C	treatment/control
T	threonine
T1/2	drug half-life

T790M	thrionie 790 methionine
TBAF	tetrabutylammonium fluoride
TBDMS	*tert*-butyldimethylsilyl ethers
TEA	triethylamine
tert	tertiary
***tert*-BuO$_2$H**	*tert*-butanoic acid
***tert*-BuOK**	potassium *tert-butoxide*
TFA	trifluoroacetic acid
TGI	tumor growth inhibition
THF	tetrahydrofuran
THPP	tetrahydropyrido [4,3-d]pyrimidine
Ti(OisoPr)$_4$	titanium isopropoxide
TiCl$_4$	titanium tetrachloride
TKI	tyrosine kinase inhibitor
TLC	thin-layer chromatography
TMdel	epidermal growth factor receptor (T790M/del746-750)
TMLR	epidermal growth factor receptor (T790M/L858R)
TMS	trimethylsilyl
Tyk2	tyrosine kinase 2
UPLC	ultra performance liquid chromatography
Vobs	apparent volume of distribution
WSC	water-soluble carbodiimide
WT EGFR	wild-type epidermal growth factor receptor
XO	xanthine oxidase
XPhos	2-dicyclohexylphosphino-2′,4′,6′-triisopropylbiphenyl
Zn(CN)$_2$	zinc cyanide
Zn	zinc
μmol/L	micromole/liter
μW	microwave

CHAPTER 1

Introduction

Lung cancer remains one of the leading causes of cancer-related mortality, reflecting the limited efficacy of conventional cytotoxic chemotherapies in patients with this disease [1]. Lung cancer is categorized into small ncell lung cancer (SCLC, 15% of all cases) and nonsmall cell lung cancer (NSCLC, 85% of all cases). NSCLC comprises of adenocarcinoma (AC) (about 40%–50% of all lung cancers), squamous cell carcinoma (SQCC), and large cell carcinoma (LCLC) [2].

The role of the epidermal growth factor receptor (EGFR) in nonsmall cell lung cancer (NSCLC) is well known, and considerable therapeutic progress in the treatment of this disease was made over the past 10 years [3, 4]. The EGFR was established as one of the most important therapeutic targets for NSCLC [5, 6].

The EGFR is composed of an extracellular ligand binding region, a transmembrane region, and an intracellular tyrosine kinase domain. Ligand binding to the extracellular region of EGFR induces a conformational change in the receptor that promotes dimerization. Dimerization of the receptor results in autophosphorylation of the intracellular domain, which in turn recruits proteins involved in downstream signaling. The resultant signal transduction cascade activates pathways including the RAS/RAF/MEK and PI3K/AKT pathways leading to cell proliferation and survival, as given in Fig. 1.1 [7].

Two small molecule inhibitors of *EGFR*, gefitinib and erlotinib, have been approved by the US Food and Drug Administration (FDA) for the treatment of NSCLC [8–11]. Gefitinib and erlotinib, as the first-generation, ATP-competitive, and reversible *EGFR* inhibitors, were effective for NSCLC patients harboring somatic *EGFR* mutations *L858R* and delE746_A750, which account for 90% of all *EGFR* mutations in NSCLC [12]. Unfortunately, due to acquired secondary mutations such as *T790M* missense in *EGFR* that leads to drug resistance in roughly 50% NSCLC patients, the ability of these *EGFR* inhibitors to effectively treat NSCLC patients is short lived (10–12 months) to later treatments [13, 14]. *EGFR T790M* mutation restores the affinity for ATP similar to that of wild-type (WT) EGFR and prevents the reversible inhibitors from binding at higher ATP concentrations [15–17].

Third-Generation EGFR Inhibitors
https://doi.org/10.1016/B978-0-08-102661-8.00001-9

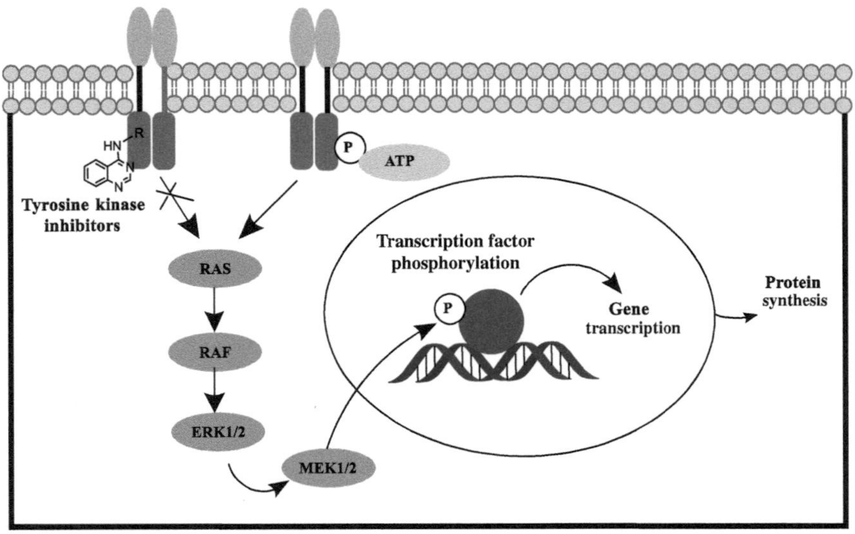

Fig. 1.1 Tyrosine kinase intracellular signaling pathway activation and inhibition by TKs blocker.

Second-generation EGFR inhibitors, including afatinib (BIBW2992) and dacomitinib (PF00299804), are irreversible EGFR inhibitors that bind to Cys797 and covalently modifies it. They have been shown in preclinical experiments to effectively inhibit EGFR with activating mutations (Exon 19 deletion or L858R) as well as those with the T790M resistance mutation [18–20]. However, their activity in patients with erlotinib-resistant cancers harboring T790M has been minimal [21, 22]. The discordance between laboratory and clinical results is likely due to a poor therapeutic window. These drugs are equally potent against wild-type EGFR and EGFR T790M, thus the toxicity resulting from inhibiting wild-type EGFR (rash and diarrhea) precludes the use of doses that would be needed to effectively suppress T790M [23].

More recently, third-generation EGFR TKIs including WZ4002, PF-06747775, nazartinib (EGF816), naquotinib (ASP8273), rociletinib (CO1686), olmutinib (HM61713), and osimertinib (AZD9291) have been developed to target mutant EGFR harboring T790M [24–26]. This class of inhibitor also binds covalently to Cys797 and largely spares WT EGFR, thereby decreasing toxicity and permitting the use of doses that fully suppress T790M. This large therapeutic window likely underlies the greater than 50% response rates observed in clinical trials with rociletinib (CO1686) and osimertinib (AZD-9291) in erlotinib resistant, T790M-positive NSCLCs (Fig. 1.2) [27].

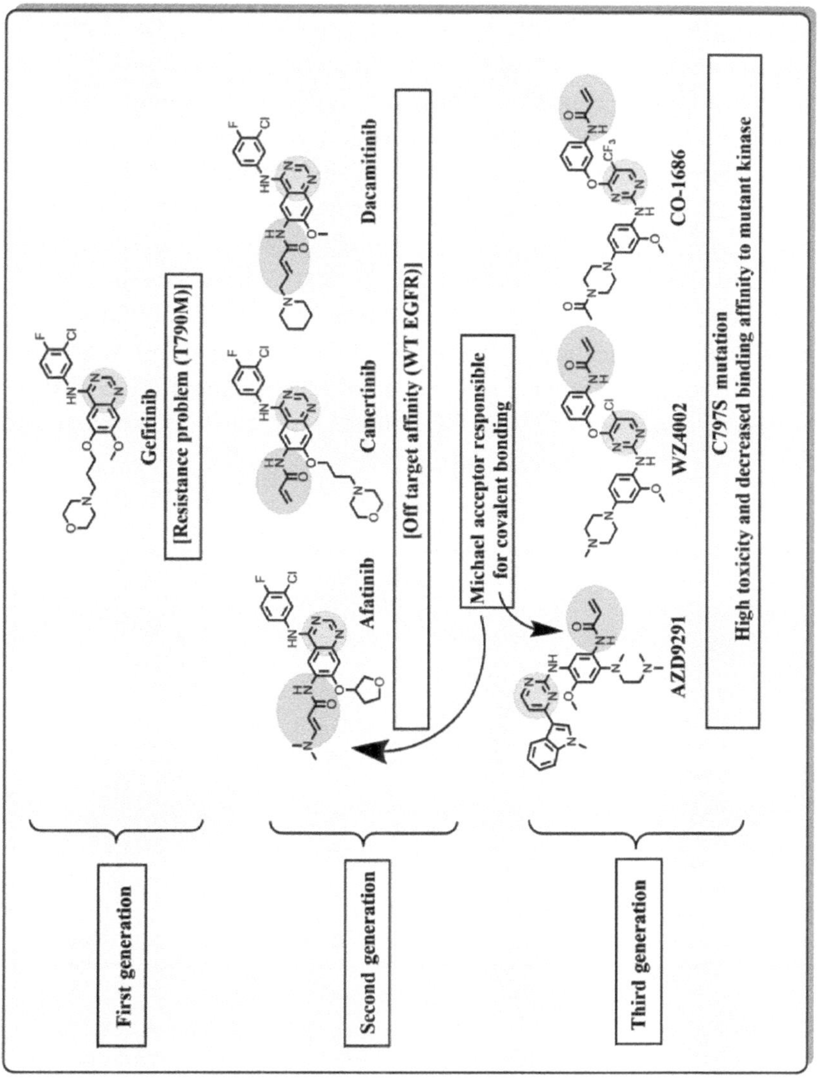

Fig. 1.2 First-, second-, and third-generation EGFR tyrosine kinase inhibitors.

Based on these promising results, both drugs have received FDA "breakthrough therapy designation," and this class of inhibitors is on the verge of becoming widely implemented for the treatment of this patient population [28]. However, as with first- and second-generation compounds, resistance has emerged. There have been recent reports of C797S mutation or the loss of T790M mutation in cell-free plasma DNA samples from patients who have developed resistance, and there is a second separate report of C797S mutation in biopsy samples from a single patient, as shown in Fig. 1.3 [29, 30]. Additional studies with third generation cell lines have shown that the allelic context of the activating gatekeeper and C797S mutations affects the sensitivity to three generations of inhibitors with no EGFR TKIs alone or in combination able to suppress activity when mutation are in cis as given in Fig. 1.3 [31]. These data suggest that there is a need for drugs that can overcome EGFR (C797S) resistance obstacles in NSCLC. This book summarizes the third-generation inhibitors, their mechanism of resistance, and the latest developments in the discovery of fourth-generation EGFR TKIs and U to Y allosteric strategies to combat the C797S EGFR resistance problem.

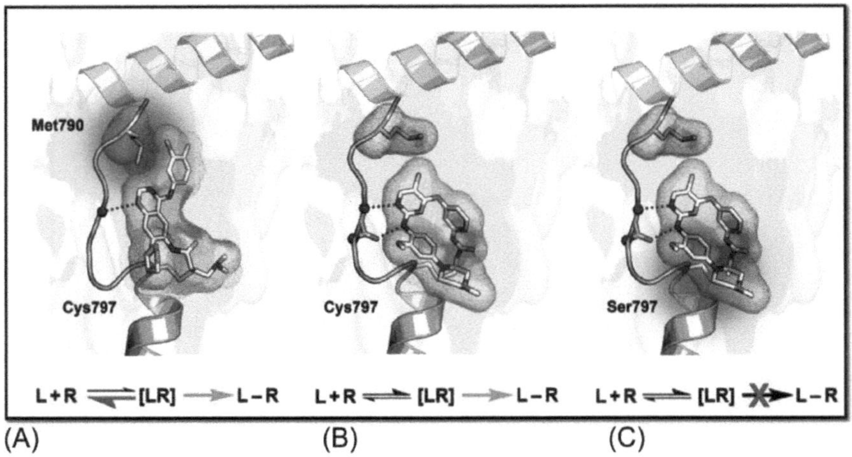

Fig. 1.3 Binding mode of covalent EGFR tyrosine kinase inhibitors. (A) Second-generation (Afatinib) bind to the T790M EGFR. (B) Third-generation inhibitors (WZ4002) avoid the steric conflict with Met790 and therefore achieve complete receptor occupancy. (C) The C797S mutation mediate resistance to irreversible drugs, as the less nucleophilic serine side chain cannot undergo covalent bond formation at physiological condition. *(Reprinted (adapted) with permission from J. Engel, J. Lategahn, D. Rauh, Hope and disappointment: covalent inhibitors to overcome drug resistance in non-small cell lung cancer, ACS Med. Chem. Lett. (2016), Copyright 2016, American Chemical Society.)*

1.1 KINASE INHIBITOR BINDING SITES

Protein kinases are defined by their ability to catalyze the transfer of the terminal phosphate of ATP to substrates that usually contain a serine, threonine, or tyrosine residue. They typically share a conserve arrangement of secondary structure elements that are arranged into 12 subdomains that fold into a bilobed catalytic core structure with ATP binding in a deep cleft located between the lobes [32–34]. ATP binds in the cleft with the adenine ring forming hydrogen bonds with the kinase "hinge"; that is, the segment that connects the amino- and carboxy-terminal kinase domains. The ribose and triphosphate groups of ATP bind in a hydrophilic channel extending to the substrate binding site that features conserved residues essential to catalysis. All kinases have a conserved activation loop, which is important in regulating kinase activity and is marked by conserved DFG and APE motifs at the start and end of the loop, respectively. The activation loop can assume a large number of conformations, with the extremes being a conformer that is catalytically competent and usually phosphorylated and an "inactive" conformer in which the activation loop blocks the substrate binding site.

Most kinase inhibitors discovered to date are ATP competitive and present one to three hydrogen bonds to the amino acids located in the hinge region of the target kinase, thereby mimicking the hydrogen bonds that are normally formed by the adenine ring of ATP as given in Fig. 1.4 [35].

Kinase inhibitors can be broadly classified into three major classes (i.e., first-, second-, and third-generation) based on their mechanism of action [35, 36]. While the first-generation kinase inhibitors are reversible in nature,

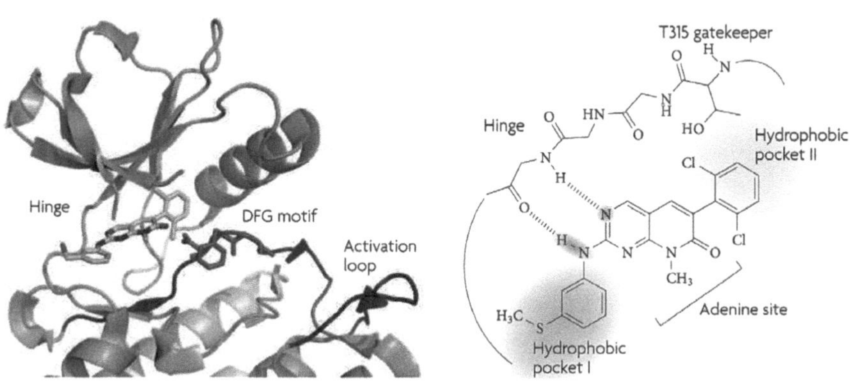

Fig. 1.4 Type I EGFR tyrosine kinase inhibitor. *(Reprinted (adapted) with permission from J. Zhang, P.L. Yang, N.S. Gray, Targeting cancer with small molecule kinase inhibitors, Nat. Rev. Cancer. (2009), Copyright 2009, Nature Publishing Group.)*

the second- and third-generation inhibitors are both irreversible and varying in their selectivity toward the wild-type kinase domain. All the kinase inhibitors vary in their binding mode with the ATP binding domain of the kinases and can be categorized into the four classes, type I, II, III, and covalent [35, 37].

1.1.1 Type I Inhibitors

Type I inhibitors are ATP-competitive inhibitors that recognize the active conformation of the kinase. They bind to the active conformation of the kinase with the aspartate residue of the DFG motif (Asp-Phe-Gly motif) pointing into the ATP binding pocket [35, 38]. Most of the inhibitors mimicking the ATP structure show type I binding mode, as shown in Fig. 1.4.

1.1.2 Type II Inhibitors

Type II inhibitors bind and stabilize the inactive conformation of the kinase with the flipped aspartate residue facing outward of the binding pocket. This conformation, recognized by type II inhibitors, is referred to as DFG-out, owing to the rearrangement of this motif. Alteration in the activation loop to the DFG-out conformation reveals an additional hydrophobic binding site directly adjacent to the ATP binding site, as shown in Fig. 1.5 [35–39].

1.1.3 Type III or Allosteric Inhibitors

Type III or allosteric inhibitors modulate kinase activity by binding outside the ATP-binding site to an allosteric pocket that is adjacent to the ATP-binding

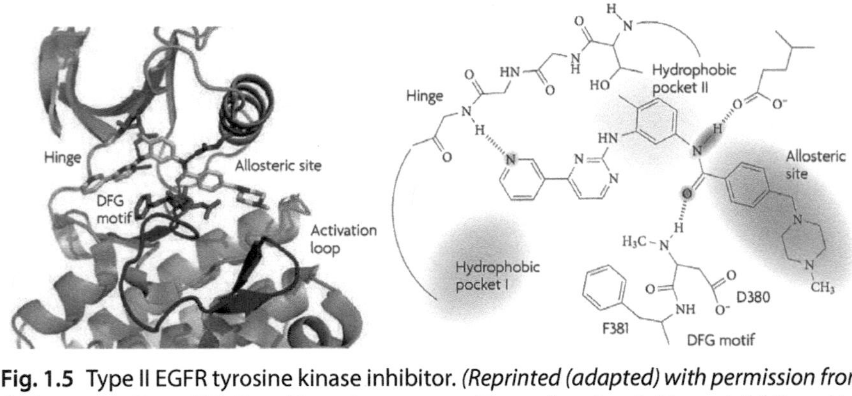

Fig. 1.5 Type II EGFR tyrosine kinase inhibitor. *(Reprinted (adapted) with permission from J. Zhang, P.L. Yang, N.S. Gray, Targeting cancer with small molecule kinase inhibitors, Nat. Rev. Cancer. (2009), Copyright 2009, Nature Publishing Group.)*

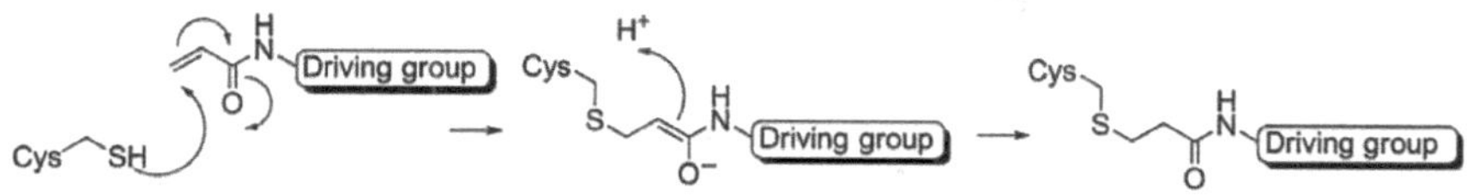

Fig. 1.6 Reaction mechanism of acrylamide toward cysteine.

pocket but does not overlap with it. Inhibitors of this class show the highest degree of kinase selectivity as they exploit a binding site unique to a particular kinase [35, 40, 41].

1.1.4 Covalent Inhibitors

Of another class are covalent inhibitors, which form covalent bonds with the specific cysteine residue in the kinase domain and inhibit it irreversibly [42, 43]. These compounds were rationally designed by appending an electrophile, which is reactive toward the electron-rich sulphur present in the cysteine residue. The resulting inhibitors undergo a Michael addition reaction, in which the solvent-exposed cysteine residue that is present in EGFR forms a covalent bond with the inhibitor, bestowing it with infinite affinity for the ATP binding site. As a result, the inhibitor irreversibly blocks the binding of ATP to the kinase, thereby rendering the kinase inactive as given in Fig. 1.6 [35, 44–47].

1.2 FIRST-GENERATION KINASE INHIBITORS (REVERSIBLE) FOR THE TREATMENT OF NSCLC

Gefitinib and erlotinib were the first EGFR tyrosine kinase inhibitors approved for the treatment of patients with NSCLC. These drugs inhibit kinase activity by competitively interacting with the ATP-binding site of EGFR, preventing autophosphorylation and consequently inhibiting downstream signaling. This inhibition leads to apoptosis in cells dependent upon EGFR signaling, such as those with EGFR mutations, as given in Table 1.1 [48–50].

1.2.1 Gefitinib

Zeneca initially synthesized getfitinib in 1994. It contains an amino-quinoline ring that fits into the adenine binding site and a chlorofluoro-substituted aromatic ring that occupies the hydrophobic pocket. It shows the type I binding mode with EGFR. Usually, getfitinib leads to dramatic clinical responses in cases of NSCLCs harboring activation of mutations in

Table 1.1 First-generation EGFR inhibitors [48, 49]

First-generation inhibitors	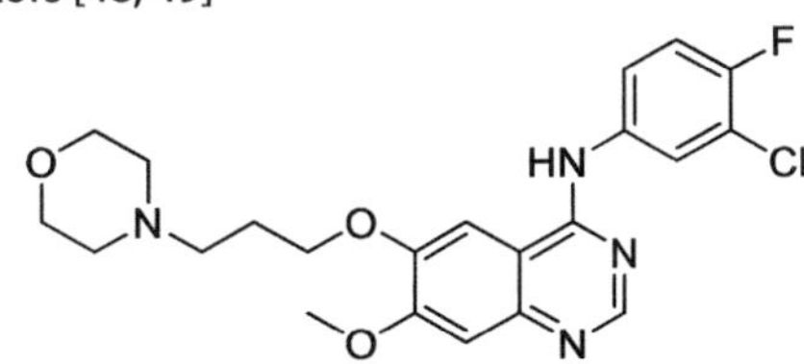 **Gefitinib**	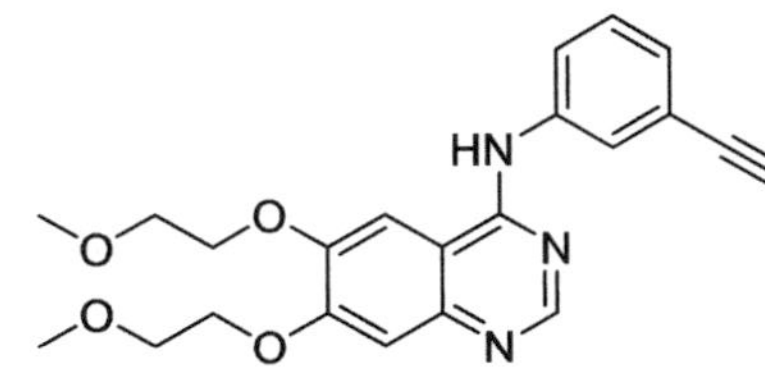 **Erlotinib**
IC_{50} against EGFR	33 nm	2 nm
Indication	US labeling limits its use to patients who, in the opinion of their treating physician, have previously benefited from treatment; in Europe, it has been approved for advanced NSCLC with an EGFR activating mutation	Treatment of patients with locally advanced or metastatic NSCLC after failure of at least one previous chemotherapy regimen; maintenance therapy after first-line chemotherapy
Recommended daily dose	One 250-mg tablet orally, once daily, with or without food	150 mg orally, once daily at least 1 h before or 2 h after the ingestion of food
Most common adverse effect	Diarrhea, rash, acne, dry skin, nausea, and vomiting	Rash and diarrhea
Serious adverse effect reported with treatment	Interstitial lung disease	Interstitial lung disease-like events; hepatic failure and hepatorenal syndrome (including fatalities); acute renal failure (including fatalities); and renal insufficiency

the EGF receptor [51]. The most suitable dose of gefitinib is 250 mg/day given both as a single agent as well as in combination; the mean bioavailability after oral administration averages around 60% with a half–life of about 48 h [52]. Therapeutic responses to gefitinib remain for as long as 2–3 years, but the mean duration of response in most cases of NSCLC is found to be only 6–8 months [53]. A single secondary mutation, T790M, within the EGFR kinase domain has been well reported for three of six cases with recurrent disease after gefitinib therapy [10].

Gefitinib's efficacy was first evaluated in two single-arm phase 2 studies in patients with non-NSCLC, who had received prior chemotherapy. The success of these trials led to the accelerated approval of gefitinib in 2003 [53, 54], as well as initiation of a phase III trial [55] (ISEL), which randomized patients to gefitinib versus placebo and found no difference in median survival (5.6 versus 5.1 months, respectively). The lack of an overall survival benefit in ISEL prompted the FDA to restrict gefitinib use. Notably, these early studies did not select or evaluate patients based upon EGFR mutation status.

Subsequent trials of gefitinib focused on subsets of patients more likely to respond to gefitinib. IPASS enrolled East Asian patients with adenocarcinoma who were either former light smokers or who had never smoked (the key clinical characteristics associated with response to EGFR TKI and presence of EGFR mutations) in order to receive either gefitinib or the combination of carboplatin and paclitaxel as a first-line treatment [3]. Gefitinib prolonged progression-free survival (PFS) (9.5 versus 6.3 months for gefitinib versus chemotherapy), with the positive study results largely driven by the predominance of EGFR mutations in the study population (60% of patients had tumors with EGFR mutations). Of note, patients with these clinical characteristics who did not have EGFR mutations and received initial gefitinib had a median progression-free survival of less than 2 months and a response rate of 1%. Based on IPASS and other studies the EMEA approved gefitinib for use in patients with EGFR mutant-advanced NSCLC. Gefitinib has also been studied specifically in patients with EGFR mutant lung cancers.

In both the WJOTG 3405 [4] and NEJ2002 [5] studies, patients were randomized to gefitinib or platinum-based chemotherapy. Gefitinib led to improved PFS and objective response rates (ORR) compared to cytotoxic chemotherapy in these patients with EGFR mutations [48, 49].

1.2.2 Erlotinib

Patented by Pfizer in the mid-1990s, erlotinib contains an aminoquinoline moiety similar to gefitinib but varies in substitutions; for example, alkoxy chains are used for solvent regions and cyano-substituted phenyl rings are used for hydrophobic pockets [56]. It binds as a type I inhibitor and was previously recommended for second- or third-line therapy in NSCLC [57]. The suggested dosage is 150 mg/day on an empty stomach at least 2 h before or after a meal [58]. For erlotinib, its use in therapy is not dependent on mutation status, and it is based on a placebo-controlled, phase III trial in unselected relapsed or refractory NSCLC patients [2]. Further studies indicated that the activity of erlotinib in second-/third-line therapy is mainly limited to patients with activating EGFR mutations. In a phase III trial, erlotinib and ocetaxel were compared as second-line therapy in patients with NSCLC and wild-type GFR, where erlotinib was found to had poor progression-free survival than docetaxel [48, 49], and it showed improved results in the subgroup of patients with SQCC [59]. In another phase III trial within an EGFR-unselected patient population, erlotinib failed to show any improvement in progression-free survival versus docetaxel [60].

1.3 PROBLEMS ASSOCIATED WITH FIRST-GENERATION KINASE INHIBITORS

The first-generation kinase inhibitors initially proved to be significant in the cancer therapy; however, within a few years, these somewhat "magical" kinase inhibitors lost their charm. The major problem associated with first-generation kinase inhibitors, other than toxicity or a lack of selectivity for one specific isoform of the kinase, was revealed to be an inevitable development of resistance in the patient subjected to the therapy of these class inhibitors. One of the most common resistance mechanisms defined to date (50–60% of patients) is the T790M missense mutation (49%) in exon 20 (T790M) [61, 62].

The secondary mutation of a single amino acid in the catalytic domain of EGFR (i.e., the conversion of gatekeeper threonine 790 to methionine; T790M), is the most common mechanism of an acquired resistance to the reversible tyrosine kinase inhibitors, as shown in Fig. 1.7 [63]. The increased ATP affinity shown by T790M mutants with respect to the L858R mutant of EGFR, which prevents reversible inhibitors from binding at cellular high ATP concentrations.

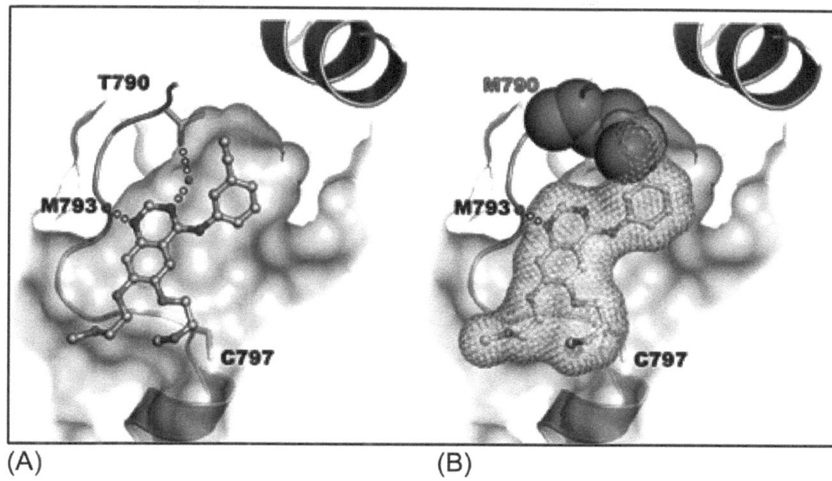

Fig. 1.7 Crystal structure of EGFR complexed with the reversible ATP competitive drug erlotinib (PDB entry 1M17). (A) Key hydrogen bonds formed between the quinazoline core of the inhibitor; T790 and M793 are indicated by dotted lines. (B) Drug resistance mutation T790M is modeled and displayed in a sphere representation (magenta) and highlights the steric clash with the acetylene moiety of erlotinib. The Van der Waals radii of the inhibitor are shown as mesh surface. Due to the T or M mutation the water-mediated hydrogen bond between N3 of the quinazoline core and the side chain of T790 is lost. *(Reprinted (adapted) with permission from J. Engel, J. Lategahn, D. Rauh, Hope and disappointment: covalent inhibitors to overcome drug resistance in non-small cell lung cancer, ACS Med. Chem. Lett. (2016), Copyright 2016, American Chemical Society.)*

1.4 SECOND-GENERATION KINASE INHIBITORS (IRREVERSIBLE)

Acquired resistance to first-generation EGFR TKIs has prompted the clinical development of second-generation EGFR tyrosine kinase inhibitors, as given in Table 1.2. Second-generation EGFR TKIs may be able to overcome some of the mechanisms of resistance to first-generation EGFR TKIs and have the potential to be more effective than erlotinib or gefitinib as first EGFR TKI. Most of the second-generation TKIs form irreversible covalent attachments to the EGFR kinase domain. In addition, due to their covalent binding, they may have more activity against T790M or other secondary mutations, unlike first-generation TKIs [48, 49].

1.4.1 Afatinib

Synthesized by the Boehringer Ingelheim Corp. in 2008, afatinib consists of an aminoquinazoline nucleus to bind with the adenine pocket; a 3-chloro-4-fluoro aniline ring, which fits into the hydrophobic cavity; and a Michael

Table 1.2 Second-generation EGFR inhibitors and their clinical trial status

Drug	Target kinase	Company	Cancer	Development phase
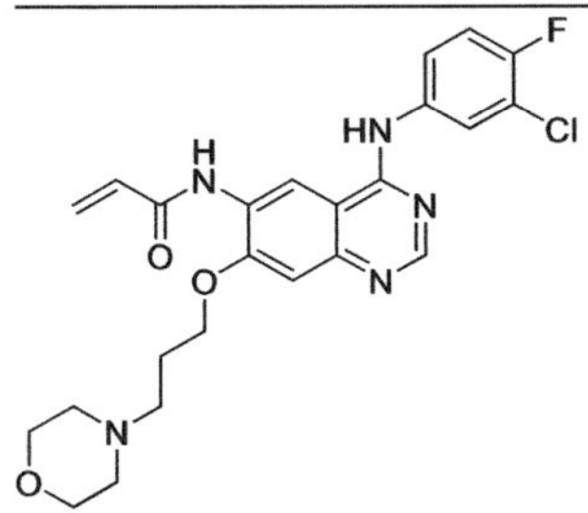 Canertinib	EGFR, HER2-4	Pfizer	NSCLC Breast cancer	Phase II Phase II
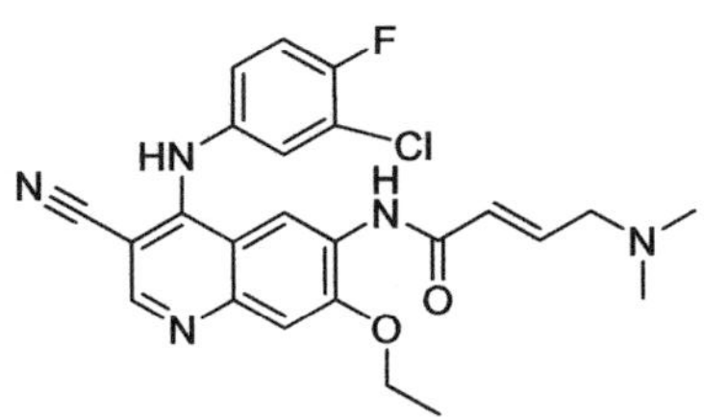 Pelitinib	EGFR, HER2-2	Wyeth/Pfizer	NSCLC Colorectal cancer	Phase II Phase II
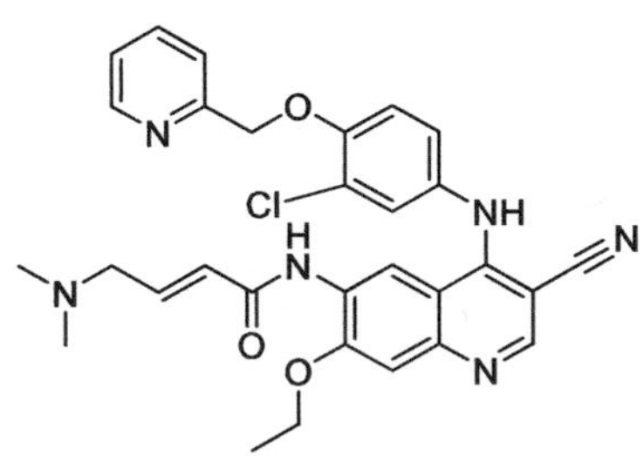 Neratinib	EGFR, HER2-2	Wyeth/Pfizer	NSCLC Breast cancer Solid tumors	Phase II Phase III Phase II

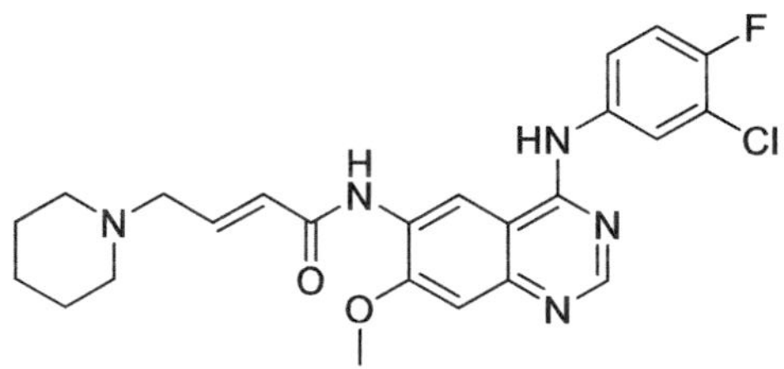 Dacomitinib	EGFR, HER2-4	Pfizer	NSCLC	Phase III
			Gastric cancer	Phase I/II
			Head and neck cancer	Phase I/II
			Glioblastoma	Phase I/II
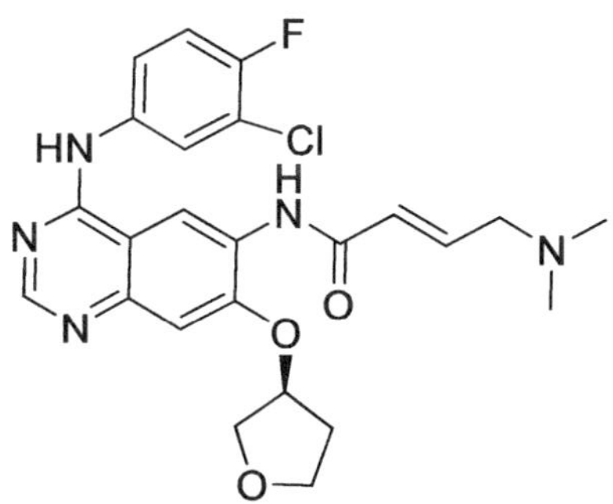 Afatinib	EGFR, HER2-2	Boehringer Ingelheim	NSCLC	Phase III
			Breast cancer	Phase III
			Head and neck cancer	Phase III
			Prostate cancer	Phase II
			Esophagogastric cancer	Phase II
			Colorectal cancer	Phase II
			Glioma	Phase II
			Glioblastoma	Phase I

acceptor to form a covalent linkage with a cysteine residue. It is the first clinically approved irreversible kinase inhibitor that covalently binds to EGFR, HER2, and ErbB4. It covalently alkylate a specific cysteine residue (Cys797 in EGFR) close to the ATP-binding site of the receptor [48, 50].

Afatinib results in the potent and prolonged inhibition of kinase activity in comparison to the reversible first-generation EGFR-TKIs due to obvious reasons (i.e., irreversible inhibition) [19]. Afatinib is an approved drug for the first-line treatment of metastatic NSCLC patients, whose tumors have EGFR possess exon 19 deletions or exon 21 substitution mutations as detected by an FDA-approved test. It is approved for patients with NSCLC in the United States and over 40 other countries, including the European Union, Japan, Taiwan, and Canada [64].

In a phase I trial, which established the MTD of 50 mg orally daily, the dose-limiting toxicities were rash and dyspnea secondary to pneumonitis 73. Common side effects of afatinib include nausea, vomiting, fatigue, and rash. Extensive clinical studies of afatinib have been completed and other studies are in progress in both the first line setting in patients with EGFR mutant tumors and in the setting of acquired resistance to first-generation EGFR TKIs.

In patients with advanced lung adenocarcinoma who have EGFR mutations and have not received prior EGFR TKI, three studies with afatinib were performed or are in progress, including a single-arm phase 2 study and a pair of randomized studies that compare afatinib to chemotherapy in this population of patients (NCT01121393, NCT00949650). In the phase 2 study (LUX Lung2), 129 patients with advanced EGFR mutant lung adenocarcinoma were treated with afatinib 74. The objective response rate was 66% with a median OS and PFS of 24 and 14 months, respectively. To validate these findings and compare the efficacy of afatinib in this population with that of cytotoxic chemotherapy, two randomized phase III trials were performed. The trial of cisplatin/gemcitabine versus afatinib is ongoing with no preliminary results available. The preliminary results of the trial randomizing patients with previously untreated lung adenocarcinoma to afatinib or cisplatin/pemetrexed chemotherapy (LUX Lung3) were recently presented. In the overall population, the median PFS was 11 months for those treated with afatinib compared to 7 months with cisplatin and pemetrexed chemotherapy 75. When only patients with the most common EGFR sensitizing mutations were included, the median PFS for afatinib treatment was 14 months and 7 months for cisplatin and pemetrexed [48, 49].

Three prospective studies of afatinib have been conducted in patients with advanced NSCLC who have progressed on first-generation EGFR TKIs. The first reported trial (LUX Lung1) was a phase 3 study that randomized 585 patients with advanced lung adenocarcinoma who had progression of the disease after platinum-based chemotherapy and at least 3 months of gefitinib or erlotinib 2:1 to afatinib or placebo 76. PFS (3 months versus 1 month) and response rate (7% versus 0%) were increased with afatinib, but the primary endpoint of overall survival was not different. Only 37% of patients met the standard criteria for acquired resistance, which includes previous treatment with a single agent EGFR TKI, and either a known sensitizing EGFR mutation or prior clinical benefit with an EGFR TKI, as well as documented systemic progression on EGFR TKI 77. A post hoc subset analysis of this subgroup showed no differences from the overall study results.

To evaluate the hypothesis that continue EGFR inhibition is important after progression of disease, there is a phase 3 trial, evaluating continued afatinib plus chemotherapy versus chemotherapy alone (Lux Lung 5) in those patients, who have progressed on single agent afatinib. This study is ongoing with a primary endpoint of overall survival (NCT01085136) [48, 49].

There is preclinical data to suggest that a dual targeting of EGFR with both a second generation EGFR TKI and an EGFR monoclonal antibody may be especially effective in shrinking EGFR mutant tumors that have developed and are resistant to erlotinib or gefitinib by way of acquired EGFR T790M 79. An earlier phase 1/2 study of erlotinib and cetuximab did not show any significant activity in patients with acquired resistance to erlotinib 79. A recently completed study looked at the combination of afatinib and cetuximab in patients with advanced NSCLC who have progressed on erlotinib or gefitinib (NCT0109001). Confirmed partial responses were seen in 35% of patients, including 11/35 patients with EGFR T790M+ NSCLC 80. Ninety percent of patients had PR or SD. This is the first targeted therapy in the setting of EGFR-acquired resistance that has shown significant activity in T790M+ EGFR mutant tumors.

1.4.2 Ceritinib

Ceritinib contains an aminopyrimidine nucleus as a core substructure occupying an adenine pocket along with a sulphonyl-substituted aniline ring that occupies a hydrophobic pocket. It shows the type I binding mode and is a second-generation ALK inhibitor for the treatment of NSCLC in the patient developing crizotinib resistance due to rearrangements of the ALK gene [65]. Reported data show that ceritinib efficiently overcomes ALK (one of the main kinase targeted in lung cancer) possessing L1196M, G1269A, I1171T, and S120Y mutations, but ALK with G1202R and F1174C mutations are still not targeted. Cocrystallized ALK–ceritinib revealed that the hydrogen bonds between the 2,4-diaminopyrimidine core and hinge residue Met1199 were found to be conserved. The isopropylsulfonylphenyl side chain binds deep inside the hydrophobic pocket, while the anchor-shaped terminal 2-isopropoxy-3-(piperidin-4-yl)phenyl group resides at the interface between the solvent and the ATP-binding pocket [66].

1.4.3 Dacomitinib

Dacomitinib contains aminoquinazoline as basic moiety for the adenine pocket, a chlorofluoro-substituted phenyl that fits into a hydrophobic pocket and a Michael acceptor to form a covalent bond with cysteine. It is an irreversible inhibitor acting on EGFR, HER2, and HER4. It prevents cell signaling via all ErbB homodimers and heterodimers. Preclinical studies disclosed that dacomitinib inhibits EGFR signaling in tumors/cells that are harboring several different EGFR mutations, including T790M, but show very limited activity against tumors expressing KRAS mutations [20]. Recently a phase II study revealed that dacomitinib delivers a median PFS of 18.2 months in patients treated earlier for NSCLC with activating EGFR mutations [67]. However, further studies with dacomitinib in the relapsed/refractory setting failed to solve their primary objectives. In the phase III ARCHER1009 study, which compared dacomitinib with erlotinib, the median PFS was found to be 2.6 months [48, 49]. Again in phase III, no significant difference between the two treatment arms was found, even though PFS was significantly longer in patients who received dacomitinib [68].

1.4.4 Neratinib

Neratinib consists of an aminoquinoline ring for binding in the adenine pocket, a pyridine substitution to occupy the solvent regions, and a Michael acceptor as a necessary structural feature for covalent inhibitors [48, 49]. Neratinib leads to the irreversible blockade of EGFR and HER2. Preclinical studies showed that it suppressed the growth of bronchoalveolar cancer cells possessing both L858R and T790M mutations [69]. But very low response rates were obtained during the phase II evaluation of Neratinib, involving a study of patients with advanced NSCLC and who had previously received a minimum of 12 weeks of therapy with gefitinib or erlotinib. No patients with T790M achieved an objective response (OR), and only 3% of patients with activating EGFR mutations showed OR [70]. Adding to the woes, 50% of the patients developed grade 3 diarrhea at the earlier determined maximum tolerated dose (MTD) of 320 mg/day, finally leading to decline in the phase II trial dose to 240 mg/day [71].

1.4.5 Canertinib

It consist of the aminoquinazoline ring that occupies the adenine pocket, the oxazine ring that occupies the solvent region and an acrylamide substitution as a Michael acceptor to form a covalent bond with cysteine residue (cysteine 773 in EGFR and relevant cysteines in analogues kinases, such as cysteines 784 and 778 of ErbB2 and ErbB4, respectively) [72]. A phase II trial of canertinib at three different dose levels (oral administration of 50, 150, and 450 mg daily doses) against advanced NSCLC patients, which were divided into three different groups on the basis of a response to platinum-based chemotherapy; a (complete response/partial response (CR/PR), SD, or progressive disease (PD)) was performed. No difference in the PFS was observed at all the three dose levels [48, 49]. However, there were adverse events such as diarrhea and rash that surfaced, hinting at a dose-dependent toxicity. Adding to the despair the observed RRs were 2%, 2%, and 4%, respectively, for the three dose levels. Thus the overall phase II study confirmed that canertinib is not suitable to be pursued as a treatment in NSCLC [73].

1.4.6 Pelitinib

Pelitinib contains an aminoquinoline nucleus fulfilling the pharmacophoric features for the adenine pocket and a Michael acceptor for covalent bond formation with cysteine [48, 49]. Thus it is an irreversible selective inhibitor of EGFR, although at higher concentrations, it also shows an affinity for the ErbB2 receptor. After a phase I study the MTD was found to be 75 mg per day, while the dose-limiting toxicity (DTL) was observed to be grade 3 diarrhea. Pelitinib was subjected to a clinical trial in combination therapy with capecitabine for colorectal cancer, but it did not yield many fruitful results. Additionally, results of a phase I trial using pelitinib and tensirolimus as control in 48 patients with advanced solid tumors have been reported recently [74]. Phase II trials have been completed with pelitinib as a single agent in advanced colorectal cancer (NCT00072748) and in advanced NSCLC (NCT00067548) but the results are not yet completely evaluated and so further trials are on hold for this molecule [75].

1.5 PROBLEMS ASSOCIATED WITH SECOND-GENERATION KINASE INHIBITORS

Second-generation irreversible kinase inhibitors like afatinib and dacomitinib were at first found potent against untreated EGFR-mutated lung cancer. However, as in monotherapy, they have failed to overcome T790M mediated resistance in patients [76] because concentrations at which these irreversible inhibitors overcome T790M activity is preclinically not achievable in humans due to the dose-limiting toxicity related to the nonselective inhibition of wild-type EGFR [77]. Furthermore, these inhibitors procure T790M resistance, providing supportive evidence that they have lower potency against T790M [78]. Thus there remains a significant unsatisfied need for an inhibitor that can more effectively target T790M tumors while sparing wild-type EGFR. This has led to the development of "third-generation" inhibitors.

REFERENCES

[1] R.L. Siegel, K.D. Miller, A. Jemal, Cancer statistics, 2017, CA Cancer J. Clin. 67 (1) (2017) 7–30.

[2] F.A. Shepherd, J. Rodrigues Pereira, T. Ciuleanu, E.H. Tan, V. Hirsh, S. Thongprasert, D. Campos, S. Maoleekoonpiroj, M. Smylie, R. Martins, M. van Kooten, M. Dediu, B. Findlay, D. Tu, D. Johnston, A. Bezjak, G. Clark, P. Santabarbara, L. Seymour, Erlotinib in previously treated non-small-cell lung cancer, N. Engl. J. Med. 353 (2005) 123–132.

[3] T. Mok, Y. Wu, S. Thongprasert, C. Yang, D. Chu, N. Saijo, P. Sunpaweravong, B. Han, B. Margono, Y. Ichinose, Y. Nishiwaki, Y. Ohe, J. Yang, B. Chewaskulyong, H. Jiang, E. Duffield, C. Watkins, A. Armour, M. Fukuoka, Gefitinib or carboplatin–paclitaxel in pulmonary adenocarcinoma, N. Engl. J. Med. 361 (2009) 947–957.

[4] T. Mitsudomi, S. Morita, Y. Yatabe, S. Negoro, I. Okamoto, J. Tsurutani, T. Seto, M. Satouchi, H. Tada, T. Hirashima, K. Asami, N. Katakami, M. Takada, H. Yoshioka, K. Shibata, S. Kudoh, E. Shimizu, H. Saito, S. Toyooka, K. Nakagawa, M. Fukuoka, Gefitinib versus cisplatin plus docetaxel in patients with non-small-cell lung cancer harbouring mutations of the epidermal growth factor receptor (WJTOG3405): an open label, randomised phase 3 trial, Lancet Oncol. 11 (2010) 121–128.

[5] M. Maemondo, A. Inoue, K. Kobayashi, S. Sugawara, S. Oizumi, H. Isobe, A. Gemma, M. Harada, H. Yoshizawa, I. Kinoshita, Y. Fujita, S. Okinaga, H. Hirano, K. Yoshimori, T. Harada, T. Ogura, M. Ando, H. Miyazawa, T. Tanaka, Y. Saijo, K. Hagiwara, S. Morita, T. Nukiwa, Gefitinib or chemotherapy for non–small-cell lung cancer with mutated EGFR, N. Engl. J. Med. 362 (2010) 2380–2388.

[6] R. Rosell, E. Carcereny, R. Gervais, A. Vergnenegre, B. Massuti, E. Felip, R. Palmero, R. Garcia-Gomez, C. Pallares, J. Sanchez, R. Porta, M. Cobo, P. Garrido, F. Longo, T. Moran, A. Insa, F. De Marinis, R. Corre, I. Bover, A. Illiano, E. Dansin, J. de Castro, M. Milella, N. Reguart, G. Altavilla, U. Jimenez, M. Provencio, M. Moreno, J. Terrasa, J. Muñoz-Langa, J. Valdivia, D. Isla, M. Domine, O. Molinier, J. Mazieres, N. Baize, R. Garcia-Campelo, G. Robinet, D. Rodriguez-Abreu, G. Lopez-Vivanco, V. Gebbia, L. Ferrera-Delgado, P. Bombaron, R. Bernabe, A. Bearz, A. Artal, E. Cortesi, C. Rolfo, M. Sanchez-Ronco, A. Drozdowskyj, C. Queralt, I. de Aguirre, J. Ramirez, J. Sanchez,

M. Molina, M. Taron, L. Paz-Ares, Erlotinib versus standard chemotherapy as first-line treatment for European patients with advanced EGFR mutation-positive non-small-cell lung cancer (EURTAC): a multicentre, open-label, randomised phase 3 trial, Lancet Oncol. 13 (2012) 239–246.

[7] N.E. Hynes, H.A. Lane, ERBB receptors and cancer: the complexity of targeted inhibitors, Nat. Rev. Cancer. 5 (2005) 341–354.

[8] L. Sequist, J. Yang, N. Yamamoto, K. O'Byrne, V. Hirsh, T. Mok, S. Geater, S. Orlov, C. Tsai, M. Boyer, W. Su, J. Bennouna, T. Kato, V. Gorbunova, K. Lee, R. Shah, D. Massey, V. Zazulina, M. Shahidi, M. Schuler, Phase III study of afatinib or cisplatin plus pemetrexed in patients with metastatic lung adenocarcinoma with EGFR mutations, J. Clin. Oncol. 31 (2013) 3327–3334.

[9] K. Ohashi, L. Sequist, M. Arcila, C. Lovly, X. Chen, C. Rudin, T. Moran, D. Camidge, C. Vnencak-Jones, L. Berry, Y. Pan, H. Sasaki, J. Engelman, E. Garon, S. Dubinett, W. Franklin, G. Riely, M. Sos, M. Kris, D. Dias-Santagata, M. Ladanyi, P. Bunn, W. Pao, Characteristics of lung cancers harboring NRAS mutations, Clin. Cancer Res. 19 (2013) 2584–2591.

[10] S. Kobayashi, T. Boggon, T. Dayaram, P. Jänne, O. Kocher, M. Meyerson, B. Johnson, M. Eck, D. Tenen, B. Halmos, EGFR mutation and resistance of non–small-cell lung cancer to gefitinib, N. Engl. J. Med. 352 (2005) 786–792.

[11] W. Pao, V. Miller, K. Politi, G. Riely, R. Somwar, M. Zakowski, M. Kris, H. Varmus, Acquired resistance of lung adenocarcinomas to gefitinib or erlotinib is associated with a second mutation in the EGFR kinase domain, PLoS Med. 2 (2005) e73.

[12] M. Niederst, J. Engelman, Bypass mechanisms of resistance to receptor tyrosine kinase inhibition in lung cancer, Sci. Signal. 6 (294) (2013) re6.

[13] J. Engelman, K. Zejnullahu, T. Mitsudomi, Y. Song, C. Hyland, J. Park, N. Lindeman, C. Gale, X. Zhao, J. Christensen, T. Kosaka, A. Holmes, A. Rogers, F. Cappuzzo, T. Mok, C. Lee, B. Johnson, L. Cantley, P. Janne, MET amplification leads to gefitinib resistance in lung cancer by activating ERBB3 signaling, Science 316 (2007) 1039–1043.

[14] K. Takezawa, V. Pirazzoli, M. Arcila, C. Nebhan, X. Song, E. de Stanchina, K. Ohashi, Y. Janjigian, P. Spitzler, M. Melnick, G. Riely, M. Kris, V. Miller, M. Ladanyi, K. Politi, W. Pao, HER2 amplification: a potential mechanism of acquired resistance to EGFR inhibition in EGFR-mutant lung cancers that lack the second-site EGFR T790M mutation, Cancer Discov. 2 (2012) 922–933.

[15] H. Yu, M. Arcila, N. Rekhtman, C. Sima, M. Zakowski, W. Pao, M. Kris, V. Miller, M. Ladanyi, G. Riely, Analysis of tumor specimens at the time of acquired resistance to EGFR-TKI therapy in 155 patients with EGFR-mutant lung cancers, Clin. Cancer Res. 19 (2013) 2240–2247.

[16] E. De Bruin, C. Cowell, P. Warne, M. Jiang, R. Saunders, M. Melnick, S. Gettinger, Z. Walther, A. Wurtz, G. Heynen, D. Heideman, J. Gomez-Roman, A. Garcia-Castano, Y. Gong, M. Ladanyi, H. Varmus, R. Bernards, E. Smit, K. Politi, J. Downward, Reduced NF1 expression confers resistance to EGFR inhibition in lung cancer, Cancer Discov. 4 (2014) 606–619.

[17] L. Sequist, B. Waltman, D. Dias-Santagata, S. Digumarthy, A. Turke, P. Fidias, K. Bergethon, A. Shaw, S. Gettinger, A. Cosper, S. Akhavanfard, R. Heist, J. Temel, J. Christensen, J. Wain, T. Lynch, K. Vernovsky, E. Mark, M. Lanuti, A. Iafrate, M. Mino-Kenudson, J. Engelman, Genotypic and histological evolution of lung cancers acquiring resistance to EGFR inhibitors, Sci. Transl. Med. 3 (75) (2011) 75ra26.

[18] M. Niederst, L. Sequist, J. Poirier, C. Mermel, E. Lockerman, A. Garcia, R. Katayama, C. Costa, K. Ross, T. Moran, E. Howe, L. Fulton, H. Mulvey, L. Bernardo, F. Mohamoud, N. Miyoshi, P. VanderLaan, D. Costa, P. Jänne, D. Borger, S. Ramaswamy, T. Shioda, A. Iafrate, G. Getz, C. Rudin, M. Mino-Kenudson, J. Engelman, RB loss in resistant EGFR mutant lung adenocarcinomas that transform to small-cell lung cancer, Nat. Commun. 6 (2015) 6377.

[19] D. Li, L. Ambrogio, T. Shimamura, S. Kubo, M. Takahashi, L. Chirieac, R. Padera, G. Shapiro, A. Baum, F. Himmelsbach, W. Rettig, M. Meyerson, F. Solca, H. Greulich, K. Wong, BIBW2992, an irreversible EGFR/HER2 inhibitor highly effective in preclinical lung cancer models, Oncogene 27 (2008) 4702–4711.
[20] J. Engelman, K. Zejnullahu, C. Gale, E. Lifshits, A. Gonzales, T. Shimamura, F. Zhao, P. Vincent, G. Naumov, J. Bradner, I. Althaus, L. Gandhi, G. Shapiro, J. Nelson, J. Heymach, M. Meyerson, K. Wong, P. Janne, PF00299804, an irreversible pan-ERBB inhibitor, is effective in lung cancer models with EGFR and ERBB2 mutations that are resistant to gefitinib, Cancer Res. 67 (2007) 11924–11932.
[21] K. Reckamp, G. Giaccone, D. Camidge, S. Gadgeel, F. Khuri, J. Engelman, M. Koczywas, A. Rajan, A. Campbell, D. Gernhardt, A. Ruiz-Garcia, S. Letrent, J. Liang, I. Taylor, J. O'Connell, P. Jänne, A phase 2 trial of dacomitinib (PF-00299804), an oral, irreversible pan-HER (human epidermal growth factor receptor) inhibitor, in patients with advanced non small cell lung cancer after failure of prior chemotherapy and erlotinib, Cancer 120 (8) (2014) 1145–1154.
[22] V. Miller, V. Hirsh, J. Cadranel, Y. Chen, K. Park, S. Kim, C. Zhou, W. Su, M. Wang, Y. Sun, D. Heo, L. Crino, E. Tan, T. Chao, M. Shahidi, X. Cong, R. Lorence, J. Yang, Afatinib versus placebo for patients with advanced, metastatic non-small-cell lung cancer after failure of erlotinib, gefitinib, or both, and one or two lines of chemotherapy (LUX-lung 1): a phase 2B/3 randomised trial, Lancet Oncol. 13 (2012) 528–538.
[23] D. Camidge, W. Pao, L. Sequist, Acquired resistance to TKIs in solid tumours: learning from lung Cancer, Nat. Rev. Clin. Oncol. 11 (8) (2014) 473–481.
[24] W. Zhou, D. Ercan, L. Chen, C.H. Yun, D. Li, M. Capelletti, A.B. Cortot, L. Chirieac, R.E. Iacob, R. Padera, J.R. Engen, K.K. Wong, M.J. Eck, N.S. Gray, P.A. Janne, Novel mutant-selective EGFR kinase inhibitors against EGFR T790M, Nature 462 (7276) (2009) 1070–1074.
[25] A.O. Walter, R.T. Sjin, H.J. Haringsma, K. Ohashi, J. Sun, K. Lee, A. Dubrovskiy, M. Labenski, Z. Zhu, Z. Wang, M. Sheets, T.S. Martin, R. Karp, D.V. Kalken, P. Chaturvedi, D. Niu, M. Nacht, R.C. Petter, W. Westlin, K. Lin, S. Jaw-Tsai, M. Raponi, T.V. Dyke, J. Etter, Z. Weaver, W. Pao, J. Singh, A.D. Simmons, T.C. Harding, A. Allen, Discovery of a mutant-selective covalent inhibitor of EGFR that overcomes T790M-mediated resistance in NSCLC, Cancer Discov. 3 (12) (2013) 1404–1415.
[26] D.A. Cross, S.E. Ashton, S. Ghiorghiu, C. Eberlein, C.A. Nebhan, P.J. Spitzler, J.P. Orme, M.R.V. Finlay, R.A. Ward, M.J. Mellor, G. Hughes, A. Rahi, V.N. Jacobs, M.R. Brewer, E. Ichihara, J. Sun, H. Jin, P. Ballard, K. Al-Kadhimi, R. Rowlinson, T. Klinowska, G.H.P. Richmond, M. Cantarini, D.W. Kim, M.R. Ranson, W. Pao, AZD9291, an irreversible EGFR TKI, overcomes T790M-mediated resistance to EGFR inhibitors in lung cancer, Cancer Discov. 4 (9) (2014) 1046–1061.
[27] L.V. Sequist, J.C. Soria, J.W. Goldman, H.A. Wakelee, S.M. Gadgeel, A. Varga, V. Papadimitrakopoulou, B.J. Solomon, G.R. Oxnard, R. Dziadziuszko, D.L. Aisner, R.C. Doebele, C. Galasso, E.B. Garon, R.S. Heist, J. Logan, J.W. Neal, M.A. Mendenhall, S. Nichols, Z. Piotrowska, A.J. Wozniak, M. Raponi, C.A. Karlovich, S. Jaw-Tsai, J. Isaacson, D. Despain, S.L. Matheny, L. Rolfe, A.R. Allen, D.R. Camidge, Rociletinib in EGFR-mutated non- small-cell lung cancer, N. Engl. J. Med. 372 (2015) 1700–1708.
[28] P.A. Janne, J.C. Yang, D.W. Kim, D. Planchard, Y. Ohe, S.S. Ramalingam, M.J. Ahn, S.W. Kim, W.C. Su, L. Horn, D. Haggstrom, E. Felip, J.H. Kim, P. Frewer, M. Cantarini, K.H. Brown, P.A. Dickinson, S. Ghiorghiu, M. Ranson, AZD9291 in EGFR inhibitor-resistant non-small-cell lung cancer, N. Engl. J. Med. 372 (2015) 1689–1699.
[29] K.S. Thress, C.P. Pawelctz, E. Felip, B.C. Cho, D. Stetson, B. Dougherty, Z. Lai, A. Markovets, A. Vivancos, Y. Kuang, D. Ercan, S.F. Matthews, M. Cantarini, J.C. Barrett, P.A. Janne, G.R. Oxnard, Acquired EGFR C797S mutation mediates resistance to AZD9291 in non-small cell lung cancer harboring EGFR T790M, Nat. Med. 21 (6) (2015) 560–562.

[30] J. Woyach, R. Furman, T. Liu, H. Ozer, M. Zapatka, A. Ruppert, L. Xue, D. Li, S. Steggerda, M. Versele, S. Dave, J. Zhang, A. Yilmaz, S. Jaglowski, K. Blum, A. Lozanski, G. Lozanski, D. James, J. Barrientos, P. Lichter, S. Stilgenbauer, J. Buggy, B. Chang, A. Johnson, J. Byrd, Resistance mechanisms for the Bruton's tyrosine kinase inhibitor ibrutinib, N. Engl. J. Med. 370 (2014) 2286–2294.

[31] M.J. Niederst, H. Hu, H. Mulvey, E.L. Lockerman, A.R. Garcia, Z. Piotrowska, L.V. Sequist, J.A. Engelman, The allelic context of the C797S mutation acquired upon treatment with third generation EGFR inhibitors impacts sensitivity to subsequent treatment strategies, Clin. Cancer Res. 21 (17) (2015) 3924–3933.

[32] L.N. Johnson, E.D. Lowe, M.E. Noble, D.J. Owen, The Eleventh Datta Lecture. The structural basis for substrate recognition and control by protein kinases, FEBS Lett. 430 (1998) 1–11.

[33] G. Manning, D.B. Whyte, R. Martinez, T. Hunter, S. Sudarsanam, The protein kinase complement of the human genome, Science 298 (2002) 1912–1934.

[34] P. Traxler, P. Furet, Strategies toward the design of novel and selective protein tyrosine kinase inhibitors, Pharmacol. Ther. 82 (1999) 195–206.

[35] J. Zhang, P.L. Yang, N.S. Gray, Targeting cancer with small molecule kinase inhibitors, Nat. Rev. Cancer 09 (2009) 28–39.

[36] L.F. Hennequin, et al., N.-(5-chloro-1,3-benzodioxol-4-yl)-7-[2-(4-methylpiperazin-1-yl)ethoxy]-5-(tetrahydro-2H-pyran-4-yloxy)quinazolin-4-amine, a novel, highly selective, orally available, dual-specific c-Src/Abl kinase inhibitor, J. Med. Chem. 49 (2006) 6465–6488.

[37] P.W. Manley, S.W. Cowan-Jacob, J. Mestan, Advances in the structural biology, design and clinical development of Bcr-Abl kinase inhibitors for the treatment of chronic myeloid leukaemia, Biochim. Biophys. Acta 1754 (2005) 3–13.

[38] P.T. Wan, et al., Mechanism of activation of the RAF–ERK signaling pathway by oncogenic mutations of B-RAF, Cell 116 (2004) 855–867.

[39] J.F. Ohren, et al., Structures of human MAP kinase kinase 1 (MEK1) and MEK2 describe novel noncompetitive kinase inhibition, Nat. Struct. Mol. Biol. 11 (2004) 1192–1197.

[40] S.F. Barnett, et al., Identification and characterization of pleckstrin-homology-domain-dependent and isoenzyme-specific Akt inhibitors, Biochem. J. 385 (2005) 399–408.

[41] C.W. Lindsley, et al., Allosteric Akt (PKB) inhibitors: discovery and SAR of isozyme selective inhibitors, Bioorg. Med. Chem. Lett. 15 (2005) 761–764.

[42] D.W. Fry, et al., A specific inhibitor of the epidermal growth factor receptor tyrosine kinase, Science 265 (1994) 1093–1095.

[43] J.V. Heymach, M. Nilsson, G. Blumenschein, V. Papadimitrakopoulou, R. Herbst, Epidermal growth factor receptor inhibitors in development for the treatment of non-small cell lung cancer, Clin. Cancer Res. 12 (2006) 4441s–4445s.

[44] T.A. Baillie, Future of toxicology—metabolic activation and drug design: challenges and opportunities in chemical toxicology, Chem. Res. Toxicol. 19 (7) (2006) 889–893.

[45] D.C. Liebler, Protein damage by reactive electrophiles: targets and consequences, Chem. Res. Toxicol. 21 (2008) 117–128.

[46] J. Singh, R.C. Petter, T.A. Baillie, A. Whitty, The resurgence of covalent drugs, Nat. Rev. Drug Discov. 10 (2011) 307–317.

[47] J. Singh, R.C. Petter, A.F. Kluge, Targeted covalent drugs of the kinase family, Curr. Opin. Chem. Biol. 14 (2010) 475–480.

[48] H.A.Y. Gregory, J. Riely, Second generation epidermal growth factor receptor tyrosine kinase inhibitors in lung cancers, J. Natl. Compr. Cancer Netw. 11 (2013) 161–169.

[49] P.K. Singh, H. Singh, O. Silakari, Kinases inhibitors in lung cancer: from benchside to bedside, Biochim. Biophys. Acta 1866 (2016) 128–140.

[50] T.J. Lynch, D.W. Bell, R. Sordella, et al., Activating mutations in the epidermal growth factor receptor underlying responsiveness of non-small-cell lung cancer to gefitinib, N. Engl. J. Med. 350 (2004) 2129–2139.
[51] T. Kawaguchi, M. Ando, K. Asami, Y. Okano, M. Fukuda, H. Nakagawa, H. Ibata, T. Kozuki, T. Endo, A. Tamura, Randomized phase III trial of erlotinib versus docetaxel as secondor third-line therapy in patients with advanced non–small-cell lung cancer: Docetaxel and Erlotinib Lung Cancer Trial (DELTA), J. Clin. Oncol. 32 (2014) 1902–1908.
[52] J.T. Hartmann, J.P. Pintoffl, H. Kröning, C. Bokemeyer, M. Holtmann, T. Hoehler, Gefitinib in combination with oxaliplatin and 5-fluorouracil in irinotecan-refractory patients with colorectal cancer: a phase I study of the Arbeits gemeinschaft Internistische Onkologie (AIO), Oncol. Res. Treat. 31 (2008) 237–241.
[53] M. Fukuoka, S. Yano, G. Giaccone, T. Tamura, K. Nakagawa, J.-Y. Douillard, Y. Nishiwaki, J. Vansteenkiste, S. Kudoh, D. Rischin, Multi-institutional randomized phase II trial of gefitinib for previously treated patients with advanced non–small-cell lung cancer, J. Clin. Oncol. 21 (2003) 2237–2246.
[54] M. Kris, R.B. Natale, R. Herbst, et al., Efficacy of gefitinib, an inhibitor of the epidermal growth factorreceptor tyrosine kinase, in symptomatic patients with non-small cell lung cancer, JAMA 290 (2003) 2149–2158.
[55] N. Thatcher, A. Chang, P. Parikh, et al., Gefitinib plus best supportive care in previously treated patients with refractory advanced non-small-cell lung cancer: results from a randomized, placebo controlled, multicentre study (Iressa survival evaluation in lung cancer), Lancet 366 (2005) 1527–1537.
[56] D.E. Thurston, Chemistry and pharmacology of anti-cancer drugs, Br. J. Cancer 97 (2007) 1713.
[57] D.S. Ettinger, W. Akerley, G. Bepler, M.G. Blum, A. Chang, R.T. Cheney, L.R. Chirieac, T.A. D'Amico, T.L. Demmy, A.K.P. Ganti, Non–small cell lung cancer, J. Natl. Compr. Cancer Netw. 8 (2010) 740–801.
[58] J.T. Hartmann, M. Haap, H.-G. Kopp, H.-P. Lipp, Tyrosine kinase inhibitors-a review on pharmacology, metabolism and side effects, Curr. Drug Metab. 10 (2009) 470–481.
[59] M.C. Garassino, O. Martelli, M. Broggini, G. Farina, S. Veronese, E. Rulli, F. Bianchi, A. Bettini, F. Longo, L. Moscetti, Erlotinib versus docetaxel as second-line treatment of patients with advanced non-small-cell lung cancer and wild-type EGFR tumours (TAILOR): a randomised controlled trial, Lancet Oncol. 14 (2013) 981–988.
[60] T. Kawaguchi, M. Ando, K. Asami, Y. Okano, M. Fukuda, H. Nakagawa, H. Ibata, T. Kozuki, T. Endo, A. Tamura, Randomized phase III trial of erlotinib versus docetaxel as second or third-line therapy in patients with advanced non–small-cell lung cancer: Docetaxel and Erlotinib Lung Cancer Trial (DELTA), J. Clin. Oncol. 32 (2014) 1902–1908.
[61] W. Pao, V.A. Miller, K.A. Politi, G. Ricly, R. Somwar, M.F. Zakowski, M.G. Kris, H. Varmus, Acquired resistance of lung adenocarcinomas to gefitinib or erlotinib is associated with a second mutation in the EGFR kinase domain, PLoS Med. 2 (2005) 225–235.
[62] O. Tetsu, M. Hangauer, J. Phuchareon, D. Eisele, F. McCormick, Drug resistance to EGFR inhibitors in lung cancer, Chemotherapy 61 (2016) 223–235.
[63] C.H. Yun, K.E. Mengwasser, A.V. Toms, M.S. Woo, H. Greulich, K.K. Wong, The T790M mutation in EGFR kinase causes drug resistance by increasing the affinity for ATP, Proc. Natl. Acad. Sci. U. S. A. 105 (2008) 2070–2075.
[64] B. Ingelheim, Gilotrif® Prescribing Information, Cited 2013.
[65] A.T. Shaw, D.W. Kim, R. Mehra, D.S. Tan, E. Felip, L.Q. Chow, D.R. Camidge, J. Vansteenkiste, S. Sharma, T. De Pas, Ceritinib in ALK-rearranged non–small-cell lung cancer, N. Engl. J. Med. 370 (2014) 1189–1197.

[66] L. Friboulet, N. Li, R. Katayama, C.C. Lee, J.F. Gainor, A.S. Crystal, P.Y. Michellys, M.M. Awad, N. Yanagitani, S. Kim, The ALK inhibitor ceritinib overcomes crizotinib resistance in non–small cell lung cancer, Cancer Discov. 4 (2014) 662–673.
[67] P.A. Janne, S.H.I. Ou, D.W. Kim, G.R. Oxnard, R. Martins, M.G. Kris, F. Dunphy, M. Nishio, J. O'Connell, C. Paweletz, Dacomitinib as first-line treatment in patients with clinically or molecularly selected advanced non-small-cell lung cancer: a multicentre, open-label, phase 2 trial, Lancet Oncol. 15 (2014) 1433–1441.
[68] S.S. Ramalingam, P.A. Janne, T. Mok, K. O'Byrne, M.J. Boyer, J. Von Pawel, A. Pluzanski, M. Shtivelband, L.I. Docampo, J. Bennouna, Dacomitinib versus erlotinib in patients with advanced-stage, previously treated non-small-cell lung cancer (ARCHER 1009): a randomised, double-blind, phase 3 trial, Lancet Oncol. 15 (2014) 1369–1378.
[69] S.K. Rabindran, C.M. Discafani, E.C. Rosfjord, M. Baxter, M.B. Floyd, J. Golas, W.A. Hallett, B.D. Johnson, R. Nilakantan, E. Overbeek, Antitumor activity of HKI-272, an orally active, irreversible inhibitor of the HER-2 tyrosine kinase, Cancer Res. 64 (2004) 3958–3965.
[70] L.V. Sequist, B. Besse, T.J. Lynch, V.A. Miller, K.K. Wong, B. Gitlitz, K. Eaton, C. Zacharchuk, A. Freyman, C. Powell, Neratinib, an irreversible pan-ErbB receptor tyrosine kinase inhibitor: results of a phase II trial in patients with advanced non–small-cell lung cancer, J. Clin. Oncol. 28 (2010) 3076–3083.
[71] K.K. Wong, P.M. Fracasso, R.M. Bukowski, T.J. Lynch, P.N. Munster, G.I. Shapiro, P.A. Jänne, J.P. Eder, M.J. Naughton, M.J. Ellis, A phase I study with neratinib (HKI-272), an irreversible pan ErbB receptor tyrosine kinase inhibitor, in patients with solid tumors, Clin. Cancer Res. 15 (2009) 2552–2558.
[72] D.W. Fry, A.J. Bridges, W.A. Denny, A. Doherty, K.D. Greis, J.L. Hicks, K.E. Hook, P.R. Keller, W.R. Leopold, J.A. Loo, Specific, irreversible inactivation of the epidermal growth factor receptor and erbB2, by a new class of tyrosine kinase inhibitor, Proc. Natl. Acad. Sci. 95 (1998) 12022–12027.
[73] P.A. Janne, J. von Pawel, R.B. Cohen, L. Crino, C.A. Butts, S.S. Olson, I.A. Eiseman, A.A. Chiappori, B.Y. Yeap, P.F. Lenehan, Multicenter, randomized, phase II trial of CI-1033, an irreversible pan-ERBB inhibitor, for previously treated advanced non–small-cell lung cancer, J. Clin. Oncol. 25 (2007) 3936–3944.
[74] A.H. Bryce, R. Rao, J. Sarkaria, J.M. Reid, Y. Qi, R. Qin, C.D. James, R.B. Jenkins, J. Boni, C. Erlichman, Phase I study of temsirolimus in combination with EKB-569 in patients with advanced solid tumors, Investig. New Drugs 30 (2012) 1934–1941.
[75] C. Erlichman, M. Hidalgo, J.P. Boni, P. Martins, S.E. Quinn, C. Zacharchuk, P. Amorusi, A.A. Adjei, E.K. Rowinsky, Phase I study of EKB-569, an irreversible inhibitor of the epidermal growth factor receptor, in patients with advanced solid tumors, J. Clin. Oncol. 24 (2006) 2252–2260.
[76] N. Yamamoto, N. Katakami, S. Atagi, T. Hida, K. Goto, T. Horai, A. Inoue, Y. Ichinose, K. Kobayashi, K. Takeda, A phase II trial of afatinib (BIBW 2992) in patients (pts) with advanced non-small cell lung cancer previously treated with erlotinib (E) or gefitinib (G), J. Clin. Oncol. 29 (2011) 7524.
[77] F. Eskens, C. Mom, A.T. Planting, J. Gietema, A. Amelsberg, H. Huisman, L. van Doorn, H. Burger, P. Stopfer, J. Verweij, A phase I dose escalation study of BIBW 2992, an irreversible dual inhibitor of epidermal growth factor receptor 1 (EGFR) and 2 (HER2) tyrosine kinase in a 2-week on, 2-week off schedule in patients with advanced solid tumours, Br. J. Cancer 98 (2008) 80–85.
[78] J. Chmielecki, J. Foo, G.R. Oxnard, K. Hutchinson, K. Ohashi, R. Somwar, L. Wang, K.R. Amato, M. Arcila, M.L. Sos, Optimization of dosing for EGFR-mutant non–small cell lung cancer with evolutionary cancer modeling, Sci. Transl. Med. 3 (2011) 90ra59.

FURTHER READING

[79] T.A. Yap, L. Vidal, J. Adam, et al., Phase I trial of the irreversible EGFR and HER2 kinase inhibitor BIBW 2992 in patients with advanced solid tumors, J. Clin. Oncol. 28 (2010) 3965–3972.

[80] J.C. Yang, J.Y. Shih, W.C. Su, et al., Afatinib for patients with lung adenocarcinoma and epidermal growth factor receptor mutations (LUX-Lung 2): a phase 2 trial, Lancet Oncol. 13 (2012) 539–548.

[81] J.C. Yang, M.H. Schuler, N. Yamamoto, et al., LUX-Lung 3: a randomized, open-label, phase III study of afatinib versus pemetrexed and cisplatin as first-line treatment for patients with advanced adenocarcinoma of the lung harboring EGFR-activating mutations, J. Clin. Oncol. 30 (2012). abstr LBA7500.

[82] V.A. Miller, V. Hirsh, J. Cadranel, et al., Afatinib versus placebo for patients with advanced, metastatic non-small-cell lung cancer after failure of erlotinib, gefitinib, or both, and one or twolines of chemotherapy (LUX-Lung 1): a phase 2b/3 randomised trial, Lancet Oncol. (2012). (published online ahead of print March 26, 2012).

[83] D. Jackman, W. Pao, G.J. Riely, et al., Clinical definition of acquired resistance to epidermal growthfactor receptor tyrosine kinase inhibitors in non-small-cell lung cancer, J. Clin. Oncol. 28 (2010) 357–360.

[84] J.E. Chaft, G.R. Oxnard, C.S. Sima, M.G. Kris, V.A. Miller, G.J. Riely, Disease flare after tyrosine kinaseinhibitor discontinuation in patients with EGFR-mutant lung cancer and acquired resistance toerlotinib or gefitinib: implications for clinical trial design, Clin. Cancer Res. 17 (2011) 6298–6303.

[85] L. Regales, Y. Gong, R. Shen, et al., Dual targeting of EGFR can overcome a major drug resistance mutation in mouse models of EGFR mutant lung cancer, J. Clin. Investig. 119 (2009) 3000–3010.

[86] Y.Y. Janjigian, C.G. Azzoli, L.M. Krug, et al., Phase I/II trial of cetuximab and erlotinib in patients with lung adenocarcinoma and acquired resistance to erlotinib, Clin. Cancer Res. 17 (2011) 2521–2527.

CHAPTER 2

Approved and Clinical Trial Third-Generation EGFR Inhibitors

Third-generation tyrosine kinase inhibitors (TKIs) such as osimertinib (AZD9291), rociletinib (CO1686), HM61713, EGF816, and ASP8273 are mutant selective and EGFR wild-type (WT) sparing, targeting sensitizing EGFR mutations as well as T790M EGFR (Table 2.1) [1]. Furthermore, they have a very low inhibitory effect on WT EGFR, thus overcoming the toxicity limitation seen with the first- and second-generation EGFR TKIs [1]. WZ4002 was one of the earliest compounds investigated. In vitro, WZ4002 was 30–100 times more potent against EGFR T790M and 100 times less potent against WT EGFR; similar potency was seen in vivo using T790M driven murine lung models. WZ4002 is currently not in clinical development [2]. The current chapter discusses various clinical trials and approved third-generation drugs.

2.1 OSIMERTINIB (AZD9291)

Osimertinib (previously known as mereletinib or AZD9291, trade name Tagrisso) is a third-generation EGFR TKI drug developed by AstraZeneca Pharmaceuticals for mutated EGFR cancers [3–6]. It is a mono-anilino-pyrimidine compound that is structurally different from other third-generation EGFR TKIs (Table 2.1). Osimertinib irreversibly targets the cysteine-797 residue in the ATP binding site of the EGFR kinase through covalent bond formation [1, 2]. It is structurally different from the first- and second-generation EGFR TKIs. This compound is an irreversible mutant-selective EGFR TKI (exon 19 deletion EGFR $IC_{50} = 12.92$ nM, L858R/T790M EGFR $IC_{50} = 11.44$ nM, WT EGFR $IC_{50} = 493.8$ nM) [1, 2]. It is the only approved EGFR TKI currently indicated for patients with metastatic EGFR T790M mutation-positive NSCLC [1, 2]. Two circulating active metabolites, a des-methyl indole analog AZ5104 and an N-demethylated analog

Third-Generation EGFR Inhibitors
https://doi.org/10.1016/B978-0-08-102661-8.00002-0

Table 2.1 Third-generation EGFR tyrosine kinase inhibitors

Drug	Structure	Research code	Status	Company	Trade name	CAS no.	Clinical trial govt. identifier
Osimertinib	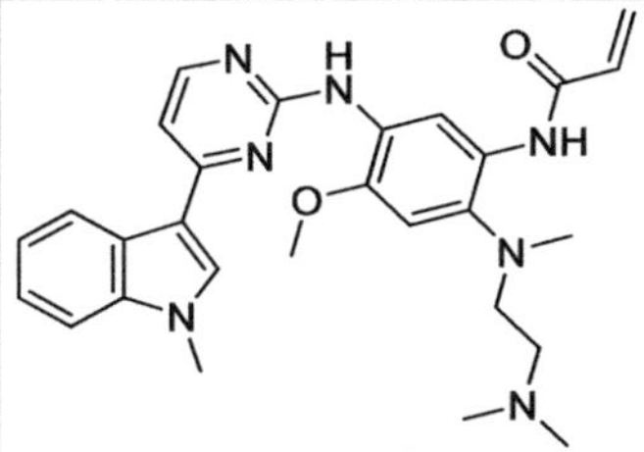	AZD9291	Approved	AstraZeneca	Tagrriso	1421373-65-0	NCT02151981
Rociletinib	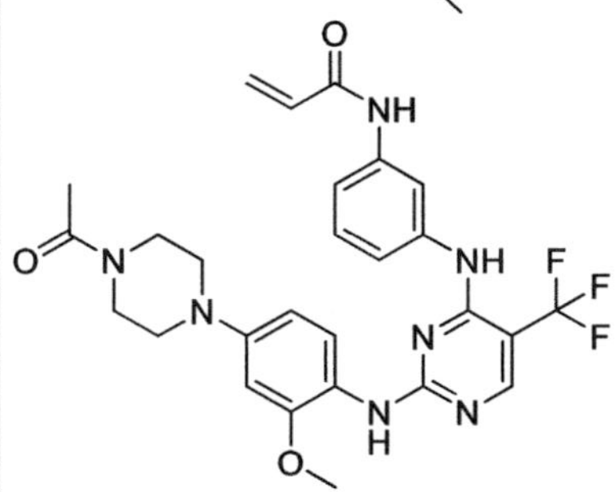	CO1686	Phase II/III Rejected by FDA April 2016	Avila Therapeutics and codeveloped by the Originator and Clovis Oncology	NA	1374640-70-6	NCT02186301
Olmutinib	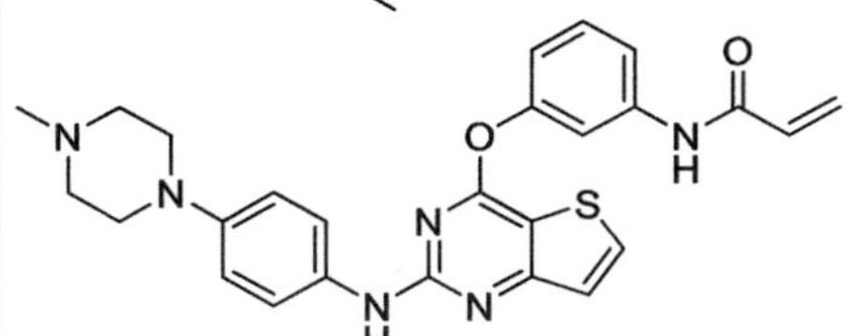	HM61713	Phase I/II clinical trial Stopped further development	Hanmi, Boehringer Ingelheim, Zai Lab	Olita	1353550-13-6	NCT01588145

Nazartinib	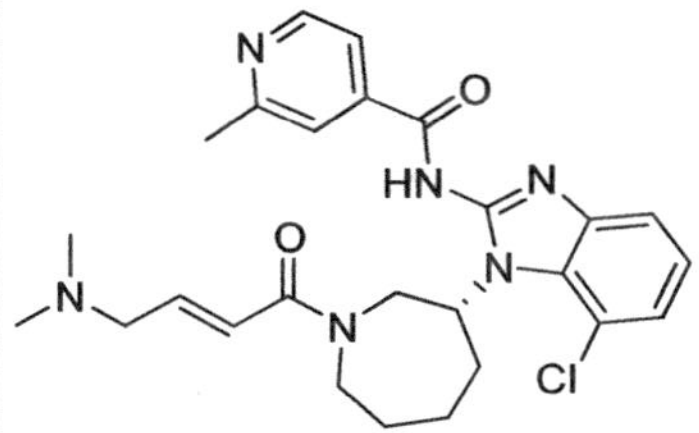	EGF816	Phase II (active)	Novartis	NA	1508250-71-2	NCT02108964
Naquotinib	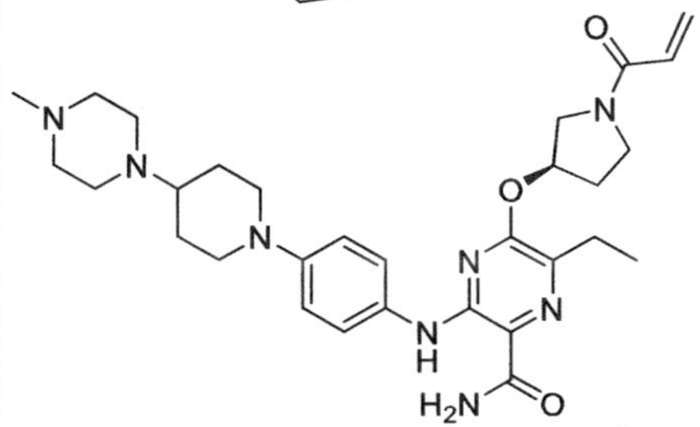	ASP8273	Phase II (active)	Astellas	NA	1197953-54-0	NCT02500297
Brigatinib	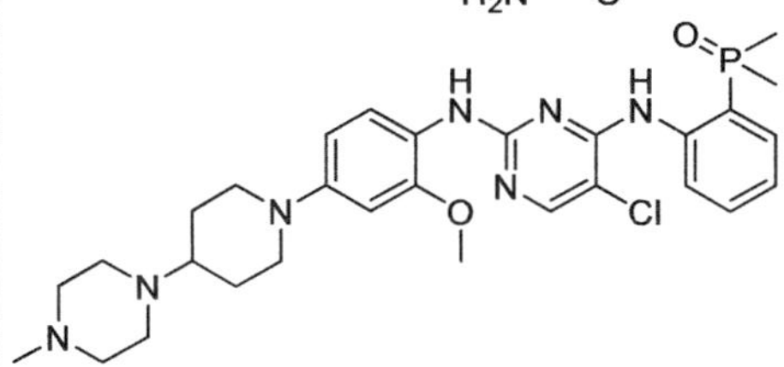	AP26113	Phase I/II	Ariad	NA	1197953-54-0	NCT01449461

Continued

Table 2.1 Third-generation EGFR tyrosine kinase inhibitors—cont'd

Drug	Structure	Research code	Status	Company	Trade name	CAS no.	Clinical trial govt. identifier
Avitinib	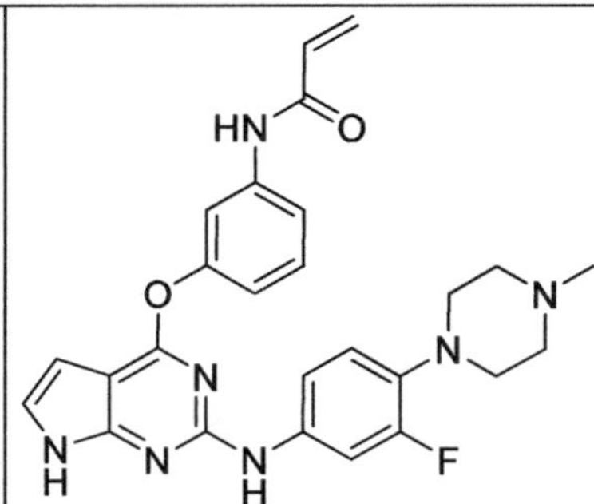	AC0010	Phase II (active)	ACEA Bioscience	NA	319460-85-0	NCT02274337
PF06747775	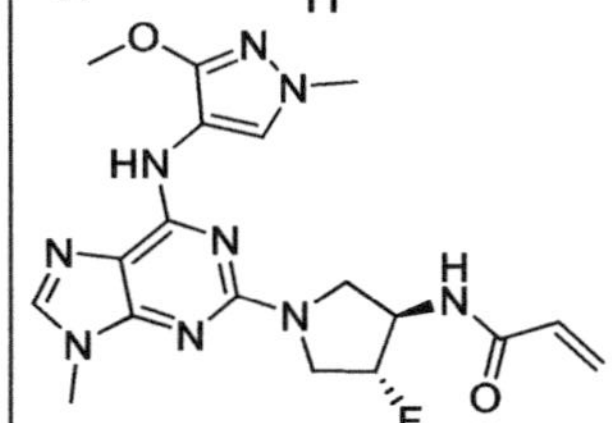	PF06747775	Phase I/II (active)	Pfizer	NA	1776112-90-3	NCT02349633

AZ7550, were observed in both preclinical as well as clinical settings [1, 2]. Compared with osimertinib, AZ5104 showed better potency against the activating mutant (IC_{50} of 2 nM in PC9 cell line) and the double mutant (IC_{50} of 2 nM in H1975 cell line), as well as the IC_{50} of 33 nM in LoVo cell line with WT EGFR. AZ7550 showed a very similar potency (IC_{50} of 26 nM in PC9 cell line and IC_{50} of 45 nM in H1975 cell line) and selectivity profile (WT EGFR; IC_{50} of 786 nM in LoVo cell line) to osimertinib [7–10]. In a pharmacokinetic study in mice, osimertinib demonstrated good bioavailability, moderate clearance, and wide tissue distribution with a half-life of 3 h following a single oral dose of 25 mg/kg. The half-lives of the circulating metabolites in plasma were similar to osimertinib, and the total exposure levels (AUC) were approximately 68% and 33% compared to parent for AZ7550 and AZ5104, respectively. It is interesting to note that the level of metabolites (AZ7550 and AZ5104) relative to osimertinib is significantly lower in humans than that in mice [11–13].

On March 30, 2017, the US Food and Drug Administration granted regular approval to osimertinib (TAGRISSO, AstraZeneca Pharmaceuticals, LP) for the treatment of patients with metastatic EGFR T790M mutation-positive NSCLC (as detected by an FDA-approved test) whose disease has progressed on or after *EGFR* TKI therapy [14]. In November 2015, osimertinib received accelerated approval for this indication based on an overall response rate (ORR) of 59% among 411 patients in 2 single-arm clinical trials. The current approval is based on AURA3 (NCT012151981), a randomized, multicenter open-label, active-controlled trial conducted in patients with metastatic EGFR T790M mutation-positive NSCLC who had progressive disease following first-line EGFR TKI therapy [15]. All patients were required to have EGFR T790M mutation-positive NSCLC identified by the cobas EGFR mutation test performed in a central laboratory. AURA3 randomized 419 patients (2:1) to receive osimertinib ($n=279$) 80 mg orally once daily or platinum-based doublet chemotherapy ($n=140$). Patients in the chemotherapy arm received either pemetrexed (500 mg with carboplatin AUC5 or 500 mg with cisplatin 75 mg) on day 1 of every 21-day cycle for up to 6 cycles, followed by pemetrexed maintenance therapy. Patients on the chemotherapy arm with radiological progression according to both investigator and blinded independent central review were offered osimertinib at progression [16]. AURA 3 demonstrated an improvement in investigator-assessed, progression-free survival (PFS), with a hazard ratio of 0.30 (95% CI: 0.23, 0.41; $P<0.001$). The estimated median PFS was 10.1 months in the osimertinib arm and 4.4 months in the chemotherapy arm. Confirmed ORR, according to investigator assessment, was 65% (95% CI: 59%, 70%) and 29% (95% CI: 21%, 37%) in the osimertinib and chemotherapy arms, respectively ($P<0.0001$). The estimated median response durations were 11 months (95% CI: 8.6, 12.6) and 4.2 months (95% CI: 3.9, 5.9) in the osimertinib and chemotherapy arms, respectively. In patients with measurable central nervous system (CNS) lesions on baseline brain scans, the confirmed CNS ORR, assessed by independent central review, was 57% (95% CI: 37%, 75%) and 25% (95% CI: 7%, 52%) in the osimertinib and chemotherapy arms, respectively. The median CNS response duration was not reached in the osimertinib arm (range: 1.4–12.5 months) and was 5.7 months (range: 1.4, 5.7 months) in the chemotherapy arm. Overall survival data are immature. Serious adverse reactions were evaluated in 833 patients receiving osimertinib. The most serious adverse reactions were interstitial lung disease/pneumonitis (3.5%), QTc interval prolongation (0.7%), cardiomyopathy (1.9%), and keratitis (0.7%). The most common adverse reactions (occurring in at least 20% of patients) were diarrhea, rash, dry skin, nail toxicity, and fatigue [14–16]. The recommended dose of osimertinib is 80 mg orally once daily, with or without food, until disease progression or unacceptable toxicity. The presence of an EGFR T790M mutation in a tumor specimen or plasma specimen (if tumor tissue is unavailable) should be confirmed by an FDA-approved test prior to initiation of treatment [14–16].

2.2 ROCILETINIB (CO1686)

Rociletinib, also known as AVL-301 and CO1686, is an orally available irreversible inhibitor of EGFR with potential antineoplastic activity. EGFR inhibitor rociletinib (CO1686) binds to and inhibits mutant forms of EGFR, including T790M, thereby leading to the cell death of resistant tumor cells. Compared to other EGFR inhibitors, rociletinib (CO1686) inhibits T790M, a secondary acquired resistance mutation, as well as other mutant EGFRs and may have therapeutic benefits in tumors with T790M-mediated resistance to other EGFR TKIs [17].

It has a 22-fold selectivity over WT EGFR. In NSCLC cell lines containing EGFR mutations, rociletinib demonstrates the following cellular pEGFR IC_{50}: 62 nM in NCI-1975 (L858R/T790M), 187 nM in HCC827 (exon 19 deletion), and 211 nM in PC9 (exon 19 deletion). In cell lines expressing WT EGFR, cellular pEGFR IC_{50} are: >4331 nM in A431, >2000 nM in NCIH1299, and >2000 nM in NCI-H358 [18].

Rociletinib hydrobromide had been filed in the United States for the treatment of EGFR mutant NSCLC, but was rejected by the FDA in April 2016. A breakthrough therapy designation was assigned by FDA for this indication in 2014 [17, 18].

Clovis has terminated enrollment in all ongoing rociletinib studies, including the phase III TIGER-3 trial, and has withdrawn its application for regulatory approval in the European Union. Rociletinib will continue to be provided to patients whose clinicians recommend continuing therapy, according to Clovis [19].

This final chapter in the rociletinib saga follows a series of negative developments for the drug, including updated data revealing lower response rates than initially reported, a negative vote from the FDA's Oncologic Drugs Advisory Committee (ODAC), and FDA approval of rociletinib's main competitor in the setting, osimertinib (Tagrisso) [20].

In April 2016, ODAC voted 12-1 against the accelerated approval of rociletinib for patients with metastatic *EGFR T790M*-mutated NSCLC who previously received an *EGFR*-targeted therapy. The panel was considering whether a pooled efficacy and safety analysis from the early-stage CO-1686-008 (TIGER-X) and CO-1686-019 (TIGER-2) trials was sufficient to recommend accelerated approval of the drug. With its negative vote, the panel recommended that the results of the now-terminated phase III CO-1686-020 trial (TIGER-3) should be submitted before the FDA made a decision on the application. The open-label, multinational TIGER-3 trial was comparing rociletinib with single-agent chemotherapy (pemetrexed, docetaxel, or gemcitabine) in patients with *EGFR* mutation-positive NSCLC with disease progression following both an *EGFR* TKI and platinum doublet chemotherapy. In the pooled TIGER-X/TIGER-2 analysis, the ORR with rociletinib (dose range: 500–750 mg twice daily) among 325 patients with *EGFR T790M*-positive metastatic NSCLC who progressed on at least 1 *EGFR* inhibitor was 30.2% (95% CI, 25.2–35.5). The ORR was 32% (95% CI, 25–40) and 23% (95% CI, 14–34) in patients receiving the 625 mg ($n=170$) and 500 mg ($n=79$) doses, respectively. The median duration of response (DOR) for the 2 treatment doses was 8.8 and 9.1 months, respectively. For its safety analysis, ODAC considered pooled data from 400 patients enrolled in either TIGER-X or TIGER-2 who were treated at either 500, 625, 750, or 1000 mg twice daily [21]. The most common all-grade adverse events (AEs), occurring in >30% of patients were diarrhea, hyperglycemia, fatigue, nausea, decreased appetite, QT prolongation, and vomiting. The most frequent (>10%) grade 3/4 AEs were hyperglycemia and QTc prolongation. Dose reductions occurred in 51% of patients, most commonly due to hyperglycemia (22%) and QTc prolongation (11%). Fifty-seven percent of patients had dose interruptions, which were most often due to hyperglycemia (22%), QTc prolongation (10%), and nausea (10%). AEs led to discontinuation for 11% of patients, most frequently due to QTc prolongation (2%) and pneumonia/pneumonitis (2%). Serious AEs were experienced by 47% of patients, with the most common being malignant neoplasm progression (16%), hyperglycemia (8%), and pneumonia (4%). Postbaseline QTc intervals of >500 ms occurred on at least one occasion in 17% of patients. One patient experienced Torsades de pointes, and there were two sudden deaths (on day 4 and day 13), according to ODAC briefing documents [19–21].

A rolling submission was originally completed for rociletinib in July 2015, which was subsequently granted a priority review by the FDA. However, at a preplanned 90-day review meeting, changes in the response rates from TIGER-X and TIGER-2 prompted the FDA to request additional data, which set off a chain of events leading to the scheduling of the ODAC hearing. In the earlier TIGER-X/TIGER-2 data submitted to the FDA, patients with *T790M* mutations ($n = 243$) experienced an ORR across all dose levels of 53% and a disease control rate of 85%. TIGER-X included 456 patients with *EGFR*-positive NSCLC who received rociletinib across four doses (range: 500–1000 mg). The median age of patients was 63 years, 10% had a prior history of diabetes, and 41% had CNS metastases. The median number of prior therapies was 2, and nearly half of patients had received more than one TKI (44%) [19–21].

At a data cutoff of April 27, 2015, the median PFS in evaluable patients with *T790M* mutations across the 500 and 625 mg doses ($n = 270$) was 8.0 months. In those without baseline CNS metastases the median PFS was 10.3 months. The company did not release new data for PFS. In the safety analysis, the most frequently occurring all-grade AEs in the 500-mg arm were hyperglycemia (35%), diarrhea (33%), fatigue (29%), decreased appetite (15%), muscle spasms (14%), weight loss (10%), and vomiting (8%). Grade 3 QTc prolongation was observed in 2.5% of patients [19–21]. No cases of interstitial lung disease were seen at the 500-mg dose level. AEs leading to treatment discontinuation were seen in 2.5% of patients with the 500-mg dose. Grade 3/4 hyperglycemia occurred in 17% of patients treated with the 500-mg dose. To adjust for this, a monitoring and treatment algorithm was implemented to detect glucose levels and initiate treatment with oral insulin sensitizing agents as needed. Prior to initiating these measures in

September 2014, the rate of grade 3/4 hyperglycemia was 22%. With proper monitoring and treatment, this rate dropped to 8% [19–21].

2.3 OLMUTINIB (HM61713)

Olmutinib (HM61713) is an irreversible kinase inhibitor and covalently binds to a cysteine residue near the kinase domain of mutant *EGFR*. Olmutinib (HM61713) has a half-life of over 24 h for *EGFR* inhibition [22–26]. Olmutinib (HM61713) caused potent inhibition in cell lines H1975 (*L858R* and *T790M*) and HCC827 (exon 19 deletion) [27–30]. Olmutinib (HM61713) has a low potency for NSCLC cell line H358 harboring WT *EGFR* (GI_{50} of 2225 nM) [27–30]. In the in vivo studies of xenograft models with grafts of H1975 and HCC827, olmutinib (HM61713) was active against the tumors without showing any side effects [27–30]. Olmutinib was approved by the Korean Ministry of Food and Drug Safety on May 2016 [31, 32]. It was originally discovered by Hanmi, which then licensed to Boehringer Ingelheim the development and global commercialization rights, except South Korea, China, and Hong Kong in July 2015, and licensed to ZAI Lab the exclusive rights in China (including Hong Kong and Macau) in November 2015 [33]. Olmutinib is an *EGFR* TKI for the treatment of patients with locally advanced or metastatic epidemial growth factor receptor (EGFR) T790M mutation-positive NSCLC [34, 35].

Olmutinib is an oral *EGFR* mutant-specific TKI active against mutant *EGFR* isoforms, including *T790M*, while sparing WT *EGFR*. A phase I/II trial HM-EMSI-101 (NCTO 1588145) was conducted to evaluate the safety tolerability, pharmacokinetics, and preliminary activity of olmutinib in Korean patients with *EGFR* TKI-pretreated NSCLC [36]. Patients received olmutinib at doses ranging from 75 to 1200 mg/day. The ORR was 58.8% in the 34 patients who received olmutinib with a dose >650 mg. The most common DLTs involved gastrointestinal symptoms and increased aspartate aminotransferase, alanine aminotransferase, amylase, and lipase levels. The recommended phase II dose was 800 mg/day. In part II of the study, 76 patients with centrally confirmed *T790M*+ NSCLC were enrolled, 70 of whom were evaluable for response. The ORR was 61% and median PFS (n=76) was 6.9 months [95% Cl, 5.36–9.49] [37]. The most common drug-related AEs (all grades) were diarrhea (59%), pruritus (42%), rash (41%), and nausea (39%). These data validate previous preliminary trial results presented at the European Society for Medical Oncology

Asia Congress in December 2015 [38]. These results were the basis for the Breakthrough Therapy Designation granted by the FDA in 2015 and the first approval for the treatment of patients with *EGFR T790M+* NSCLC in South Korea in 2016. Following promising early clinical data, Boehringer Ingelheim launched the ELUXA clinical trial program to investigate olmutinib as a monotherapy in different settings as well as in combination with other anticancer treatments. Nevertheless, due to an unexpected increase in grade 3/4 skin toxicity (epidermolysis) in previous trials, Boehringer decided to definitively stop the development of this drug [36–38].

2.4 NAZARTINIB (EGF816)

Nazartinib, also known as EGF816, is an orally available, irreversible, third-generation, mutant-selective *EGFR* inhibitor, with potential antineoplastic activity [39]. Nazartinib (EGF816) covalently binds to and inhibits the activity of mutant forms of *EGFR*, including the *T790M EGFR* mutant, thereby preventing *EGFR*-mediated signaling [40]. This may both induce cell death and inhibit tumor growth in *EGFR*-overexpressing tumor cells. Nazartinib (EGF816) preferentially inhibits mutated forms of *EGFR* including *T790M*, a secondarily acquired resistance mutation, and may have therapeutic benefits in tumors with *T790M*-mediated resistance when compared to other *EGFR* TKIs [41]. Nazartinib (EGF816) target profiles suggest that it represents an alternative and better therapy option against *T790M* mutations [1, 2].

Nazartinib (EGF816) (Novartis Pharmaceuticals) has preclinical activity targeting sensitizing *EGFR* mutants as well as *T790M* mutants with 60-fold selectivity over *WT EGFR* [1, 2]. EGF-816 is in phase II clinical trials by Novartis for the treatment of NSCLC in combination with nivolumab, and it is also in phase I/II clinical trials for the treatment of solid tumors [42–45].

Clinical trial phase I/II dealt with the human study of nazartinib in a patient with EGFR-mutated locally advanced or metastatic NSCLC [46]. Updated results from the phase I dose-escalation part were recently presented. Patients were assigned to receive once-daily nazartinib, with doses ranging from 75 to 350 mg. At the cutoff date of January 29, 2016, 152 patients had been treated across seven cohorts. Among them, 147 patients were evaluable for response. The confirmed ORR was 46.9% [95% CI, 38.7%–55.3%] and the DCR was 87.1% [95% CI, 80.6%–92.0%]. The estimated median PFS across all dose levels was 9.7 months [95% CI, 7.3–11.1] [47].

Among 69 patients with confirmed responses at the cutoff date the estimated median DOR was 9.5 months [95% CI, 9.2–14.7]. The most common toxicities (all grades) were rash (54%), diarrhea (37%), and pruritus (34%). Interestingly, the rashes observed in the study tended to have a different pattern, location, and histology than those seen with other *EGER* TKIs that target WT *EGFR*. Diarrhea was the most common grade 3/4 AE (16%); of note, incidences of both diarrhea and rash tended to increase along with rising nazartinib doses [46, 47]. The phase II part, performed in six cohorts, is ongoing. In addition, the drug is being investigated in association with INC280, a specific MET inhibitor (based on the potential escape pathway for third-generation *EGFR* TKIs) in an ongoing phase Ib/II trial in patients with advanced *EGFR* mutant NSCLC (NCT02335944), and with nivolumab, an anti-PD-1 monoclonal antibody in a phase II trial in *EGFR* mutant/*T790M*+ NSCLC patients who have progressed on first-line *EGFR* TKI (NCT02323126) [46, 47].

2.5 NAQUOTINIB (ASP8273)

Naquotinib (ASP8273) is another small molecule, irreversible TKI inhibitor that inhibits the kinase activity of EGFR mutations including T790M, with limited activity against WT EGFR NSCLC [1]. In the in vitro enzymatic and cell-based assays, naquotinib (ASP8273) was evaluated against EGFR mutants (L858R, exon 19 deletion, L858R/T790M, and del19/T790M) and WT EGFR [2]. Naquotinib (ASP8273) was found by mass spectrometry to covalently bind to a mutant EGFR (L858R/T790M) via cysteine residue 797 in the kinase domain of EGFR with a long-lasting inhibition of EGFR phosphorylation for 24 h. In the NSCLC cell lines harboring the above EGFR mutations, naquotinib (ASP8273) had IC_{50} values of 8–33 nM toward EGFR mutants, more potently than that of WT EGFR (IC_{50} value of 230 nM). In mouse xenograft models, naquotinib (ASP8273) induced a complete regression of the tumors after 14 days of treatment [48]. Naquotinib (ASP8273) was further shown to suppress the signaling pathway through ERK and Akt. Naquotinib (ASP8273) even showed activity in the mutant EGFR cell line, which is resistant to other EGFR TKIs including osimertinib (AZD9291) and rociletinib (CO1686) [49, 50].

Naquotinib (ASP8273) was evaluated in an open-label phase I/II study (NCT02192697) for safety and efficacy [51]. Thirty Japanese patients were enrolled in the phase I dose-escalation cohorts across seven dose levels (25–600 mg/day), and 15 patients were enrolled in the response expansion cohorts across four dose levels (100–400 mg/day). *T790M* status was 49% positive, 13% negative, and 38% unknown, respectively. Responses were observed in patients enrolled in ≥100 mg/day cohorts. Partial responses were achieved in 50% (18/36) of all evaluable patients and 80% (12/15) of patients with *T790M*+ NSCLC (including confirmed and unconfirmed) [52, 53]. The most common AEs (all grades) were diarrhea (56%), nausea (31%), vomiting (31%), and thrombocytopenia (31%). Based on tolerability and preliminary antitumor activity the recommended phase II dose selected was 300 mg once daily [52,53]. The safety, tolerability, and antitumor activity for naquotinib (ASP8273) 300 mg/day in patients with NSCLC EGFR mutation-positive and previously treated with an EGFR TKI were recently presented in a total of 63 patients, including 7 treated in the dose-escalation part, 18 in the response expansion, 19 in recommended phase II dose part, and 19 from the food effect cohort. The majority of tumors (>90%) were positive for the T790M mutation based on local testing. All but one patient (98%) had been previously treated with an EGER TKI, with erlotinib the most common inhibitor. Among the 63 patients treated with naquotinib (ASP8273) 300 mg, the ORR was 30% [95% CI, 19.2–43.0%] and the median PFS was 6.0 [95% Cl, 4.1–9.8] months. For the subgroups with T790M+ tumors the ORRs, assessed by local or central testing, were similar: 31% [95% CI, 19.5%–44.5%] and 29% [95% CI, 13.2%–48.7%], respectively. Median PFS for T790M+ patients (local testing) was 6.0 months [95% CI, 5.3–9.8] and 6.8 months [95% CI, 5.5] months to not evaluable for T790M+ patients (central testing). The most frequent drug-related AEs (all grades) were diarrhea (48%), nausea (27%), hyponatremia (19%), paresthesia (14%), and vomiting (13%). Six patients (10%) discontinued treatment due to treatment-related toxicity [52,53]. Based on this study, the dose of naquotinib (ASP8273) 300 mg daily was selected for a recently initiated phase III study (SOLAR) to compare the clinical efficacy and safety/tolerability of naquotinib (ASP8273) with erlotinib or gefitinib as an initial treatment of advanced IGFR mutant NSCLC (NCT02588261) [54].

2.6 PF-06747775 (PFIZER)

PF-06747775 is another small molecule inhibitor of EGFR T790M. This molecule is being studied in phase I/II clinical trial (NCT02349633) in advanced NSCLC patients with EGFR mutations (del19 or L858R ± T790M) [55]. The agent will be administered as continuous daily dosing in 21-day cycles. The starting dose of PF-06747775 will be 25 mg PO daily [56, 57]; results are not yet available.

2.7 AVITINIB (AC0010)

Avitinib (AC0010) is in phase I/II clinical trials by ACEA Biosciences (Hangzhou) for the treatment of nonsmall cell lung cancer (NSCLC) [58]. Avitinib, also known as AC0010 or AC0010MA, is an orally available, irreversible, EGFR mutant-selective inhibitor, with potential antineoplastic activity [58]. Upon oral administration, avitinib covalently binds to and inhibits the activity of mutant forms of EGFR, including the drug-resistant T790M EGFR mutant, which prevents signaling mediated by mutant forms of EGFR. This may both induce cell death and inhibit tumor growth in EGFR-mutated tumor cells. EGFR, a receptor tyrosine kinase that is mutated in a variety of cancers, plays a key role in tumor cell proliferation and tumor vascularization [59]. As this agent is selective toward mutant forms of EGFR, its toxicity profile may be reduced when compared to nonselective EGFR inhibitors, which also inhibit WT EGFR [59].

Parallel clinical trials were initiated in China and the United States using Avitinib as second-line therapy in NSCLC patients progressing on first-generation EGFR TKIs and who have acquired the gatekeeper T790M mutation. Two trials evaluating the safety, tolerability, pharmacokinetics and antitumor activity of Avitinib are ongoing; a phase I/II trial (NCT02274337) in advanced NSCLC patients progressing on prior therapy with an EGFR TKI agent and a phase I trial (NTO233O367) designed to determine the MTD and/or recommend ending the phase 2 dose in previously treated mutant EGFR NSCLC patients with a T790M resistant mutation [60].

2.8 BRIGATINIB (AP26113)

Brigatinib, also known as AP26113, is an orally active, potent, reversible dual ALK/*EGFR* inhibitor [61]. It is an investigational, small molecule-targeted cancer therapy in development by ARIAD Pharmaceuticals. Brigatinib has exhibited activity as a potent dual inhibitor of anaplastic lymphoma kinase (ALK) and epidermal growth factor receptor (*EGFR*) [62]. On April 28, 2017, it was granted an Accelerated Approval from the US FDA for metastatic NSCLC; as a second-line therapy for ALK-positive NSCLC. In 2016, brigatinib was granted orphan drug status by the FDA for the treatment of NSCLC [63]. Brigatinib could overcome resistance to osimertinib conferred by the EGFR C797S mutation if it is combined with an anti-EGFR antibody such as cetuximab or panitumumab [64]. Brigatinib is a dual inhibitor of ALK and mutated EGFR. ALK was first identified as a chromosomal rearrangement in anaplastic large cell lymphoma (ALCL). Genetic studies indicate that the abnormal expression of ALK is a key driver of certain types of NSCLC and neuroblastomas, as well as ALCL. As ALK is generally not expressed in normal adult tissues, it represents a highly promising molecular target for cancer therapy [61–64]. EGFR is another validated target in NSCLC. Additionally, the T790M "gatekeeper" mutation is linked to approximately 50% of patients who grow resistant to first-generation EGFR inhibitors. While second-generation EGFR inhibitors are in development, clinical efficacy has been limited due to toxicity thought to be associated with inhibiting the native (endogenous or unmutated) EGFR. The therapeutic agents, which target the T790M EGFR by sparing the WT-EGFR is another promising approach for cancer therapy [61–64].

Brigatinib has been filed by Ariad in the United States for the treatment of patients with ALK-positive locally advanced or metastatic NSCLC and who have been resistant to crizotinib [65–69].

On April 28, 2017, the US-FDA granted accelerated approval to brigatinib for the treatment of patients with metastatic ALK-positive NSCLC, who have progressed on or are intolerant to crizotinib [60].

Approval was based on a noncomparative, two-arm, open-label, multicenter clinical trial that demonstrated a clinically meaningful and durable ORR in patients with locally advanced or metastatic ALK-positive NSCLC who had progressed on crizotinib (the ALTA Trial; NCT02094573). All patients had tumors with a documented ALK rearrangement based on an FDA-approved test or a different test with adequate archival tissue to confirm ALK arrangement by the Vysis ALK Break-Apart fluorescence in situ hybridization Probe Kit test. A total of 222 patients were randomized to brigatinib orally either 90 mg once daily ($n=112$) or 180 mg once daily, following a 7-day lead-in at 90 mg once daily ($n=110$) [60].

ORR was assessed by an independent review committee according to Response Evaluation Criteria in Solid Tumors (RECIST) 1.1. ORR was 48% (95% CI: 39%, 58%) in the 90 mg arm and 53% (95% CI: 43%, 62%) in the 180 mg arm. After a median follow-up duration of 8 months, median DOR was 13.8 months in both arms. In patients with measurable brain metastases at baseline, intracranial ORR was 42% (95% CI: 23%, 63%) in the 90 mg arm ($n=26$) and 67% (95% CI: 41%, 87%) in the 180 mg arm ($n=18$). Median intracranial DOR was not estimable in the 90 mg arm but was 5.6 months in the 180 mg arm. Among patients who exhibited an intracranial response, 78% of patients in the 90 mg arm and 68% of patients in the 180 mg arm maintained an intracranial response for at least 4 months [60].

Safety was evaluated in 219 patients who received at least one dose of brigatinib in the ALTA trial. The most common adverse reactions, occurring in at least 25% of patients taking brigatinib, were nausea, diarrhea, fatigue, cough, and headache. The most common serious adverse reactions were pneumonia and ILD/pneumonitis. Fatal adverse reactions occurred in 3.7% of patients and consisted of pneumonia (two patients), sudden death, dyspnea, respiratory failure, pulmonary embolism, bacterial meningitis, and urosepsis (one patient each). Visual disturbances also occurred in patients receiving brigatinib. Adverse reactions leading to the permanent discontinuation of brigatinib occurred in 2.8% and 8.2% of patients receiving 90 and 180 mg, respectively. Patients receiving brigatinib should be monitored for new or worsening respiratory symptoms, hypertension, bradycardia, visual symptoms, and elevations in amylase, lipase, blood glucose, and creatine phosphokinase. The recommended dosing regimen of brigatinib is 90 mg orally once daily for the first 7 days; then, if tolerated, it should be increased to 180 mg orally once daily [60].

REFERENCES

[1] H. Cheng, S.K. Nair, B.W. Murray, Recent progress on third generation covalent EGFR inhibitors, Bioorg. Med. Chem. Lett. 26 (8) (2016) 1861–1868.

[2] S. Wang, S. Cang, D. Liu, Third-generation inhibitors targeting EGFR T790M mutation in advanced non-small cell lung cancer. J. Hematol. Oncol. 9 (34) (2016), https://doi.org/10.1186/s13045-016-0268-z.

[3] M.F. Baevsky, K. Lee, D. Niu, R.C. Petter, J. Singh, Heterocyclic Compounds and Uses Thereof, WO Patent 2012061299 A1, 2012.

[4] A. Thomas, S.V. Liu, D.S. Subramaniam, G. Giaccone, Lung cancer: diverse *EGFR* mutations explain AZD9291 resistance, Nat. Rev. Clin. Oncol. 12 (2015) 438.

[5] E. Avizienyte, R. Ward, A. Garner, Comparison of the EGFR resistance mutation profiles generated by EGFR-targeted tyrosine kinase inhibitors and the impact of drug combinations, Biochem. J. 415 (2) (2008) 197–206.

[6] Y.Y. Janjigian, E.F. Smit, H.J. Groen, S. Gettinger, D.R. Camidge, G.J. Riely, B. Wang, Y. Fu, V.K. Chand, V.A. Miller, W. Pao, Dual inhibition of EGFR with afatinib and cetuximab in kinase inhibitor resistant EGFR mutant lung cancer with and without T790M mutations, Cancer Discov. 4 (9) (2014) 1036–1045.

[7] C.A. Eberlein, D. Stetson, A.A. Markovets, K.J. Al-Kadhimi, Z. Lai, P.R. Fisher, C.B. Meador, P. Spitzler, E. Ichihara, S.J. Ross, M.J. Ahdesmaki, A. Ahmed, L.E. Ratcliffe, E.L. O'Brien, C.H. Barnes, H. Brown, P.D. Smith, J.R. Dry, G. Beran, K.S. Thress, B. Dougherty, W. Pao, D.A. Cross, Acquired resistance to the mutant-selective EGFR inhibitor AZD9291 is associated with increased dependence on RAS signaling in preclinical models, Cancer Res. 75 (12) (2015) 2489–2500.

[8] T. Jiang, C. Zhou, Clinical activity of the mutant-selective EGFR inhibitor AZD9291 in patients with EGFR inhibitor-resistant non-small cell lung cancer, Transl. Lung Cancer Res. 3 (6) (2014) 370–372.

[9] R. Mah, J.R. Thomas, C.M. Shafer, Drug discovery considerations in the development of covalent inhibitors, Bioorg. Med. Chem. Lett. 24 (1) (2014) 33–39.
[10] D.A. Cross, S.E. Ashton, S. Ghiorghiu, C. Eberlein, C.A. Nebhan, P.J. Spitzler, J.P. Orme, M.R.V. Finlay, R.A. Ward, M.J. Mellor, G. Hughes, A. Rahi, V.N. Jacobs, M.R. Brewer, E. Ichihara, J. Sun, H. Jin, P. Ballard, K. Al-Kadhim, R. Rowlinson, T. Klinowska, G.H.P. Richmond, M. Cantarini, D.W. Kim, M.R. Ranson, W. Pao, AZD9291, an irreversible *EGFR* TKI, overcomes *T790M*-mediated resistance to *EGFR* inhibitors in lung cancer, Cancer Discov. 4 (9) (2014) 1046–1061.
[11] D. Ayeni, K. Politi, S.B. Goldberg, Emerging agents and new mutations in EGFR-mutant lung cancer, Clin. Cancer Res. 21 (17) (2015) 3818–3820.
[12] C.S. Tan, D. Gilligan, S. Pacey, Treatment approaches for EGFR-inhibitor-resistant patients with non-small-cell lung cancer, Lancet Oncol. 16 (9) (2015) e447–e459.
[13] B. Sam, M.R.V. Finlay, A.W. Richard, V.K. Kadambar, R.C. Chandrashekar, A. Murugan, H.M. Redfearn, 2-(2,4,5-Substituted-anilino)pyrimidine derivatives as EGFR modulators useful for treating cancer, WO Patent 2013014448 A1, 2013.
[14] P.A. Janne, J.C.H. Yang, D.W. Kim, D. Planchard, Y. Ohe, S.S. Ramalingam, M.J. Ahn, S.W. Kim, W.C. Su, L. Horn, D. Haggstrom, F. Enriqueta, J.H. Kim, P. Frewer, M. Cantarini, K.H. Brown, P.A. Dickinson, S. Ghiorghiu, M. Ranson, AZD9291 in EGFR inhibitor-resistant non-small-cell lung cancer, N. Engl. J. Med. 372 (18) (2015) 1689–1699.
[15] J. Yang, S.S. Rarnalingam, P.A. Janne, M. Cantarini, T. Mitsudomi, LBA2-PR: osirnertinib (AZD9291) in pre-treated pts with T790M-positive advanced NSCLC: updated phase 1 (P1) and pooled phase 2 (P2) results, J. Thorac. Oncol. 11 (2016) S152–S153.
[16] S. Ramalingam, J.C.H. Yang, C.K. Lee, T. Kurata, D.W. Kim, LBA1-PR: osimertinib as first-line treatment for EGFR mutation-positive advanced NSCLC: updated efficacy and safety results from two phase I expansion cohorts, J. Thorac. Oncol. 11 (2016) S152.
[17] S.V. Liu, D. Subramaniam, G.C. Cyriac, F.J. Abdul-Khalek, G. Giaccone, Emerging protein kinase inhibitors for non-small cell lung cancer, Expert Opin. Emerg. Drugs 19 (1) (2013) 51–65.
[18] N.V.D. Steen, C. Caparello, C. Rolfo, P. Pauwels, G. Peters, E. Giovannetti, New developments in the management of non-small-cell lung cancer focus on rociletinib: what went wrong? Onco Targets Ther. 9 (2016) 6065–6074.
[19] A.O. Walter, R.T.T. Sjin, H.J. Haringsrna, K. Ohashi, J. Sun, K. Lee, et al., Discovery of a mutant-selective covalent inhibitor of EGFR that overcomes T790M-mediated resistance in NSCLC, Cancer Discov. 3 (2013) 1404–1415.
[20] L.V. Sequist, J.C. Soria, J. Goldman, H.A. Wakelee, S.M. Gadgeel, A. Varga, V. Papadimitrakopoulou, B.J. Solomon, G.R. Oxnard, R. Dziadziuszko, D.L. Aisner, R.C. Doebele, C. Galasso, E.B. Garon, R.S. Heist, J. Logan, J.W. Neal, M.A. Mendenhall, S. Nichols, Z. Piotrowska, A.J. Wozniak, M. Raponi, C.A. Karlovich, S. Jaw-Tsai, J. Isaacson, D. Despain, S.L. Matheny, L. Rolfe, A.R. Allen, D.R. Camidge, Rociletinib in EGFR-mutated non-small-cell lung cancer, N. Engl. J. Med. 372 (2015) 1700–1709.
[21] L.V. Sequist, J.C. Soria, D.R. Camidge, Update to rociletinib data with the RECIST confirmed response rate, N. Engl. J. Med. 374 (2016) 2296–2297.
[22] K. Park, J.Y. Han, D.W. Kim, L. Bazhenova, S.H. Ou, Y.K. Pang, H. Hin, O. Juan, J. Son, P. Janne, 190TiP: ELUXA 1: phase II study of BI 1482694 (HM61713) in patients (pts) with *T790M*-positive non-small cell lung cancer after treatment with an epidermal growth factor receptor tyrosine kinase inhibitor (*EGFR* TKI), J. Thorac. Oncol. 11 (2016) S139.
[23] K. Park, J.S. Lee, J.Y. Han, K. Lee, J.H. Kim, E. Cho, J. Cho, Y. Min, J.S. Kim, D.W. Kim, 1300: efficacy and safety of BI 1482694 (HM61713), an *EGFR* mutant-specific inhibitor, in *T790M*-positive NSCLC at the recommended phase II dose, J. Thorac. Oncol. 11 (4 Suppl) (2016) S113.
[24] A. Chi, S. Remick, W. Tse, *EGFR* inhibition in non-small cell lung cancer: current evidence and future directions, Biomark. Res. 1 (1) (2013) 2.

[25] C.S. Tan, B.C. Cho, R.A. Soo, Next-generation epidermal growth factor receptor tyrosine kinase inhibitors in epidermal growth factor receptor-mutant non-small cell lung cancer, Lung Cancer 93 (2016) 59–68.
[26] H.N. Song, K.S. Jung, K.H. Yoo, J. Cho, J.Y. Lee, S.H. Lim, H.S. Kim, J.M. Sun, S.H. Lee, J.S. Ahn, K. Park, Y.L. Choi, W. Park, M.J. Ahn, Acquired C797S mutation upon treatment with a T790M-specific third-generation EGFR inhibitor (HM61713) in non-small cell lung cancer, J. Thorac. Oncol. 11 (4) (2016) e45–e47.
[27] R. Pirker, Third-generation epidermal growth factor receptor tyrosine kinase inhibitors in advanced nonsmall cell lung cancer, Curr. Opin. Oncol. 28 (2) (2016) 115–121.
[28] S.H. Ou, R.A. Soo, Dacomitinib in lung cancer: a "lost generation" EGFR tyrosine-kinase inhibitor from a bygone era? Drug Des. Dev. Ther. 9 (2015) 5641–5653.
[29] A. Tartarone, R. Lerose, Clinical approaches to treat patients with non-small cell lung cancer and epidermal growth factor receptor tyrosine kinase inhibitor acquired resistance, Ther. Adv. Respir. Dis. 9 (2015) 242–250.
[30] B.C. Liao, C.C. Lin, J.C. Yang, Second and third-generation epidermal growth factor receptor tyrosine kinase inhibitors in advanced nonsmall cell lung cancer, Curr. Opin. Oncol. 27 (2) (2015) 94–101.
[31] C.E. Steuer, F.R. Khuri, S. Ramalingam, The next generation of epidermal growth factor receptor tyrosine kinase inhibitors in the treatment of lung cancer, Cancer 121 (8) (2015) E1–E6.
[32] S. Peters, S. Zimmermann, A.A. Adjei, Oral epidermal growth factor receptor tyrosine kinase inhibitors for the treatment of non-small cell lung cancer: comparative pharmacokinetics and drug-drug interactions, Cancer Treat. Rev. 40 (8) (2014) 917–926.
[33] B.C. Liao, C.C. Lin, J.H. Lee, J.C. Yang, Update on recent preclinical and clinical studies of T790M mutant-specific irreversible epidermal growth factor receptor tyrosine kinase inhibitors, J. Biomed. Sci. 23 (1) (2016) 86.
[34] A. Passaro, E. Guerini-Rocco, A. Pochesci, D. Vacirca, G. Spitaleri, C.M. Catania, A. Rappa, M. Barberis, F. de Marinis, Targeting EGFR T790M mutation in NSCLC: from biology to evaluation and treatment, Pharmacol. Res. 117 (2017) 406–415.
[35] M.Y. Cha, S.J. Kang, M.R. Kim, J.Y. Lee, J.Y. Jeon, M.G. Jo, E.J. Kwak, K.O. Lee, T.H. Ha, K.H. Suh, M. Kim, Novel fused pyrimidine derivatives for inhibition of tyrosine kinase activity, WO Patent 2011162515 A2, 2011.
[36] K.O. Lee, M. Cha, M. Kim, J. Song, J.H. Lee, Y. Kim, Discovery of HM61713 as an orally available and mutant EGFR selective inhibitor, Clin. Cancer Res. 7 (2014) LB-100.
[37] K. Park, J.S. Lee, K.H. Lee, J.H. Kim, Y.J. Mm, J.Y. Cho, Updated safety and efficacy results from phase I/II study of HM61713 in patients (pts) with EGFR mutation positive non-small cell lung cancer (NSCLC) who failed previous EGFR-tyrosine kinase inhibitor (TKI), J. Clin. Oncol. 33 (2015). abstract 8084.
[38] K. Park, J.S. Lee, K. Lee, J.I. Kim, B.C. Cho, Y.J. Mm, Olmutinib (131 1482694, HM61713), an EGFR mutant-specific inhibitor, in T790M+ NSCLC: efficacy and safety at the RP2D, J. Clin. Oncol. 34 (2016). abstract 9055.
[39] Y. Jia, J. Juarez, J. Li, M. Manuia, M.J. Niederst, C. Tompkins, N. Timple, M.T. Vaillancourt, A.C. Pferdekamper, E.L. Lockerman, C. Li, J. Anderson, C. Costa, D. Liao, E. Murphy, M. DiDonato, B. Bursulaya, G. Lelais, J. Barretina, M. McNeill, R. Epple, T.H. Marsilje, N. Pathan, J.A. Engelman, P.Y. Michellys, P. McNamara, J. Harris, S. Bender, S. Kasibhatla, EGF816 exerts anticancer effects in non-small cell lung cancer by irreversibly and selectively targeting primary and acquired activating mutations in the EGF receptor, Cancer Res. 76 (6) (2016) 1591–1602.
[40] J. Singh, E. Evans, M. Hagel, M. Labinski, A. Dubrovskiy, M. Nacht, R.C. Petter, A. Prasad, M. Sheets, T. St Martin, R. Tjin Tham Sjin, W. Westlin, Z. Zhu, Superiority of a novel *EGFR* targeted covalent inhibitor over its reversible counterpart in overcoming drug resistance, Med. Chem. Commun. 3 (2012) 780.

[41] S.G. Wu, Y.N. Liu, M.F. Tsai, Y.L. Chang, C.J. Yu, P.C. Yang, The mechanism of acquired resistance to irreversible EGFR tyrosine kinase inhibitor-afatinib in lung adenocarcinoma patients, Oncotarget 7 (11) (2016) 12404–12413.

[42] F.Y. Niu, Y.L. Wu, Novel agents and strategies for overcoming EGFR TKIs resistance, Exp. Hematol. Oncol. 3 (1) (2014) 2.

[43] J.M. Sun, K. Park, Can we define the optimal sequence of epidermal growth factor receptor tyrosine kinase inhibitors for the treatment of epidermal growth factor receptor-mutant nonsmall cell lung cancer? Curr. Opin. Oncol. 29 (2017) 89–96.

[44] Y. Kobayashi, T. Mitsudomi, Not all epidermal growth factor receptor mutations in lung cancer are created equal: perspectives for individualized treatment strategy, Cancer Sci. 107 (9) (2016) 1179–1186.

[45] W. Ji, C.M. Choi, J.K. Rho, S.J. Jang, Y.S. Park, S.M. Chun, W.S. Kim, J.S. Lee, S.W. Kim, D.H. Lee, J.C. Lee, Mechanisms of acquired resistance to EGFR-tyrosine kinase inhibitor in Korean patients with lung cancer, BMC Cancer 13 (2013) 606.

[46] G. Lelais, R. Epple, T.H. Marsilje, Y.O. Long, M. McNeill, B. Chen, W. Lu, J. Anumolu, S. Badiger, B. Bursulaya, M. DiDonato, R. Fong, J. Juarez, J. Li, M. Manuia, D.E. Mason, P. Gordon, T. Groessl, K. Johnson, Y. Jia, S. Kasibhatla, C. Li, J. Isbell, G. Spraggon, S. Bender, P.Y. Michellys, Discovery of EGF816, a novel, potent, and WT sparing covalent inhibitor of oncogenic (L858R, ex19del) and resistant (T790M) EGFR mutants for the treatment of EGFR mutant non-small-cell lung cancers, J. Med. Chem. 59 (2016) 6671–6689.

[47] D.S.W. Tan, J.C.H. Yang, N.B. Leigh, G. Riely, I.N. Sequist, E. Felip, Updated results of a phase 1 study of EGF8I6, a third-generations mutant-selective EGFR tyrosine kinase inhibitor (TKI) in advanced non-small cell lung cancer (NSCLC) harboring T790M, J. Clin. Oncol. 34 (2016). abstract 9044.

[48] H. Sakagami, S. Konagai, H. Yamamoto, H. Tanaka, T. Matsuya, M. Mon, Abstract 1728: ASP8273, a novel mutant-selective irreversible *EGFR* inhibitor, inhibits growth of non-small cell lung cancer (NSCLC) cells with *EGFR* activating and *T790M* resistance mutations, Cancer Res. 74 (2014) 1728.

[49] Y. Goto, H. Nokihara, H. Murakami, T. Shimizu, T. Seto, A.P. Krivoshik, A.T. Keating, K. Uegaki, K. Takeda, K. Komatsu, S. Morita, M. Fukuoka, K. Nakagawa, in: ASP8273, a mutant-selective irreversible EGFR inhibitor in patients (pts) with NSCLC harboring EGFR activating mutations: preliminary results of first-in-human phase I study in Japan, ASCO Annual Meeting, 2015. abstract #8014.

[50] H.A. Yu, G.R. Oxnard, A.I. Spira, L. Horn, J. Weiss, Y. Feng, H.J. West, G. Giaccone, T.L. Evans, R.J. Kelly, T. Fleege, S. Poondru, F. Jie, K. Aoyama, M.A. Foley, D. Whitcom, A.T. Keating, A.P. Krivoshik, in: Phase I dose escalation study of ASP8273, a mutant-selective irreversible EGFR inhibitor, in subjects with EGFR mutation positive NSCLC, ASCO Annual Meeting, 2015. abstract #8083.

[51] H. Sakagami, S. Konagai, H. Yamamoto, H. Tanaka, T. Matsuya, M. Mon, ASP8273, a novel mutant-selective irreversible EGFR inhibitor, inhibits growth of non-small cell lung cancer (NSCLC) cells with EGFR activating and T790M resistance mutations, Cancer Res. 74 (2014). abstract 1728.

[52] S. Konagal, H. Sakagami, H. Yamamoto, H. Tanaka, T. Matsuya, S. Mimasu, ASP8273 selectively inhibits mutant EGFR signal pathway and induces tumor shrinkage in EGFR mutated tumor models, Cancer Res. 75 (2015). abstract 2586.

[53] Y. Goto, H. Nokihara, H. Murakami, T. Shimizu, T. Seto, A. Krivoshik, ASP8273, a mutant-selective irreversible EGFR inhibitor in patients (pts) with NSCLC harboring EGFR activating mutations: preliminary results of first-in-human phase I study in Japan, J. Clin. Oncol. 33 (2015). abstract 8014.

[54] H.A. Yu, A. Spira, L. Horn, J. Weiss, H. Jack, G. Giaccone, Anti-ttimor activity of ASP8273 300 mg in subjects with EGIR mutation-positive non-small cell lung cancer: interim results from an ongoing phase I study, J. Clin. Oncol. 34 (2016). abstract 9050.

[55] V.M. Rivera, F. Wang, R. Anjum, S. Zhang, R. Squillace, J. Keats, D. Miller, Y. Ning, S.D. Wardwell, L. Moran, J. Miret, N. Narasimhan, D. Dalgarno, T. Clackson, W.C. Shakepeare, AP26113 is *a* dual ALK/EGFR inhibitor: characterization against EGFR T790 M in cell and mouse models of NSCLC, Cancer Res. 72 (2014). abstract #1794.

[56] D.C. Behenna, H. Cheng, S.C. Schultz, T.O. Johnson Jr, J.C. Kath, A. Nagata, S.K. Nair, S.P. Planken, Purine derivatives, US Patent 20150141402 A1, 2015.

[57] http://apisynthesisint.blogspot.com.

[58] X. Xu, Parallel phase 1 clinical trials in the US and in China: accelerating the test of Avitinib in lung cancer as a novel inhibitor selectively targeting mutated *EGFR* and overcoming *T790M*-induced resistance, Chin. J. Cancer 34 (7) (2015) 285–287.

[59] X. Xu, L. Mao, W. Xu, W. Tang, X. Zhang, B. Xi, R. Xu, X. Fang, J. Liu, C. Fang, L. Zhao, X. Wang, J. Jiang, P. Hu, H. Zhao, L. Zhang, AC0010, an irreversible EGFR inhibitor selectively targeting mutated EGFR and overcoming T790M-induced resistance in animal models and lung cancer patients, Mol. Cancer Ther. 15 (11) (2016) 2586–2597.

[60] I. Sulivan, D. Planchard, Next generation EGFR tyrosine kinase inhibitors for treating EGFR-mutant lung cancer beyond first line, Front. Med. 3 (2017) 1–13.

[61] A. Rossi, P. Maione, P.C. Sacco, A. Sgambato, F. Casaluce, M.L. Ferrara, G. Palazzolo, F. Ciardiello, C. Gridelli, ALK inhibitors and advanced non-small cell lung cancer, Int. J. Oncol. 45 (2) (2014) 499–508.

[62] I. Okamoto, E. Iwama, T. Harada, K. Takayama, Y. Nakanishi, Development of anaplastic lymphoma kinase (ALK) inhibitors and molecular diagnosis in ALK rearrangement-positive lung cancer, Onco Targets Ther. 7 (2014) 375–385.

[63] C.A. Perez, M. Velez, L.E. Raez, E.S. Santos, Overcoming the resistance to crizotinib in patients with non-small cell lung cancer harboring EML4/ALK translocation, Lung Cancer 84 (2) (2014) 110–115.

[64] B. Solomon, K.D. Wilner, A.T. Shaw, Current status of targeted therapy for anaplastic lymphoma kinase-rearranged non-small cell lung cancer, Clin. Pharmacol. Ther. 95 (1) (2013) 15–23.

[65] H.A. Yu, G.J. Riely, Second-generation epidermal growth factor receptor tyrosine kinase inhibitors in lung cancers, J. Natl. Compr. Cancer Netw. 11 (2) (2013) 161–169.

[66] M. Ceccon, L. Mologni, W. Bisson, L. Scapozza, C. Gambacorti-Passerini, Crizotinib-resistant NPM-ALK mutants confer differential sensitivity to unrelated Alk inhibitors, Mol. Cancer Res. 11 (2) (2012) 122–132.

[67] R. Katayama, T.M. Khan, C. Benes, E. Lifshits, H. Ebi, V.M. Rivera, W.C. Shakespeare, A.J. Iafrate, J.A. Engelman, A. Shaw, Therapeutic strategies to overcome crizotinib resistance in non-small cell lung cancers harboring the fusion oncogene EML4-ALK, Proc. Natl. Acad. Sci. 108 (18) (2011) 7535–7540.

[68] W. Huang, S. Liu, D. Zou, M. Thomas, Y. Wang, T. Zhou, J. Romero, A. Kohlmann, F. Li, J. Qi, L. Cai, T.A. Dwight, Y. Xu, R. Xu, R. Dodd, A. Toms, L. Parillon, X. Lu, R. Anjum, S. Zhang, F. Wang, J. Keats, S.D. Wardwell, Y. Ning, Q. Xu, L.E. Moran, Q.K. Mohemmad, H.G. Jang, T. Clackson, N.I. Narasimhan, V.M. Rivera, X. Zhu, D. Dalgarno, W.C. Shakespeare, Discovery of Brigatinib (AP26113), a phosphine oxide-containing, potent, orally active inhibitor of anaplastic lymphoma kinase, J. Med. Chem. 59 (10) (2016) 4948–4964.

[69] L.W. Rozamus, P. Sharma, Crystalline forms of 5-chloro-N4-[-2-(dimethylphosphoryl) phenyl]-N2-{2-methoxy-4-[4-(4-methylpiperazin-1-yl)piperidin-1-yl]pyrimidine-2,4-diamine, WO Patent 2016065028 A1, 2016.

CHAPTER 3

Current Approaches in the Development of Covalent Irreversible EGFR Inhibitors

The first-generation, ATP-competitive, and reversible EGFR inhibitors such as gefitinib and erlotinib are effective for patients harboring a short in-frame deletion of exon 19 (del E746_A750) and the L858R point mutation in exon 21 in the EGFR domain [1]. Unfortunately the secondary mutation in the catalytic domain of EGFR exon 20 (T790M) causes about 50% of NSCLC cases to develop resistance after repeated treatments by the first-generation EGFR inhibitors [2]. EGFR T790M mutation restores the affinity for ATP similar to that of wild-type (WT) EGFR and prevents the reversible inhibitors from binding at higher ATP concentrations.

Second-generation EGFR inhibitors, which include afatinib (BIBW2992) and dacomitinib (PF00299804), are irreversible EGFR inhibitors that block to Cys797 and were proven in preclinical trials to successfully restrain EGFR with activating mutations (Exon 19 deletion or L858R)) as well as those with T790M resistance mutation [3]; however, their action in patients with erlotinib-resistant cancers harboring T790M has been insignificant. The conflict among laboratory and clinical outcomes is probably because of a poor therapeutic window [4]. These drugs are equally intense against WT EGFR and EGFR T790M, and in this manner the toxicity attributable to inhibiting WT EGFR (rash and diarrhea) precludes the use of doses that could be successfully stifle T790M. More recently, third-generation EGFR TKIs consisting of WZ4002, CO-1686, Osimertinib (AZD9291), and EGF816 were evolved to target mutant EGFR harboring T790M [5].

This elegance of inhibitor additionally ties covalently to Cys797 and to a great extent spares WT EGFR, accordingly diminishing toxicity and allowing the utilization of doses that completely smother T790M [3, 4]. This extensive therapeutic window in all likelihood underlies the more prominent than 50% response rates observed in a clinical study with CO-1686 and AZD9291 in erlotinib-resistant, T790M-positive NSCLC. On March 30,

Third-Generation EGFR Inhibitors
https://doi.org/10.1016/B978-0-08-102661-8.00003-2

2017 the USFDA conceded regular approval to Osimertinib (TAGRISSO, AstraZeneca Pharmaceuticals) for the treatment of sufferers with metastatic EGFR T790M mutation-positive NSCLC. The current chapter focuses on recently reported covalent EGFR irreversible inhibitors [1–5].

Xu et al. reported the design, synthesis, and biological evaluation of 2-*oxo*-3,4-dihydropyrimido[4,5-*d*]pyrimidinyl derivatives as irreversible *EGFR* inhibitors with improved pharmacokinetic properties [6]. The synthesis of compound **11(a–v)** is shown in Scheme 3.1. Commercially available

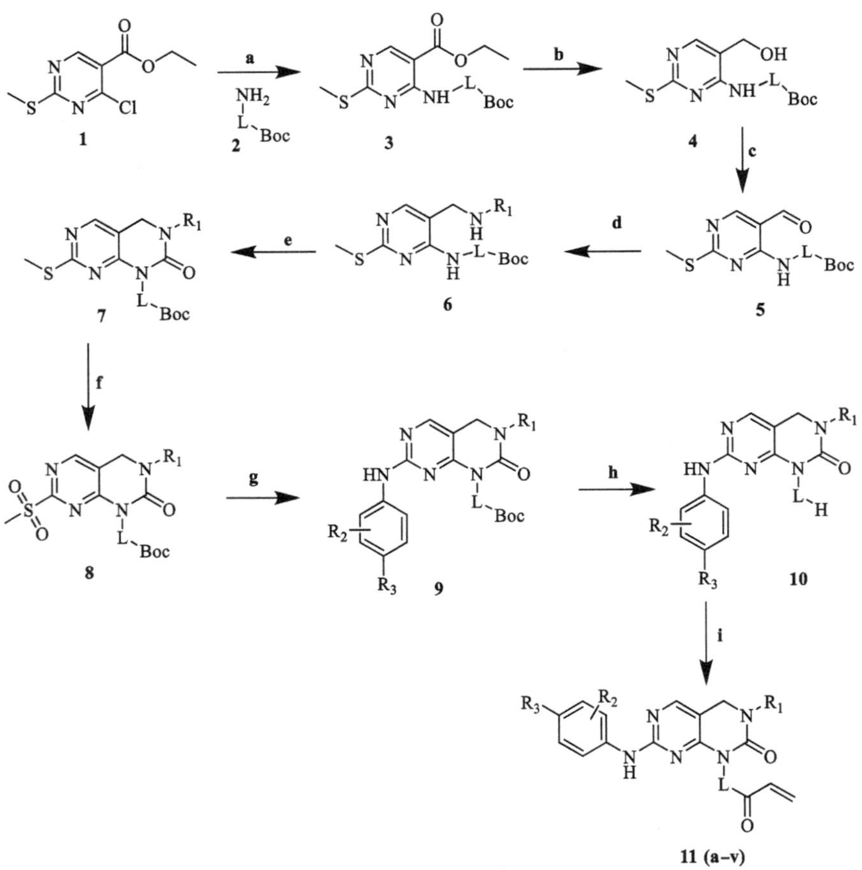

Scheme 3.1 Reagent and conditions: (a) K_2CO_3 (2 equiv.), DMF, 60°C, 85%–97%; (b) $LiAlH_4$ (2 equiv.), THF, −40°C to 0°C, 40%–58%; (c) MnO_2, CH_2Cl_2, rt, 84%–93%; (d) R_1NH_2, AcOH, $NaBH_4$, MeOH, 0°C to rt, 72%–90%; (e) triphosgene (0.34 equiv.), Et_3N (3.0 equiv.), CH_2Cl_2, 0°C to rt, 88%–96%; (f) 3-chloroperbenzoic acid (3.0 equiv.), CH_2Cl_2, rt, 80%–92%; (g) RNH_2, (1.2 equiv.), trifluoroacetic acid (1.2 equiv.), 2-BuOH, 110°C; (h) trifluoroacetic acid (10 equiv.), CH_2Cl_2, rt, 38%–60% (for two steps); and (i) acryloyl chloride (1.2 equiv.), DIPEA (2 equiv.), CH_2Cl_2, 0°C to rt, 75%–90%.

ethyl-4-chloro-2-(methylthio)-pyrimidine-5-carboxylate (**1**) was reacted with the protected aliphatic amine **2** to produce the adduct **3** in high yields. The reduction of **3** with lithium aluminum hydride gave alcohol **4**, which was subsequently oxidized with MnO_2 to obtain the aldehyde **5**. Reductive amination of aldehyde **5** with diverse amines yielded the corresponding compound **6**. Compound **8** was prepared by direct cyclization using triphosgene and subsequent oxidation with m-CPBA. Nucleophilic coupling of intermediates **8** with different substituted amines followed by a treatment with trifluoroacetic acid afforded the key intermediates **10**, which react with acryloyl chloride to produce the final products **11(a–v)**.

Structural optimization aimed at improving the aqueous solubility and pharmacokinetic properties of a series of irreversible *EGFR* inhibitors results in the most promising compounds, **11v**, having an aqueous solubility of 1265 μg/mL, which is some 300-fold greater than that of the original lead compound. Compound **11v** potently inhibits the enzymatic function of *EGFR L858R/T790M* with an IC_{50} value of 9.2 nM and suppresses the proliferation of gefitinib-resistant H1975 human NSCLC cells with an IC_{50} value of 47 nM. The compound also exhibits good pharmacokinetic properties with an oral bioavailability of 30%. Furthermore, it displayed promising antitumor efficacy in vivo in an EGFR L858R/T790M-driven human H1975 xenograft mouse model without obvious toxic signs during the treatment, as shown in Chart 3.1.

Zhang et al. designed and synthesized a novel series of 4-arylamino-6/7-substituted-5,6,7,8-tetrahydropyrido[4,3-*d*]pyrimidines and evaluated their biological activities as the potential antiproliferative agents and *EGFR* kinase inhibitors [7].

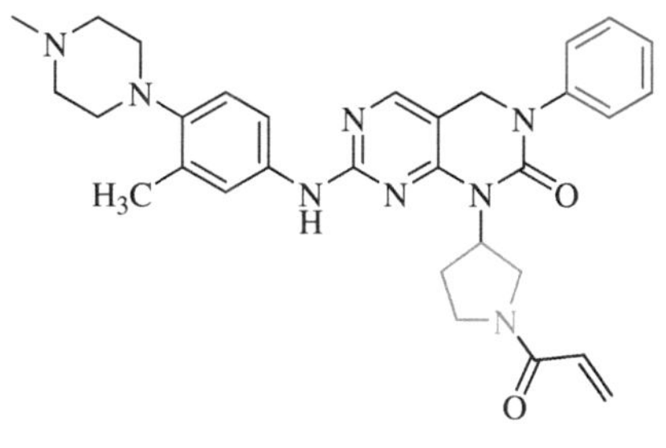

Compound 11v
Aqueous solubility = 1265 μg/mL
EGFR L858R/T790M IC_{50} = 9.2 nM
H1975 IC_{50} = 47 nM

Chart 3.1 Aqueous solubility and *EGFR T790M* inhibitory data of compound **11v**.

The designed 4-arylamino-6-substituted-5,6,7,8-tetrahydropyrido[4,3-*d*]pyrimidines (series I and II) were synthesized as shown in Schemes 3.2 and 3.3. It was critical to introduce a robust N-protecting group in tetrahydropyrido[4,3-*d*] pyrimidine (THPP) core to avoid formation of by-products occurring at arylamine in subsequent deprotection (especially for arylamines with Cl, Br, or NO_2 groups). Ethyl-1-benzyl-4-oxopiperidine-3-carboxylate (**12**) was therefore subjected to catalytic hydrogenation to give the intermediate **13**, which was protected again with CbzCl to yield intermediate **14**.

Cyclization with formamidine acetate generated **15** efficiently. After the chlorination with phosphorus oxychloride to get **16**, the optimal base and solvent were explored and identified to facilitate nucleophilic substitution with different anilines and afford the compounds **17a–q** in satisfactory yields. Deprotection in HBr and acetic acid yielded the key precursors **18a–q** cleanly.

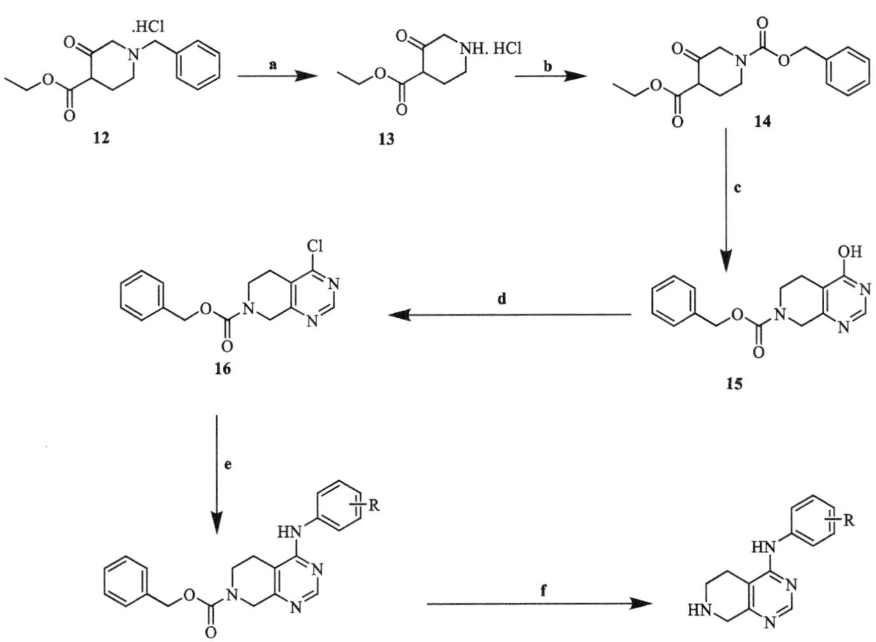

12 13 14 15 16 17a–q 18a–q

Scheme 3.2 Reagents and conditions: (a) H_2, Pd/C, 5 atm, HCl, ethanol, 40°C, 8 h; (b) CbzCl, DIPEA, DCM, rt, 2 h; (c) formamidine acetate, ethanol, $NaOCH_3$, 6 h, reflux; (d) $POCl_3$, DIPEA, CH_3CN, reflux, 6 h; (e) NH_2ArR_1, NaHMDS, THF, 0°C, 2 h; and (f) HBr, acetic acid, 50°C, 2 h.

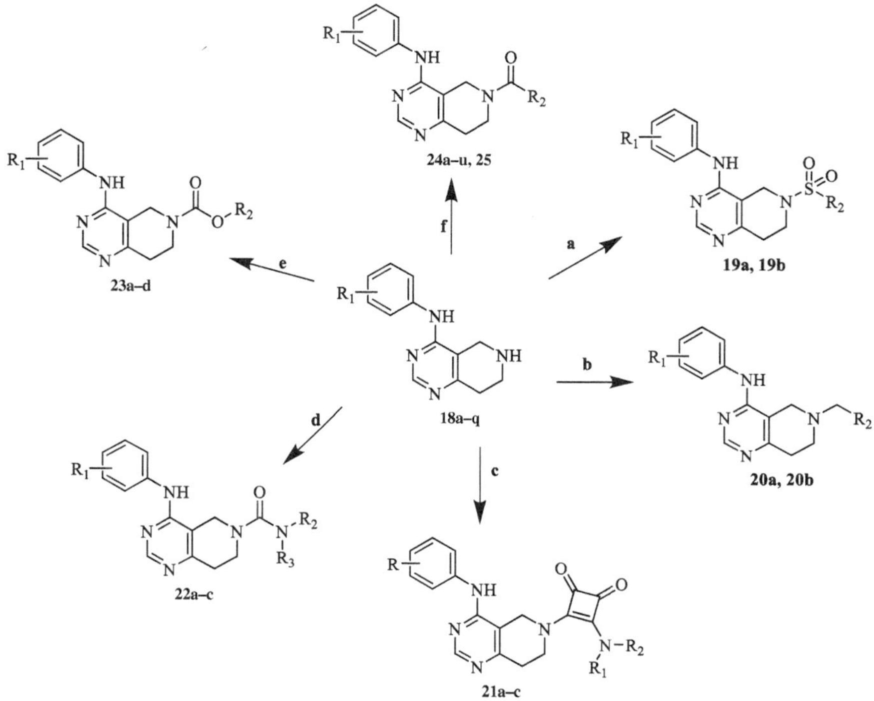

Scheme 3.3 Reagents and conditions: (a) ClO_2SR_2, DIPEA, DCM, 0°C, 2 h; (b) $ClCH_2R_2$, DIPEA, DCM, 0°C, 1 h or CH_2CHR_2, ethanol, 60°C, 8 h; (c) (i) diethyl squarate, ethanol, 60°C, 4 h and (ii) NR_1R_2, ethanol, 60°C, 4 h; (d) (i) CDI, ethanol, reflux, 4 h and (ii) NR_1R_2, ethanol, reflux, 12 h; (e) $ClOCOR_2$, DIPEA, CH_3CN, 0°C, 1 h; and (f) $ClOCR_2$, DIPEA, DCM, 0°C, 1 h or HO_2CR_2 EDCI, HOBT, DIPEA, CH_3CN, 60°C, 6 h.

The title compounds **19–25** were obtained in various conditions as depicted in Scheme 3.3. Compound **18a** reacted with alkylsulfury chloride to obtain sulfonamides **19a** and **19b**, underwent Michael addition or alkylation to get amines **20a–b**, reacted with diethyl squarate and subsequently amines to afford compounds **21a–c**, reacted with CDI and amines to prepare ureas **22a–c**, or coupled with chlorocarbonic ester or acyl chloride/aliphatic acid to afford carbamates **23a–d** or amides **24a–u** and **25**, respectively. On the other hand, in order to investigate the influence of the nitrogen atom at different positions in the THPP core, the 4-arylamino-7-alkyl substituted-5,6,7,8-tetrahydropyrido[3,4-*d*]pyrimidine compounds **32–35** and **36a–c** were synthesized by adopting similar methods as for 6-substituted compounds taking ethyl 1-benzyl-3-oxopiperidine-4-carboxylate as the starting material (Schemes 3.4 and 3.5).

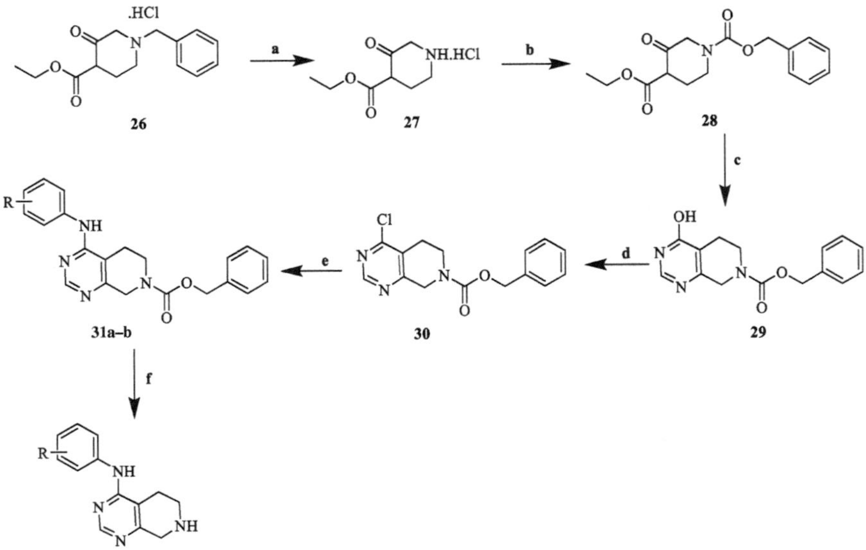

26 27 28 29 30 31a–b 32a–b

Scheme 3.4 Reagents and conditions: (a) H_2, Pd/C, 5atm, ethanol, 40°C, 8h; (b) CbzCl, DIPEA, DCM, rt, 2h; (c) formamidine acetate, ethanol, $NaOCH_3$, 6h, reflux; (d) $POCl_3$, DIPEA, DCE, reflux, 6h; (e) NH_2ArR_1, NaHMDS, THF, 0°C, 2h; and (f) HBr, acetic acid, 50°C, 2h.

Biological activity indicates that *N*-acrylamide fragment in THPPs and 4-aniline groups with substituents played key roles for the significant antiproliferative activities against four cancer cell lines (HT29, A549, H460, and H1975). The inhibitory activity of Gefitinib-resistant H1975 were especially shown as more favorable, which could be easily observed from compounds **24b**, **24c**, **24n**, **24o**, **24p**, **24r**, **24s**, **24u**, and **36c** ($IC_{50} \leq 8\,\mu M$) (Chart 3.2).

In an evaluation of inhibiting *EGFR* and HER2 kinases, seven compounds (**24b**, **24g**, **24n**, **24o**, **24p**, **24r**, and **24s**) showed stronger *EGFR* potency with $IC_{50} \leq 18\,nM$ (Chart 3.2), which could also be mechanistically understood by the preliminary docking study of **24b** with *EGFR* kinase. On the other hand, bisarylaniline derivatives (**24o**, **24p**, **24r**, and **24s**) showed obvious improvements in HER2 inhibition activities, which indicated their potential as EGFR/HER2 dual kinase inhibitors (Chart 3.2).

Moreover the carbon atom of the Michael acceptor was situated 3.73Å away from the sulfhydroyl group of Cys797, which was suitable for the formation of the covalent bond. The three key interactions might be able to explain the corresponding biological activities of the designed 4-arylamino-6-substituted-THPPs as irreversible inhibitors of *EGFR*.

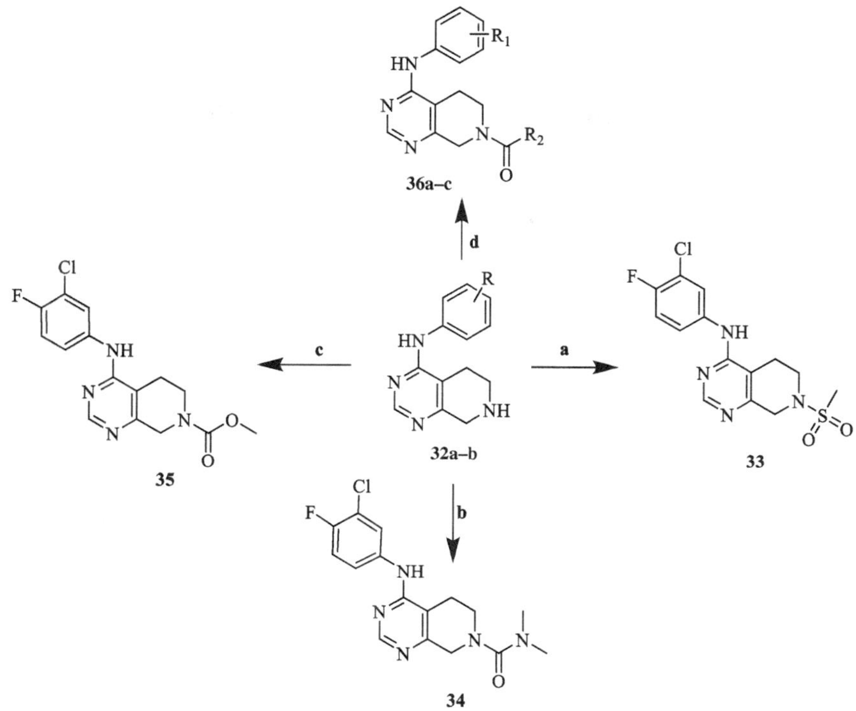

Scheme 3.5 Reagents and conditions: (a) ClO_2SCH_3, DIPEA, DCM, 0°C, 2h; (b) (i) CDI, ethanol, reflux, 4h and (ii) $N(CH_3)_2$, ethanol, reflux, 12h; (c) $ClOCOCH_3$, DIPEA, CH_3CN, 0°C, 1h; and (d) $ClOCR_2$, DIPEA, DCM, 0°C, 1h or HO_2CR_2, EDCI, HOBT, DIPEA, CH_3CN, 60°C, 6h.

Han et al. to search for potent nitric oxide (NO)-donating (*EGFR*) inhibitors, a series of phenylsulfonylfuroxan-based anilinopyrimidines **46a–h** were synthesized and biologically evaluated [8]. Compounds **46a–h** were synthesized as depicted in Scheme 3.6. Substituted phenol **40a** was prepared from hydroxycinnamic acid **37** by esterification and subsequent hydrogenation, while phenols **40b–d** were commercially available. The phenols **40a–d** were regioselectively coupled to the 4-position of 2,4,5-trichloropyrimidine to afford aryl ethers **41a–d**. On the other hand, cyclic amines **42a–c** were treated with 5-fluoro-2-nitrophenyl methyl ether to furnish nitro arylamines **43a–c**, which underwent reduction to generate corresponding arylamines **44a–c**. The coupling of **41a–d** with **44a–c** in the presence of trifluoroacetic acid under reflux offered anilinopyrimidines **45a–h**, which were condensed with diphenylsulfonylfuroxan prepared via a three-step reaction sequence as described previously to provide monophenylsulfonylfuroxans **46a–h**.

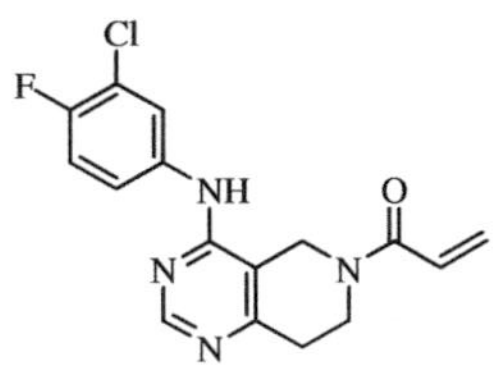

24b

H1975 = 6.32 μM/L
EGFR = 12 nM

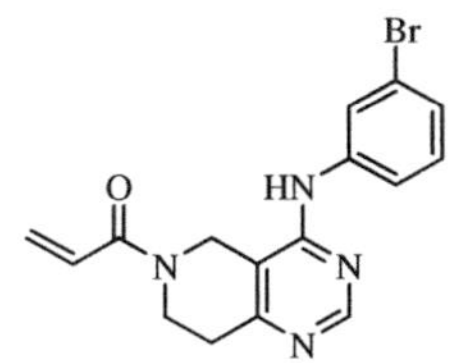

24g

H1975 = 15.32 μM/L
EGFR = 09 nM

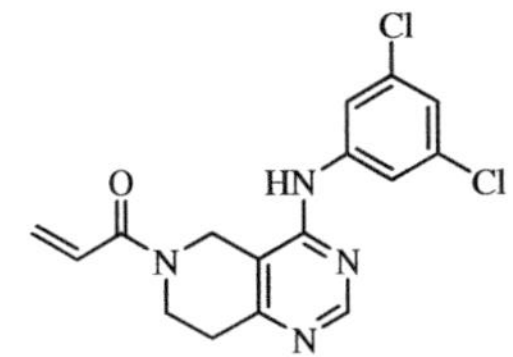

24n

H1975 = 7.00 μM/L
EGFR = 06 nM

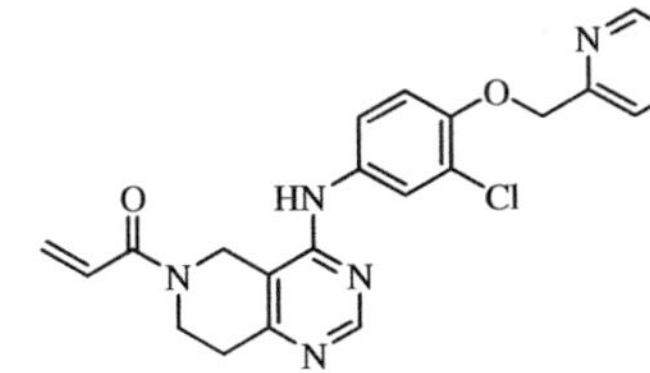

24o

H1975 = 2.79 μM/L
EGFR = 08 nM

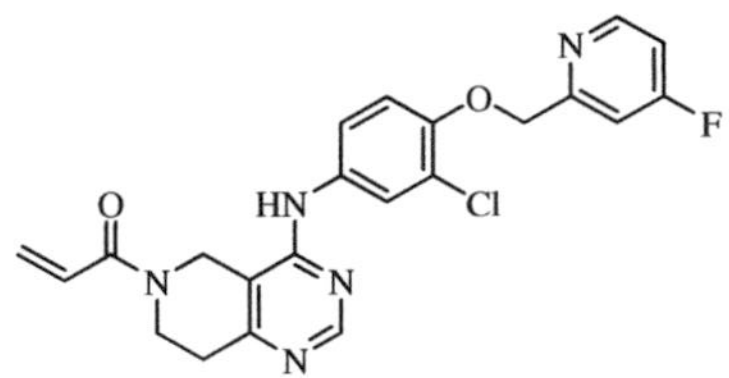

24p

H1975 = 3.34 μM/L
EGFR = 09 nM

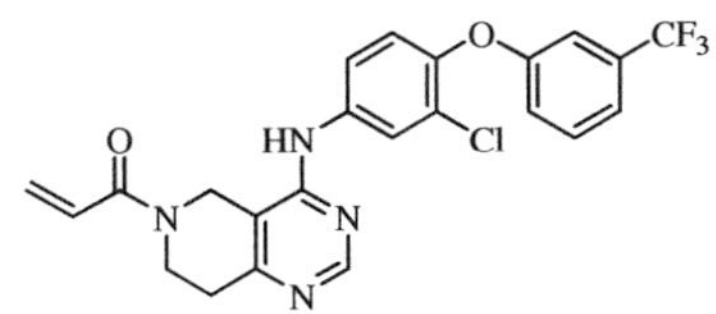

24r

H1975 = 6.93 μM/L
EGFR = 18 nM

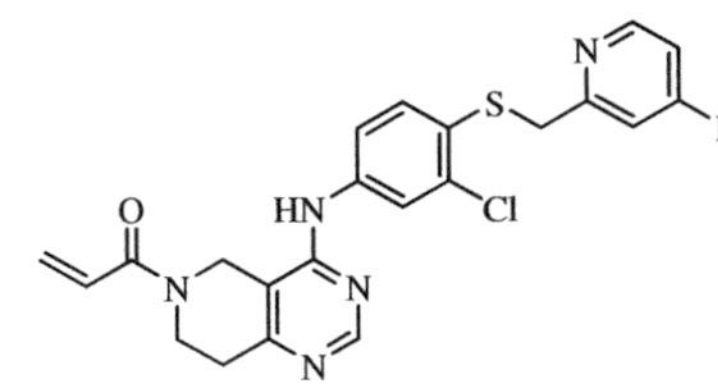

24s

H1975 = 3.96 μM/L
EGFR = 08 nM

Chart 3.2 Antiproliferative and in vitro *EGFR* inhibitory activity of potent compound.

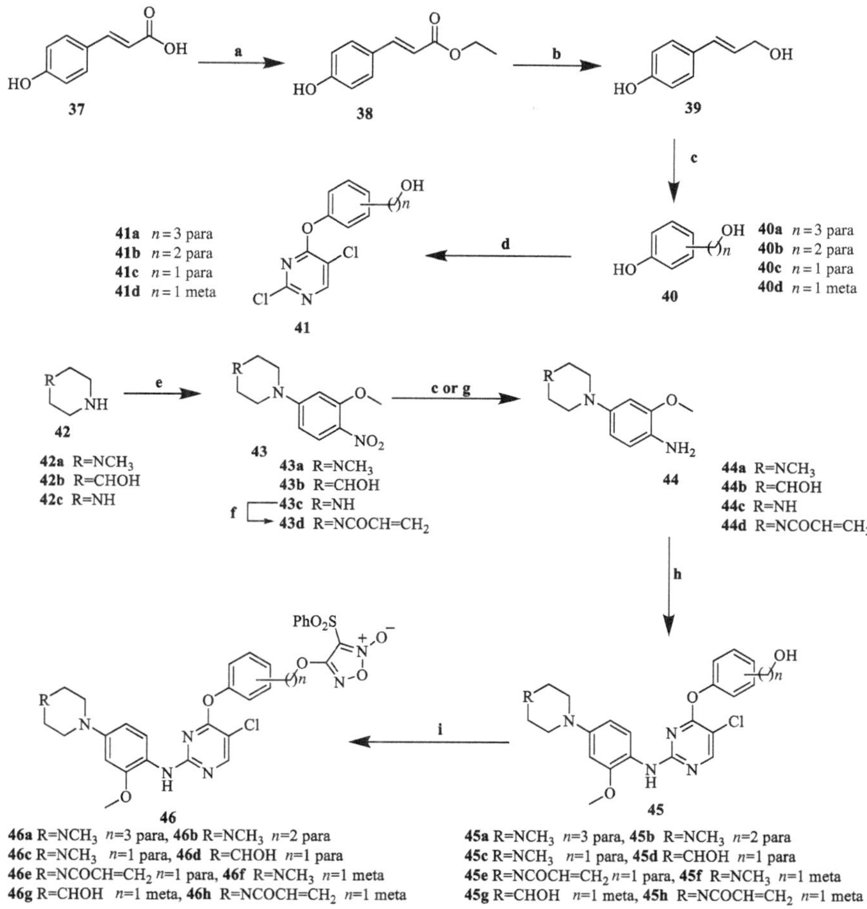

Scheme 3.6 Reagents and conditions: (a) EtOH, H_2SO_4, reflux, 4 h; (b) $LiAlH_4$, BnCl, dry THF, rt, 4 h; (c) H_2, Pd/C, MeOH, rt, overnight, (for **40a**, **44a**, and **44b**); (d) 2,4,5-trichloropyrimidine, K_2CO_3, DMF, rt, 2 h; (e) 5-fluoro-2-nitrophenyl methyl ether, K_2CO_3, DMSO, rt, overnight (for **43a** and **43b**); dioxane, 120°C, 2 h (for **43c**); (f) acrylic acid, DCC/DMAP, anhydrous CH_2Cl_2, rt, overnight; (g) iron powder, NH_4Cl, THF/H_2O (1/1, v/v), reflux, 3 h, (for **44c**); (h) **41a–d**, TFA, 2-BuOH, reflux, 3 h; and (i) 3,4-bis(phenylsulfonyl)-1,2,5-oxadiazole-2-oxide, NaH, anhydrous THF, rt, 0.5 h.

As shown in Chart 3.3, compound **46h** released much higher levels of NO in H1975 cells than in 16HBE cells, and the antiproliferative activity of **46h** in H1975 was diminished by an NO scavenger in a dose-dependent manner, suggesting that the NO released by this compound may play an important role in the antiproliferation of H1975 cells. Furthermore the *EGFR* kinase inhibitory and antiproliferative activities of **46h** could be attributed to the synergic effect of high levels of NO production and the inhibition of *EGFR* and related signaling in the cancer cells.

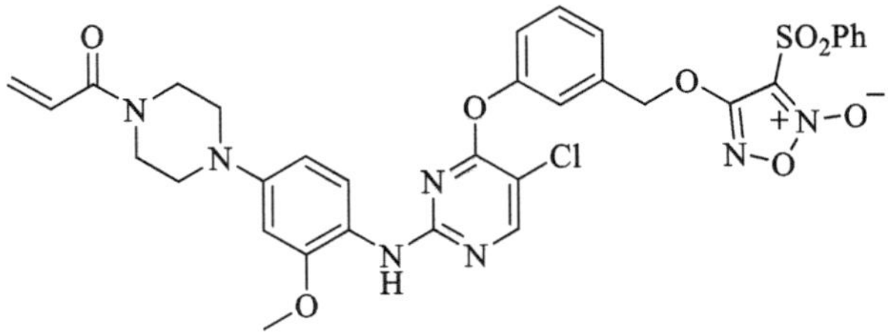

Compound 46h

H1975 = 0.055 μM	WT = 0.35 μM
HCC827 = 0.022 μM	L858R = 0.37 μM
A549 = 1.715 μM	L858R/T790M = 0.045 μM
A431 = 0.348 μM	
16HBE = 1.928 μM	

Chart 3.3 Antiproliferative activities and inhibitory activity of compounds **46h** against different types of EGFRs in vitro.

Engel et al. established seven different synthetic routes to produce pyrimidine-based compounds **50a–b**, **55**, **57l–m**, and **57a–k**, **57n–o**, **57p–q**, **58a–e**, **80**. In the case of **50a–b** (Scheme 3.7), they started with a nucleophilic aromatic substitution using 2-phenyl substituted pyrimidine **47** and 5-aminopyrazole, resulting in intermediate **48** [9]. The subsequent decoration of the pyrazole with the BOC protection group led to

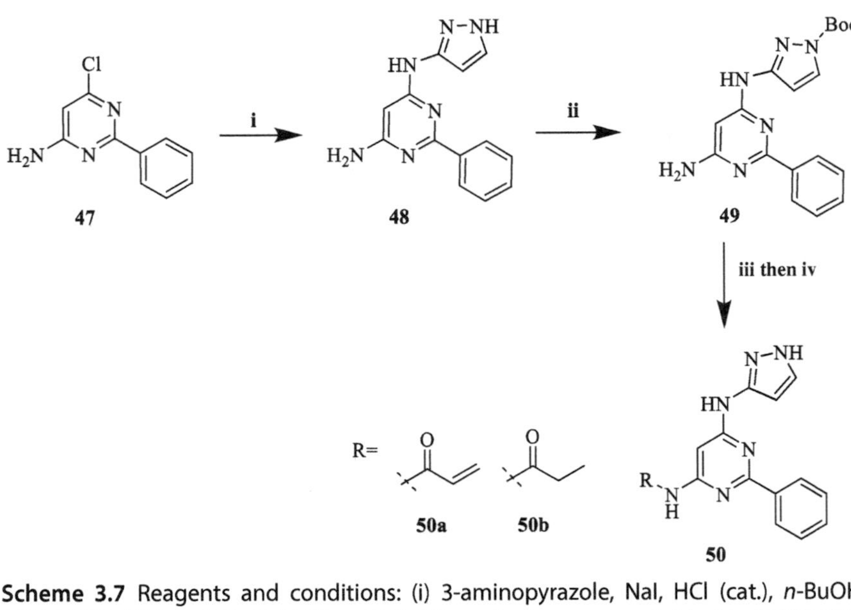

Scheme 3.7 Reagents and conditions: (i) 3-aminopyrazole, NaI, HCl (cat.), *n*-BuOH, 120°C, 72%; (ii) Boc_2O, Et_3N, DCM, rt, 21%; (iii) acryloyl or propionyl chloride, DIPEA, THF, 0°C; and (iv) TFA in DCM, rt, 58%–64% (two steps).

49, which was then converted into **50a** and its reversible counterpart **50b** by amide formation using acryloyl or propionyl chloride and a subsequent deprotection with TFA. To retain the flexibility of the pyrimidine-based inhibitors but to vary the type of substituent at the 2- and 4-positions, they synthesized several similar derivatives by the synthetic route shown in Scheme 3.8. 2,4-Dichloropyrimidine was preferentially decorated with 3-methyl-1*H*-pyrazol-5-amine at the 4-position to afford key intermediate **52**; subsequently the pyrazole moiety was protected using dihydropyran. Then, (3-nitrophenyl) boronic acid was used to introduce the 3-nitrophenyl moiety into the pyrimidine 2-position under Suzuki conditions, resulting in intermediate **54a**, which was reduced to the corresponding amine **54b**. The acrylamide Michael acceptor system was introduced by an amide formation using acryloyl chloride. THP-deprotection with TFA led to compound **55**. Starting from key intermediate **52**, they likewise generated compounds with a bridging ether at the 2-position to increase the conformational flexibility of the Michael acceptor-decorated moiety.

The nucleophilic aromatic substitution of **52** utilizing 3-nitrophenol gave intermediate **56a**, and the subsequent reduction with Pd/C and ammonium formate led to the amino equivalent **56b**. The reaction of amine **56b** with various acid chlorides yielded compounds **57l–m**.

In a third approach (Scheme 3.9) to build up 2,4,6-substituted pyrimidine compounds, **57a–k**, **57n–q**, and **77a–d**, 4,6-dichloro-2-(methylthio) pyrimidine (**58**) was oxidized to the corresponding sulfone **59** using mCPBA to generate an appropriate leaving group at the pyrimidine 2-position. Consecutive nucleophilic aromatic substitution with the ortho-, meta-, or para-substituted nitrophenol-derivatives led to intermediates **60a–c**. Following the substitution of one aromatic chlorine with either aminopyrazoles or aminoindazoles resulted in intermediates **61a–c** and **62–64**, which were then derivatized with 1-methylpiperazine at the pyrimidine 6-position, providing **65a–c** and **66–68**. The decoration of the pyrazoles or indazoles with the BOC protection group led to **69b–c** and **70–72**. The subsequent reduction of the aromatic nitro group was performed by using Pd/C and ammonium formate and resulted in the corresponding amino derivatives **73b–c** and **74–76**. Ultimately, these amines were decorated with Michael acceptors using acid chlorides or carboxylic acids under standard amide coupling conditions (EDC/HOBt), respectively. The equivalent reversible counterparts were produced accordingly utilizing propionyl chloride. Final BOC deprotection was conducted using TFA and led to compounds **57a–k**, **57n–q**, and **77a–d**.

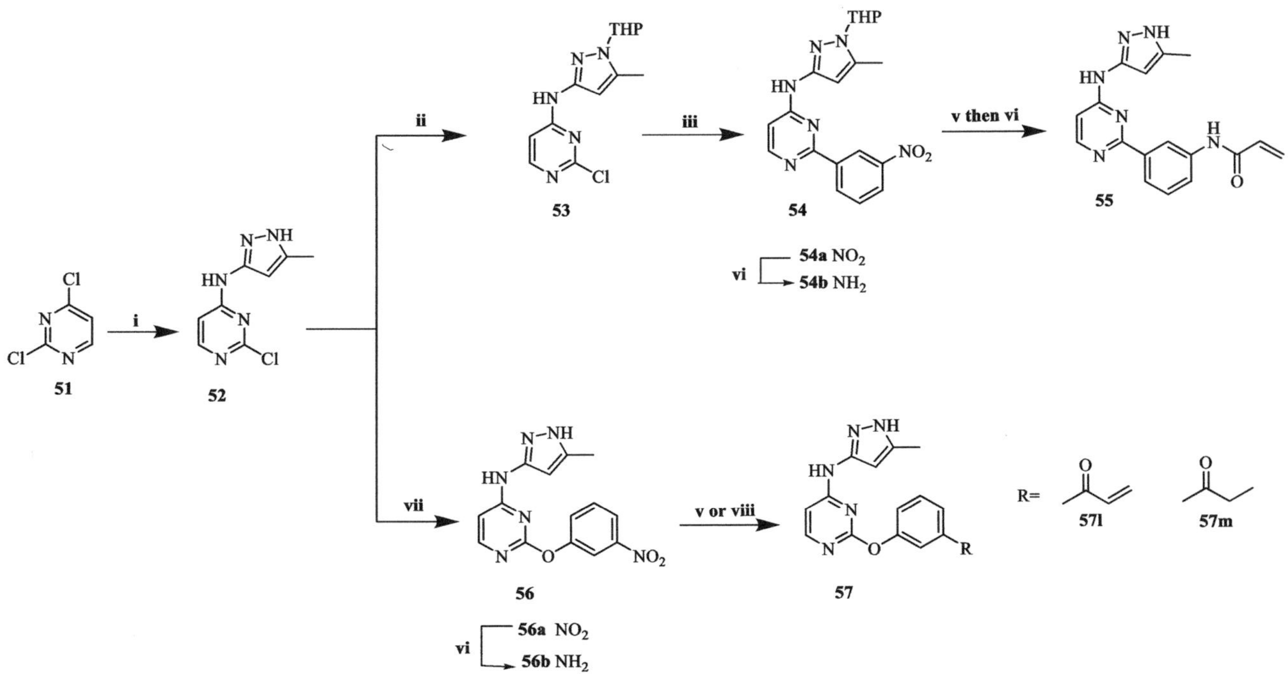

Scheme 3.8 Reagents and conditions: (i) 5-methyl-1H-pyrazol-3-amine, DIPEA, DMF, 90°C, 54%; (ii) DHP, pTsOH, THF, 90°C, 64%; (iii) (3 nitrophenyl)boronic acid, $Pd(PPh_3)_4$, Na_2CO_3, DMF/H_2O (9:1), 90°C, 70%; (iv) ammonium formate, Pd/C, EtOH, 90°C; (v) acryloyl chloride, DIPEA, THF, 0°C, 26% (**57l**); (vi) TFA in DCM, rt, 73% (2 steps); (vii) 3-nitrophenol, K_2CO_3, DMF, 140°C, 20%; and (viii) propionyl chloride, DIPEA, THF, 0°C, 27%.

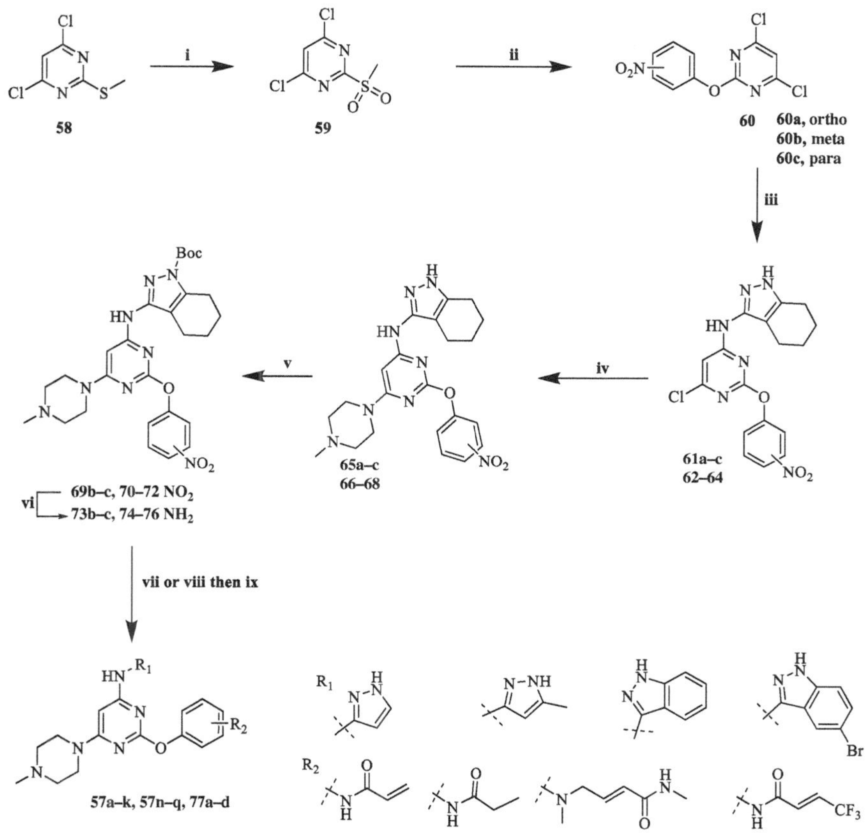

Scheme 3.9 Reagents and conditions: (i) mCPBA, DCM/THF, 0°C rt, 86%; (ii) nitrophenol, NaH, THF, 0°C rt, 29%–84%; (iii) amine, DIPEA, NaI, DMF, 85–110°C, 60%–82%; (iv) methylpiperazine, 110°C, 62%–82%; (v) Boc_2O, Et_3N, MeOH, rt or Boc_2O, Cs_2CO_3, THF, rt, 56%–95%; (vi) ammonium formate, Pd/C, EtOH or EtOAc, 90°C, 48%–87%; (vii) acryloyl or propionyl chloride, DIPEA, THF, 0°C; (viii) (*E*)-4-(dimethylamino)but-2-enoic acid or (*E*)-4,4,4-trifluorobut-2-enoic acid, EDC, HOBt, DIPEA, DCM, rt or (*E*)-4-(dimethylamino) but-2-enoic acid, oxalyl chloride, DMF (cat.), THF, rt; and (ix) TFA in DCM, rt, 33%–95% (two steps).

Engel et al. presented a structure-guided development of irreversible inhibitors to selectively target the L858R/T790M mutant form of EGFR, using the general quinazoline scaffold of **50a** as the structural basis for further rational design approaches that would result in compounds with increased specificity while retaining cellular potency. While the quinazoline-based inhibitor **77m** illustrated increasing inhibitory impact on biochemical characterizations ($IC_{50}=1.1\,\mu M$ (wt) > $0.5\,\mu M$ L858R > $0.1\,\mu M$ L858R/T790M) and thereby displaying mutant selectivity, the related pendants

77f and **77h**, differing only by the position of the attachment of the Michael acceptor, basically did not have any effect (Chart 3.4). The shortened pyrimidine scaffolds **77l** and **57l**, however, illustrated the same mutant selectivity. Notably, for the 2,4,6-substituted scaffold **57b**, the mutant selective inhibition profile was somehow inverted since the L858R mutant form of EGFR was predominantly addressed (IC_{50} = 0.4 μM (wt) > 0.03 μM (L858R) < 0.5 μM (L858R/T790M)), whereas **77f** exhibited an excellent, mutant-selective effect on EGFR mutant variants (IC_{50} = 2.2 μM (wt) > 0.5 μM L858R > 0.07 μM L858R/T790M) (Chart 3.4).

Han et al. synthesized hybrid molecule **89a–k** from phenylsulfonylfuroxan and anilinopyrimidine and biologically evaluated as EGFR inhibitors for intervention of NSCLC [10]. The synthesis of compounds **89a–k** is given in Scheme 3.10. Phenylpiperazine **79** was prepared by amination at 5-position of 5-fluoro-2-nitrophenyl methyl ether, followed by a treatment with ethyl bromoacetate to give ester **80**. Then the nitro group of **80** was reduced to yield arylamine **81**. Meanwhile, aryl ether **83** was obtained by the regioselective coupling of 3-nitrophenol **82** to the 4-position of 2,4,5-trichloropyrimidine. The subsequent reduction of nitro group **83** afforded arylamine **84**. The acylation of amino group **84** led to acrylamide **85**. The coupling of **81** with **85** in the presence of trifluoroacetic acid offered anilinopyrimidine **86**, which was converted to corresponding acid **87** by hydrolysis. Finally the condensation of **87** with various monophenylsulfonylfuroxans **88a–k** provided the target compounds **89a–k** as illustrated in Scheme 3.10.

As can be seen in Chart 3.5, all the compounds except **89c** (IC_{50} = 0.291 μM) were potent in the inhibition of HCC827 cells (IC_{50} = 0.006–0.041 μM). Compounds **89a**, **89f**, **89h**, **89i**, and **89k** (IC_{50} = 0.009, 0.007, 0.006, 0.010, and 0.007 μM, respectively) showed comparable inhibition to gefitinib (IC_{50} = 0.006 μM) and WZ4002 (IC_{50} = 0.008 μM) on HCC827 cells. Additionally, consistent with their kinase inhibition, most of these compounds displayed higher antiproliferative activity on H1975 cells (IC_{50s} = 0.029–2.017 μM) than gefitinib (IC_{50} = 6.917 μM). In general, esters (**89f–k**) exhibited better inhibitory effects than amides (**89a–e**) against H1975 cells. Among which, **89f**, **89h**, **89i**, and **89k** (IC_{50} = 0.063, 0.069, 0.043, and 0.029 μM, respectively) were as potent as WZ4002 (IC_{50} = 0.046 μM). Taken together, compounds **89f**, **89h**, **89i**, and **89k** were potent to inhibit both gefitinib-sensitive HCC827 and -resistant H1975 cell lines [10].

Finlay et al. synthesized amides of the distal piperazine nitrogen. As given in Scheme 3.11, fluoro intermediate **90** was exposed to various piperazine

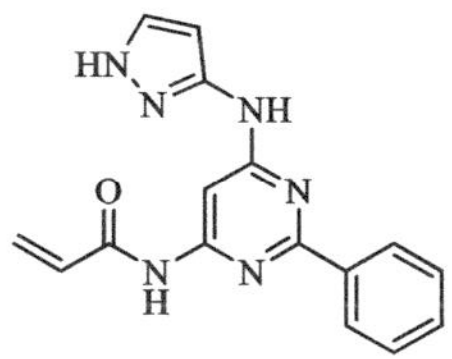

50a

WT(IC_{50}) = 0.9 μM
L858R (IC_{50}) = 0.8 μM
L858R/T790M (IC_{50}) = 0.3 μM

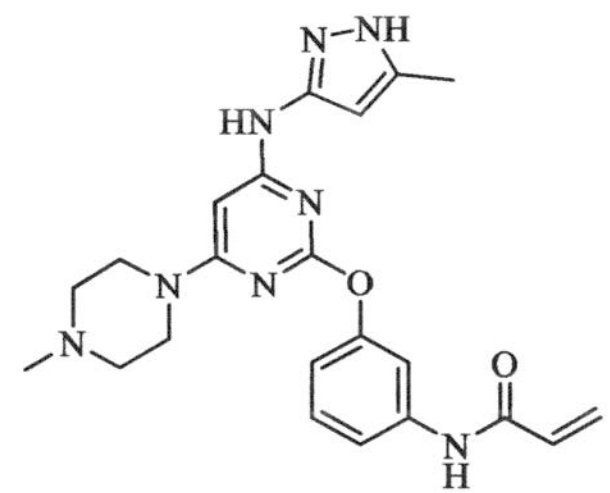

57b

WT(IC_{50}) = >10 μM
L858R (IC_{50}) = 0.03 μM
L858R/T790M (IC_{50}) = 0.5 μM

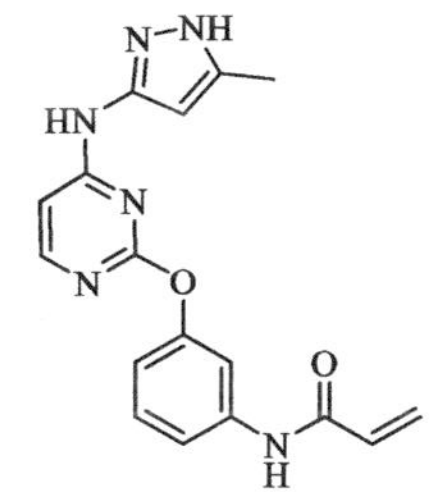

57l

WT(IC_{50}) = >10 μM
L858R (IC_{50}) = 1.18 μM
L858R/T790M (IC_{50}) = 0.9 μM

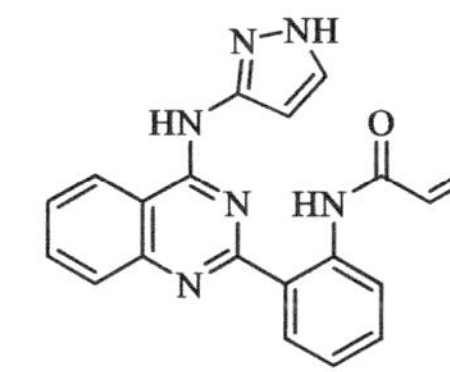

77f

WT(IC_{50}) = >10 μM
L858R (IC_{50}) = 2.7 μM
L858R/T790M (IC_{50}) = 1.1 μM

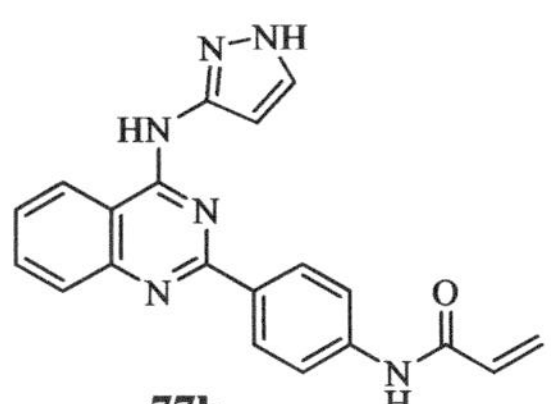

77h

WT(IC_{50}) = >10 μM
L858R (IC_{50}) = >10 μM
L858R/T790M (IC_{50}) = >10 μM

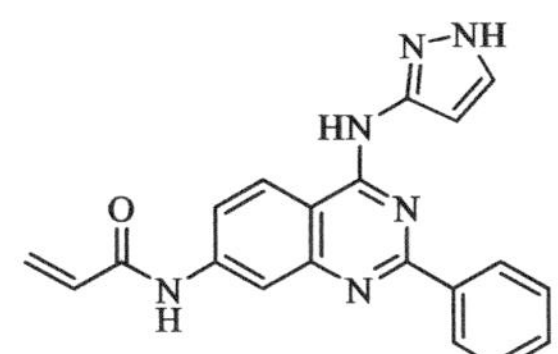

77l

WT(IC_{50}) = >10 μM
L858R (IC_{50}) = 1.9 μM
L858R/T790M (IC_{50}) = 2.2 μM

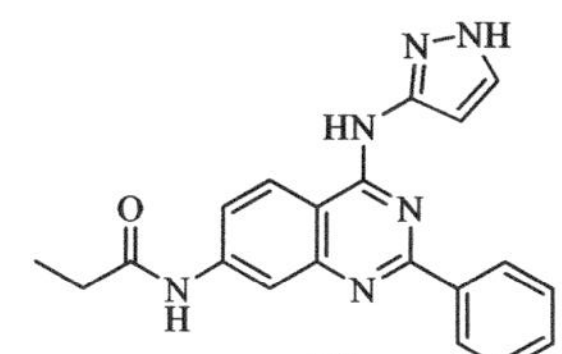

77m

WT(IC_{50}) = >10 μM
L858R (IC_{50}) = 2.1 μM
L858R/T790M (IC_{50}) = 2.2 μM

Chart 3.4 EGFR in vitro activity of potent anilinoquinazolines.

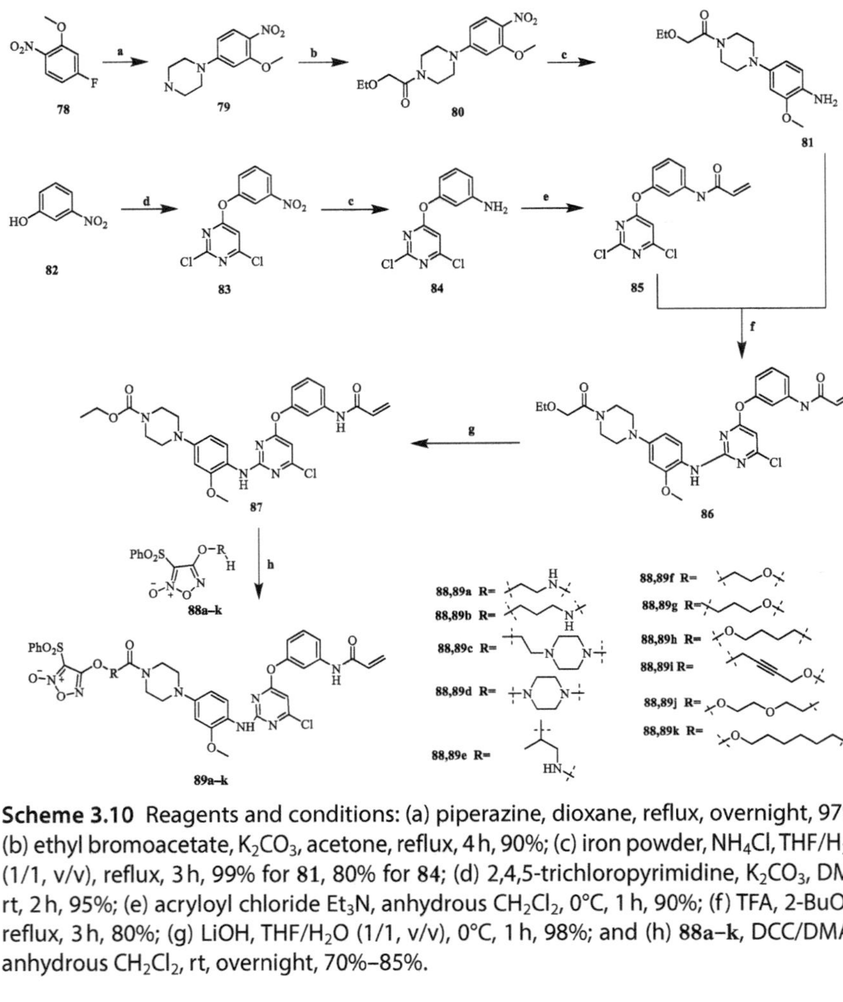

Scheme 3.10 Reagents and conditions: (a) piperazine, dioxane, reflux, overnight, 97%; (b) ethyl bromoacetate, K_2CO_3, acetone, reflux, 4 h, 90%; (c) iron powder, NH_4Cl, THF/H_2O (1/1, v/v), reflux, 3 h, 99% for **81**, 80% for **84**; (d) 2,4,5-trichloropyrimidine, K_2CO_3, DMF, rt, 2 h, 95%; (e) acryloyl chloride Et_3N, anhydrous CH_2Cl_2, 0°C, 1 h, 90%; (f) TFA, 2-BuOH, reflux, 3 h, 80%; (g) LiOH, THF/H_2O (1/1, v/v), 0°C, 1 h, 98%; and (h) **88a–k**, DCC/DMAP, anhydrous CH_2Cl_2, rt, overnight, 70%–85%.

amides bearing protected amino/hydroxyl groups [11]. The nitro group was then reduced, and the resulting aniline was acylated with acryloyl chloride. Protecting groups (either BOC or TBDMS) were then removed to afford the required compounds **92–99**. A variety of alternative diamine reagents were used in similar way as discussed in Scheme 3.11 to get the basic side chain compounds **101–103** (Scheme 3.12).

The introduction of the amine side chain was achieved via SNAr reaction; the subsequent reduction of the nitro group and a reaction with acryloyl chloride furnished the desired final compounds. To access indole compounds **111a–f**, commercial pyrimidines with $R_2 = H$, Me, and Cl were used (**104**). The introduction of the indole was achieved by deprotonation

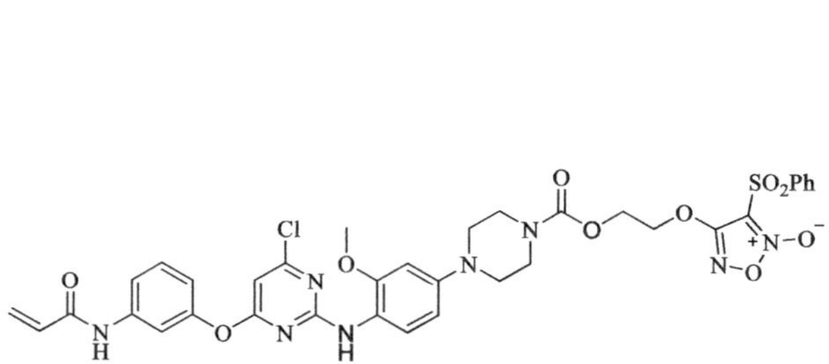

89f

H1975 (IC_{50}) = 0.063 μM
L858R (IC_{50}) = 0.083 μM
T790M/L858R (IC_{50}) = 0.017 μM
Wild type (IC_{50}) = 0.059 μM

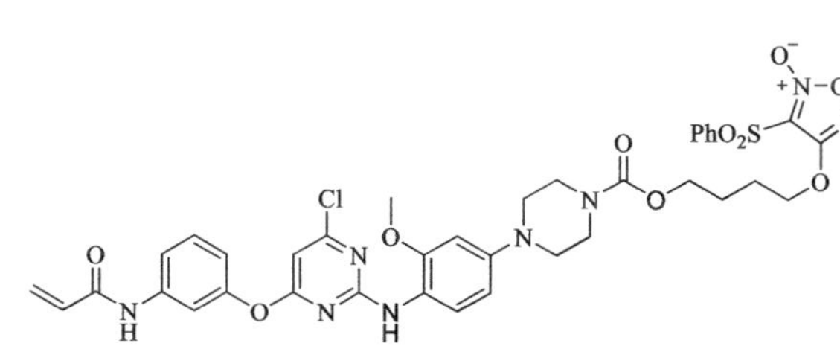

89h

H1975 (IC_{50}) = 0.069 μM
L858R (IC_{50}) = 0.189 μM
T790M/L858R (IC_{50}) = 0.038 μM
Wild type (IC_{50}) = 0.174 μM

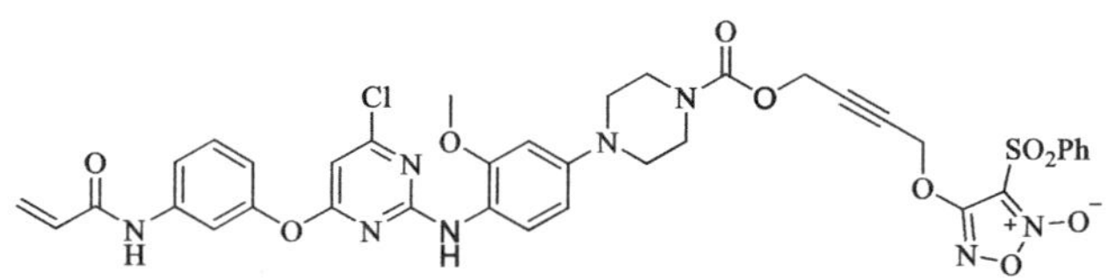

89i

H1975 (IC_{50}) = 0.043 μM
L858R (IC_{50}) = 0.064 μM
T790M/L858R (IC_{50}) = 0.016 μM
Wild type (IC_{50}) = 0.052 μM

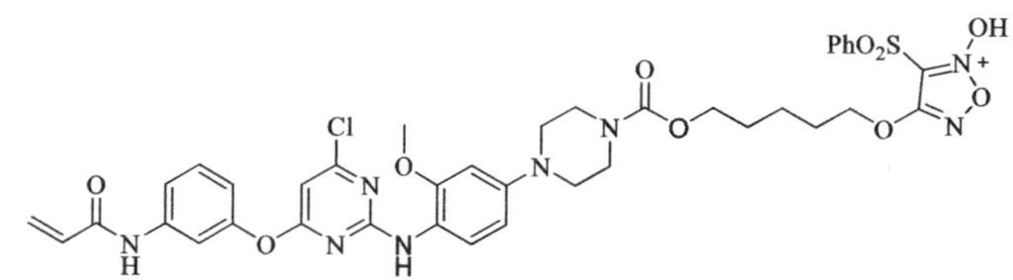

89k

H1975 (IC_{50}) = 0.029 μM
L858R (IC_{50}) = 0.305 μM
T790M/L858R (IC_{50}) = 0.047 μM
Wild type (IC_{50}) = 0.292 μM

Anti-proliferative and *in vitro EGFR* inhibitory activities

Chart 3.5 Antiproliferative and in vitro EGFR inhibitory activities.

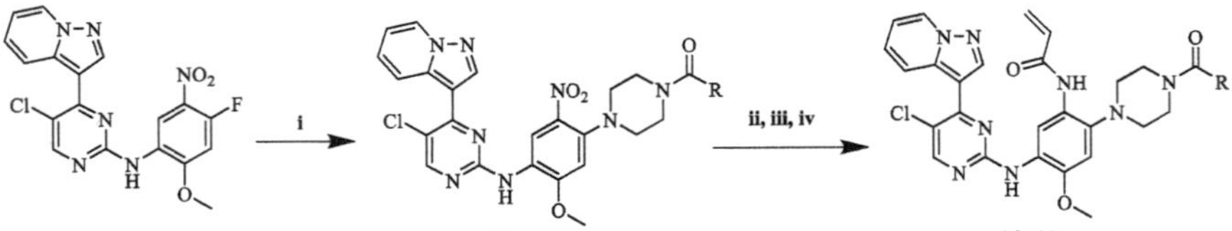

Scheme 3.11 Reagents and conditions: (i) piperazine amide, DIPEA, DMA, 140°C (microwave); (ii) iron, NH_4Cl, EtOH, water, 100°C; (iii) acryloyl chloride, DIPEA, DCM, 0°C; and (iv) functional group deprotection if required; TFA, DCM, rt; or TBAF, THF, rt.

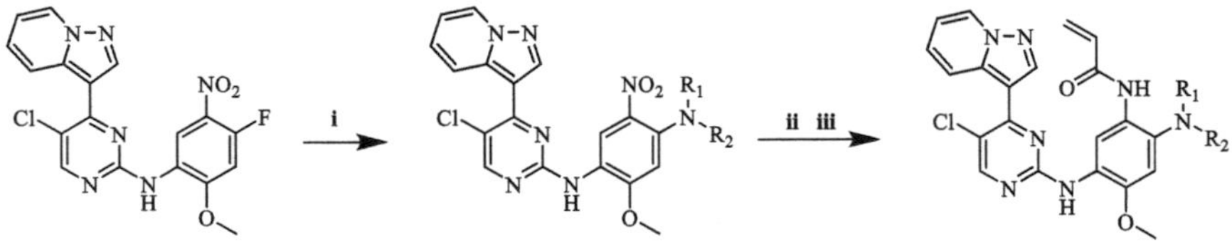

Scheme 3.12 Reagents and conditions: (i) R_1R_2NH, DIPEA, DMA, 85–100°C; (ii) iron, NH_4Cl, EtOH, water, 100°C; and (iii) acryloyl chloride, DIPEA, DCM, −15°C to 0°C.

with methyl magnesium bromide (**105**) (Scheme 3.13) and subsequent S_NAR in an adequate yield (**106**), where R_1 = Me, deprotonation of the indole NH and methylation with methyl iodide proceeded in good yield. The introduction of the aniline side chain and further elaboration into the desired compounds were achieved in a similar manner to the compounds detailed in Scheme 3.11. In the case where R_2 was cyano, this group was introduced at the penultimate step (Scheme 3.13). Although isolated in a modest yield, this approach resulted in sufficient material to deliver the desired compound **111a–f**.

Finlay et al. prepared the anilinopyrimidine with three basic side chains described above in **101–103**; additional examples were also prepared where the basic side chain was replaced with amines such as those in **97** and **99**. The presence or indeed absence of a substituent at the 5-position of the pyrimidine ring also had a differential effect on double mutant (DM) potency, IGF1R potency, and hERG binding. Compound **111f** offers a number of advantages over earlier EGFR TKI agents, including high cellular potency, superior efficacy in relevant disease models, and selectivity over the WT form of the receptor (Chart 3.6).

Qin et al. synthesized the target compounds (**118a–j** and **119a–o**) containing a hydrazone moiety, as shown in Scheme 3.14. The synthesis of the target compounds (**118a–j** and **119a–o**) containing a hydrazone moiety is shown in Scheme 3.14 [12].

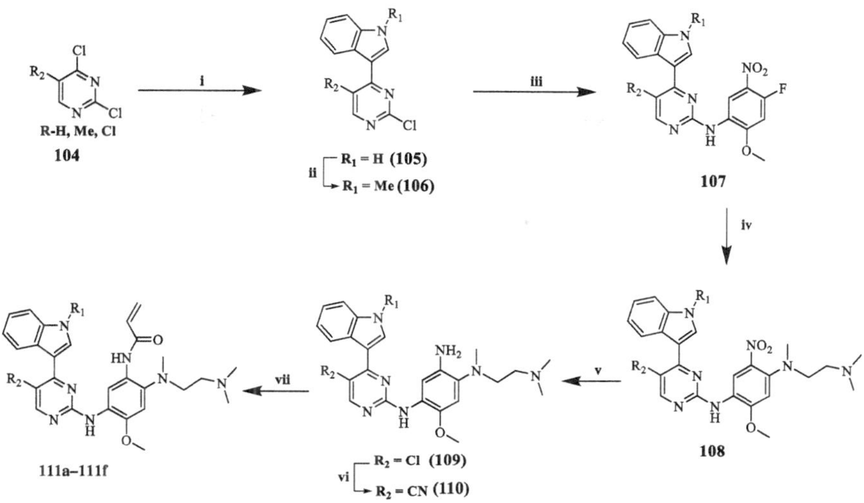

Scheme 3.13 Reagents and conditions: (i) MeMgBr, indole, THF, 0°C → 60°C; (ii) sodium hydride, methyl iodide (1.05 equiv.), THF, 0°C; (iii) 4-fluoro-2-methoxy-5-nitroaniline, tosic acid (1.1 equiv.), 2-pentanol, 125°C; (iv) *N,N,N′*-trimethylethane-1,2-diamine, DMA, 140°C; (v) iron, ammonium chloride, EtOH, water, 100°C; (vi) $Zn(CN)_2$, zinc, tris(dibenzylideneacetone) dipalladium,dicyclohexyl(2′,4′,6′-triisopropylbiphenyl-2-yl)-phosphine (XPhos), DMA, 95°C; and (vii) acryloyl chloride, DIPEA, THF, 0°C.

The intermediate **114** was prepared via a regioselective coupling of the protected benzene-1,3-diamine **113** to the 4th position of **112**. The substitution of **114** with hydrazine hydrate led to the intermediate **115**, which was reacted with an appropriate aromatic aldehyde or ketone under acid conditions to generate hydrazone **116a–o**. The key intermediate **117a–o** was obtained by the deprotection of **116a–o** in the presence of trifluoroacetic acid. The acylation of **117** with acryloyl chloride conveniently produced compounds **118a–j**. On the other hand, condensation of **117** with commercial (*E*)-4-(dimethylamino)but-2-enoic acid using HATU afforded compounds **119a–o** with good yields.

The general steps for preparing the hydrazide analogs **124a** and **124b** are outlined in Scheme 3.15. Intermediate **123** was obtained from intermediate **114** via a two-step reaction using similar conditions to those for **117** and **118**. The hydrazinolysis of ester **120** yielded the corresponding intermediate **121**, which was reacted with **123** via an SN_2 mechanism to generate the target compounds.

The structure activity relationship indicates that the substitution pattern on the terminal aryl moiety was closely related to potency (Chart 3.7).

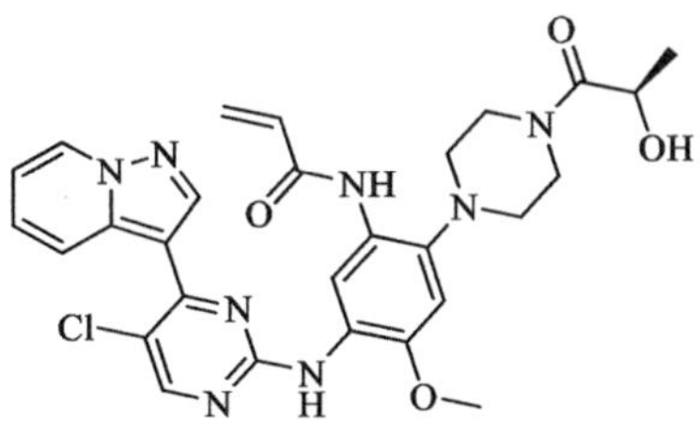

97

H1975 (IC_{50}) = 75 nM
PC9 (IC_{50}) = 189 nM
WT (IC_{50}) = 11,000 nM

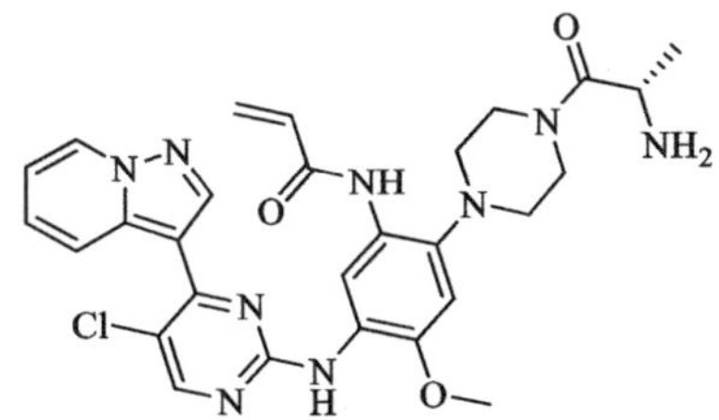

99

H1975 (IC_{50}) = 75 nM
PC9 (IC_{50}) = 189 nM
WT (IC_{50}) = 11,000 nM

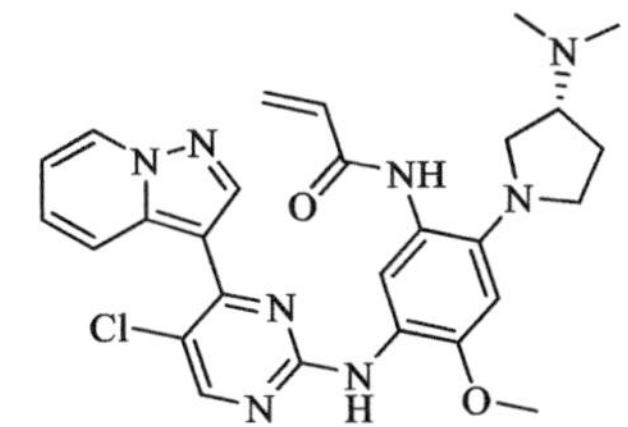

101

H1975 (IC_{50}) = 2 nM
PC9 (IC_{50}) = 16 nM
WT (IC_{50}) = 357 nM

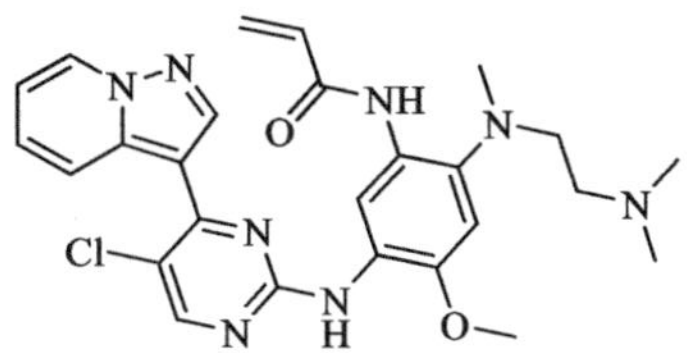

102

H1975 (IC_{50}) = 0.6 nM
PC9 (IC_{50}) = 2 nM
WT (IC_{50}) = 145 nM

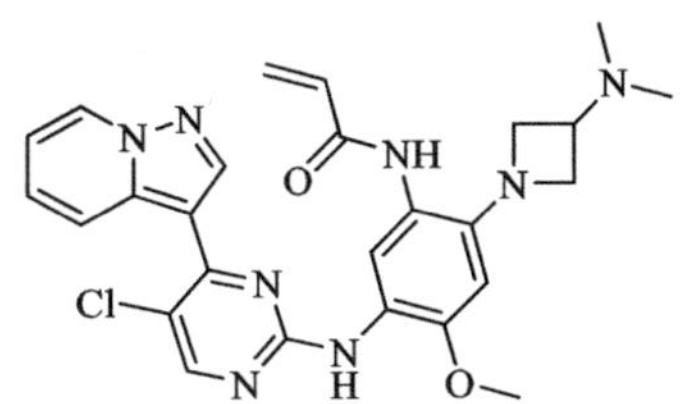

103

H1975 (IC_{50}) = 74 nM
PC9 (IC_{50}) = 21 nM
WT (IC_{50}) = 938 nM

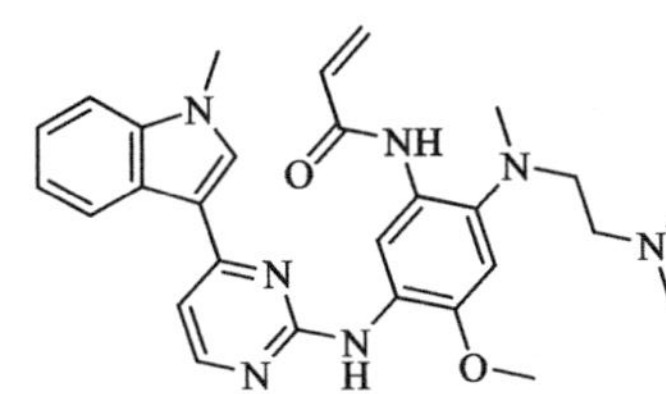

111f

H1975 (IC_{50}) = 15 nM
PC9 (IC_{50}) = 17 nM
WT (IC_{50}) = 480 nM

Chart 3.6 Antiproliferative and in vitro EGFR inhibitory activities.

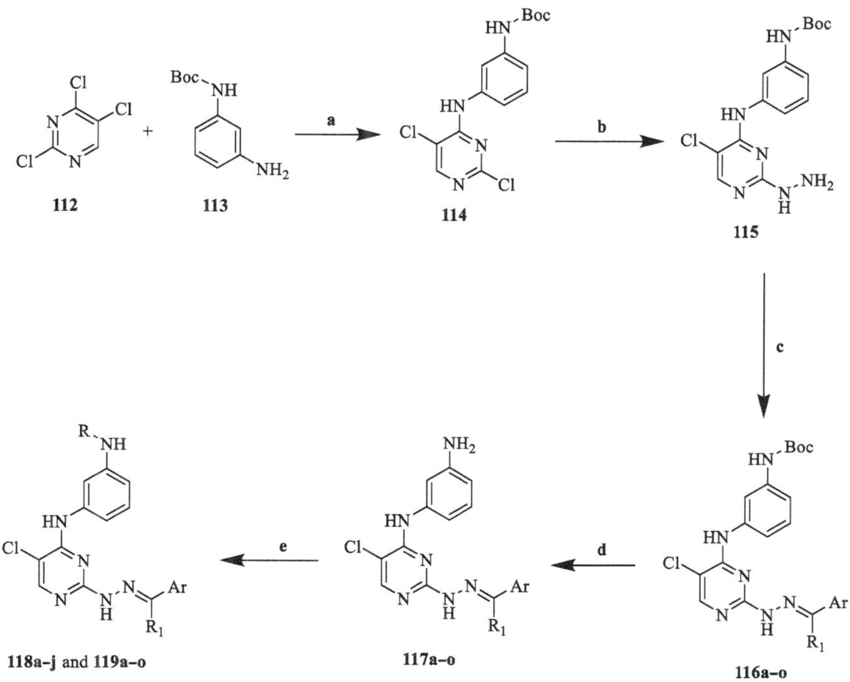

Scheme 3.14 Reagents and conditions: (a) K_2CO_3, DMF, 50°C, 3 h; (b) $NH_2NH_2{\cdot}H_2O$, MeCN, reflux, 3 h; (c) appropriate aromatic aldehyde or ketone, *i*-PrOH, reflux, 6–8 h; (d) TFA, DCM, 25°C, 2–7 h; and (e) acryloyl chloride, DIPEA, DCM, 25°C, 1 h or (*E*)-4-(dimethylamino)but-2-enoic acid, HATU, DIPEA, DCM, 25°C, 3–9 h.

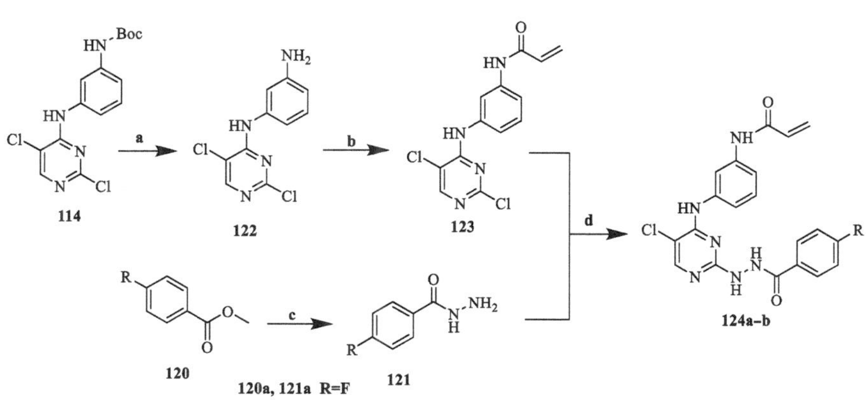

Scheme 3.15 Reagents and conditions: (a) TFA, DCM, 25°C, 1 h; (b) $NH_2NH_2{\cdot}H_2O$, EtOH, reflux, 5 h; (c) TFA, *i*-PrOH, reflux, 3 h; and (d) acryloyl chloride, DIPEA, DCM, 25°C, 1 h.

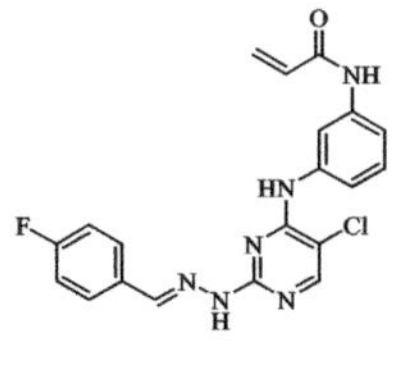

118a
H1975 (IC_{50}) = 0.43 μM
T790M/L858R(IC_{50}) = 0.86 μM
T790M(IC_{50}) = 1.12 μM
WT(IC_{50}) = >20 μM

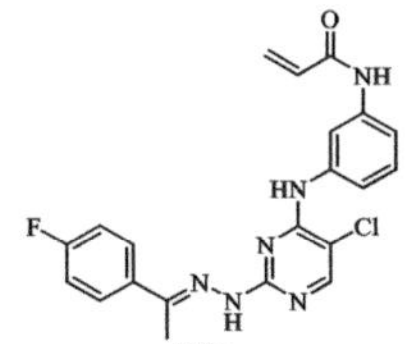

118e
H1975 (IC_{50}) = 0.39 μM
T790M/L858R(IC_{50}) = 1.68 μM
T790M(IC_{50}) = 2.53 μM
WT(IC_{50}) = >20 μM

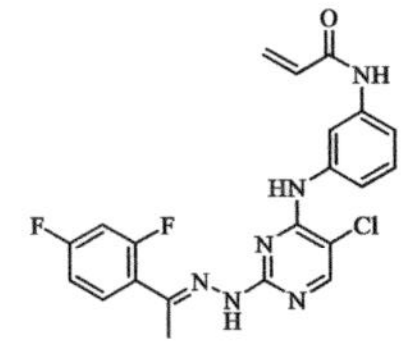

118f
H1975 (IC_{50}) = 0.30 μM
T790M/L858R(IC_{50}) = 2.27 μM
T790M(IC_{50}) = ND
WT(IC_{50}) = 11.75 μM

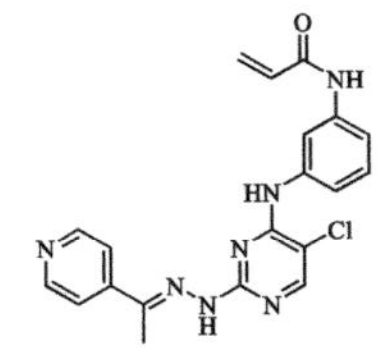

118j
H1975 (IC_{50}) = 3.69 μM
A549 (IC_{50}) = 28.71 μM
HT-29(IC_{50}) = 12.80 μM

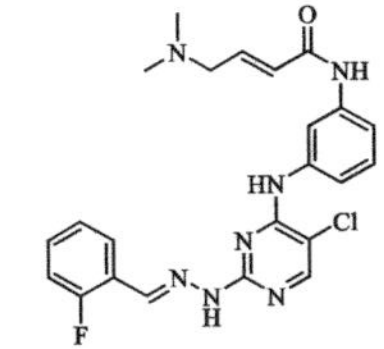

119a
H1975 (IC_{50}) = 0.44 μM
T790M/L858R(IC_{50}) = 0.79 μM
T790M(IC_{50}) = 1.56 μM
WT(IC_{50}) = >20 μM

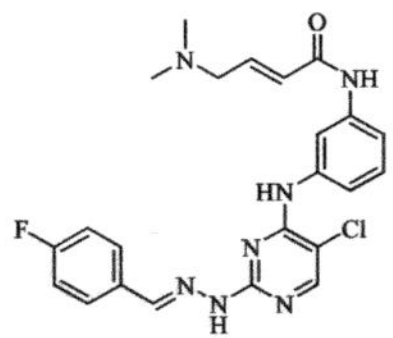

119b
H1975 (IC_{50}) = 0.59 μM
T790M/L858R(IC_{50}) = 0.68 μM
T790M(IC_{50}) = 1.49 μM
WT(IC_{50}) = 10.65 μM

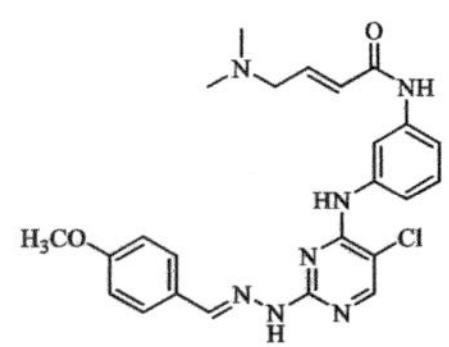

119d
H1975 (IC_{50}) = 1.58 μM
A549 (IC_{50}) = 6.24 μM
HT- 29(IC_{50}) = 4.82 μM

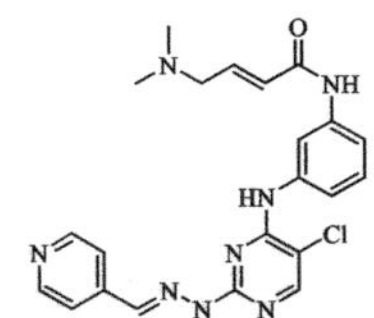

119e
H1975 (IC_{50}) = 4.13 μM
A549 (IC_{50}) = 47.68 μM
HT- 29(IC_{50}) = 25.19 μM

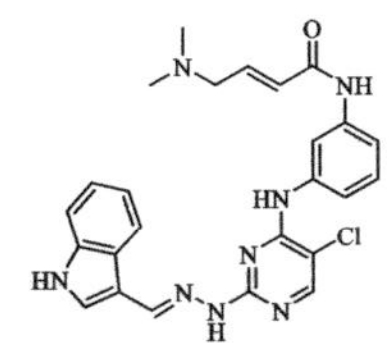

119f
H1975 (IC_{50}) = 5.45 μM
A549 (IC_{50}) = 28.68 μM
HT- 29(IC_{50}) = 9.68 μM

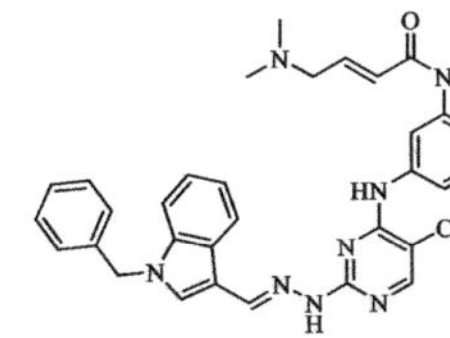

119g
H1975 (IC_{50}) = 0.26 μM
T790M/L858R(IC_{50}) = 0.88 μM
T790M(IC_{50}) = 0.79 μM
WT = >20 μM

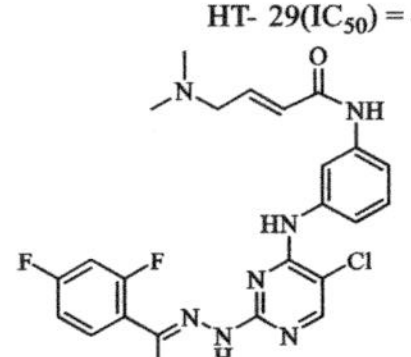

119i
H1975 (IC_{50}) = 0.48 μM
T790M/L858R(IC_{50}) = 0.72 μM
T790M(IC_{50}) = 0.82 μM
WT(IC_{50}) = >20 μM

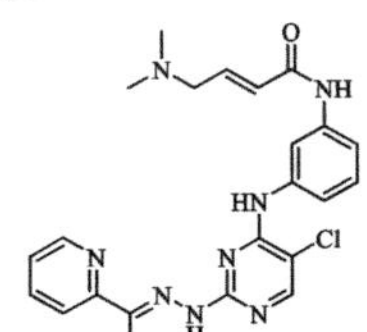

119m
H1975 (IC_{50}) = 0.077 μM
A549 (IC_{50}) = 0.83 μM
HT- 29(IC_{50}) = 0.12 μM

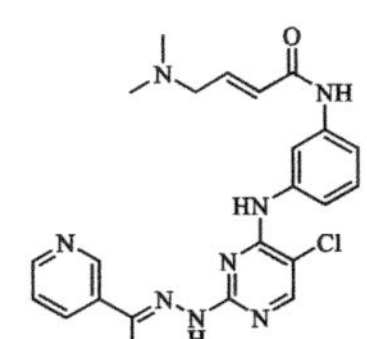

119n
H1975 (IC_{50}) = 2.35 μM
A549 (IC_{50}) = 17.29 μM
HT- 29(IC_{50}) = 9.66 μM

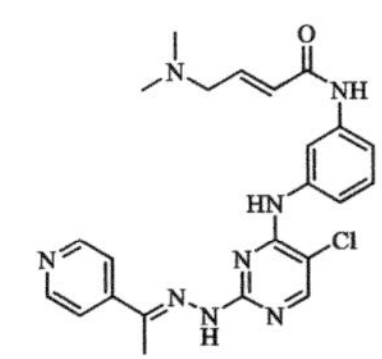

119o
H1975 (IC_{50}) = 6.24 μM
A549 (IC_{50}) = 34.86 μM
HT- 29(IC_{50}) = 10.93 μM

Chart 3.7 Antiproliferative and in vitro EGFR inhibitory activities.

Compounds **118e** and **118f**, which contained an electron-withdrawing group on the benzene ring, strongly inhibited the H1975 cells (IC_{50}, 0.39 and 0.30 μM, respectively). However, decreased activity was observed for compounds **118g** (IC_{50}, 0.74 μM) bearing an electron-donating group in the same region. Similar results were also observed when **119b** (IC_{50}, 0.59 μM) was compared with **119d** (IC_{50}, 1.58 μM).

Shifting the fluoro atom from the para position (**118a** IC_{50}, 0.43 μM) to the ortho position (**118b** IC_{50}, 0.29 μM) slightly improved the inhibition of H1975 cells, while a significant enhancement of the inhibition of A549 and HT-29 cells was also observed (Chart 3.7). The replacement of the terminal phenyl group with a pyridinyl group obviously decreased the potency, as shown with compounds **118j**, **119e**, **119n**, and **119o** (IC_{50}, 2.35–6.24 μM). A dramatically adverse result was observed with compound **119m**, which exhibited outstanding potency against H1975 cells with an IC_{50} of 0.077 μM. However, it also showed strong toxicity against HT-29 cells (IC_{50}, 0.12 μM), illustrating an off-target effect that might limit its further development.

Compound **119f** with an indole group showed a sharply reduced efficacy against H1975 cells, with an IC_{50} of 5.45 μM. Interestingly the inhibitory activity was enhanced 21-fold when an additional benzyl group was introduced on the N-position of the indole fragment (compound **119g** IC_{50}, 0.26 μM). Compound **117a** (IC_{50}, 9.92 μM) exhibited a significant loss of potency against H1975 cells compared to **118a** (IC_{50}, 0.43 μM), which demonstrated that the acrylamide fragment was essential for high activity (Chart 3.7). With an additional dimethylamino group in the acrylamide moiety, compounds (**119a–b** and **119g–i**) also possessed a high level of inhibitory activity, showing an attractive strategy for optimizing the physicochemical property of these compounds, if necessary.

Ward et al. developed a novel series of small-molecule inhibitors to target the DM form of the EGFR tyrosine kinase, which is resistant to treatment with gefitinib and erlotinib, as shown in Scheme 3.16. The installation of the 4-substituent of the pyrimidine **125** was achieved by a direct regioselective displacement of 4-chloro under basic conditions or by Suzuki coupling to give **127** [13].

Where necessary the aryl group was constructed from vinyl butyl ether intermediate **126**, which in turn was synthesized from the 4-chloropyrimidine via Heck coupling. Subsequent cyclization reactions afforded aryl pyrimidines **127**. The aniline addition to 2-chloropyrimidines **127** was achieved via Buchwald-Hartwig coupling with substituted anilines or by

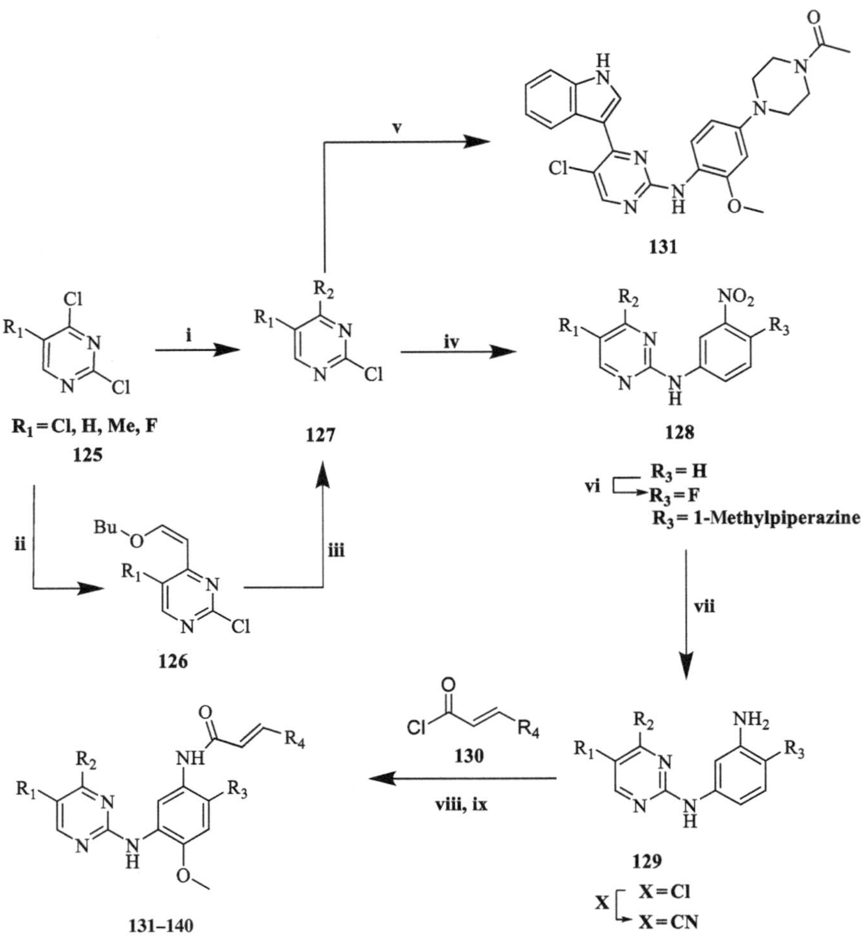

Scheme 3.16 Reagents and conditions: (i) aniline displacement: excess $NaHCO_3$, 2-pentanol or EtOH, 85°C; Suzuki coupling: Na_2CO_3, $ArB(OR)_2$, $Pd(PPh_3)_4$, MeCN, water, 100°C; indole displacement: MeMgBr (1 equiv., 3.2 M in 2-methyl THF), indole (1 equiv.), THF, 0–60°C. (ii) Et_3N (1.1 equiv.), butyl-vinyl-ether (1.1 equiv.), $Pd(OAc)_2$ (0.03 equiv.), PEG 200, 80°C. (iii) Either 1 aminopyridinium iodide (1 equiv.), K_2CO_3 (2.5 equiv.), DMA, 110°C or 2-amino-pyridine (1 equiv.), NBS (1.2 equiv.), 1,4-dioxane, water, 85°C. (iv) Either aniline (1.1 equiv.), p-TSA (1.1 equiv.), 2-pentanol, 80–120°C, or aniline (1.1 equiv.), 4 M HCl in 1,4-dioxane, TFE, 150°C (μW), or aniline (1.1 equiv.), $Pd(OAc)_2$ (0.08 equiv.), Xantphos (0.1 equiv.), Cs_2CO_3 (2 equiv.), 1,4-dioxane, 80°C. (v) Acetyl aniline (1 equiv.), p-TSA (3 equiv.), *n*-pentanol/NMP, 150°C. (vi) 1-Methylpiperazine (2.5 equiv.), TFE, 120°C. (vii) Either iron (3 equiv.), NH_4Cl (0.7 equiv.), EtOH, water, 100°C or Pt/C (1 equiv.), ZnI2 (0.1 equiv.), H_2, MeOH, EtOAc. (viii) Either acryloyl chloride (1 M, THF, 1 equiv.), DIPEA (1.1 equiv.), THF, 0°C, or (*E*)-4-(dimethylamino)but-2-enoic acid (1 equiv.), HATU (1.1 equiv.), DIPEA (2.5 equiv.), DCM, or (*E*)-4-bromobut-2-enoic acid (1.1 equiv.), HATU (1.1 equiv.), DIPEA (2.5 equiv.), 0°C then Me_2NH (20 equiv.). (ix) Deprotection (where required): either K_2CO_3 (3 equiv.), THF, MeOH or Cs_2CO_3 (3 equiv.), THF, MeOH, 0°C. (x) $Zn(CN)_2$ (0.6 equiv.), Zn (0.1 equiv.), $Pd_2(dba)_3$ (0.1 equiv.), dicyclohexyl(2′,4′,6′-triisopropylbiphenyl-2-yl)phosphine (0.2 equiv.), DMA, 90°C.

acid-catalyzed S_NAr reaction, typically by heating with p-TSA in 2-pentanol to get **128**. For compounds where $R_3 = H$, commercial 5-nitro-2-methoxy-aniline was used. Where $R_3 = F$ the aniline was not commercially available and was synthesized by the nitration of 2-methoxy-4-fluoro-aniline prior to use. Amine substitution of the fluorine by S_NAr proceeded in a fair to good yield by heating with either an excess of the amine or the DIPEA as a base. The subsequent reduction of the nitro group and reaction with acryloyl chloride ($R_4 = H$) or HATU coupling with commercial (*E*)-4-(dimethylamino)but-2-enoic acid yielded test compounds **130** in moderate yield. The substituted aniline was installed on **127** using *n*-pentanol to yield the compound **131**. The substituted aniline part was constructed from commercial starting materials, as detailed in Scheme 3.17.

Among the synthesized, compound **131** is identified as a DM-selective reversible EGFR inhibitor, as shown in Chart 3.8. Compounds **132** and **133** were found to be selective toward the DM enzyme, sparing the WT EGFR as shown in Chart 3.8. To improve the physical properties of compounds within this series, they modify the head group (Chart 3.8). Out of these head groups, the pyrollopyridine system (compound **137b**) was especially interesting, maintaining a reasonable potency in DM and activating-mutant

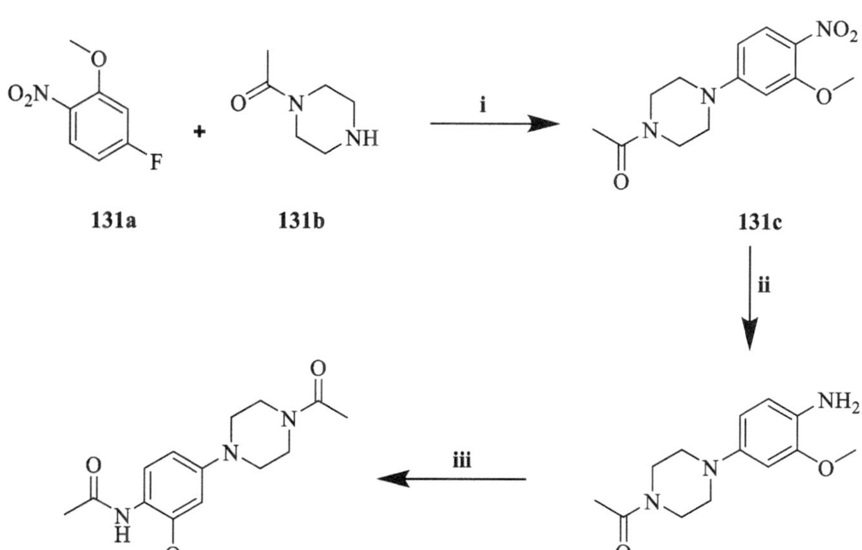

131a **131b** **131c**

131e **131d**

Scheme 3.17 Reagents and conditions: (i) 1-acetyl piperazine (1.2 equiv.), DIPEA (1 equiv.), DMA, 90°C, 18 h, 90%; (ii) platinum oxide (5 mol%), EtOH, hydrogen (1.3 bar), 3 h, 99%; and (iii) acetic anhydride (6.6 equiv.), acetic acid (4.5 equiv.), 1 h, 73%.

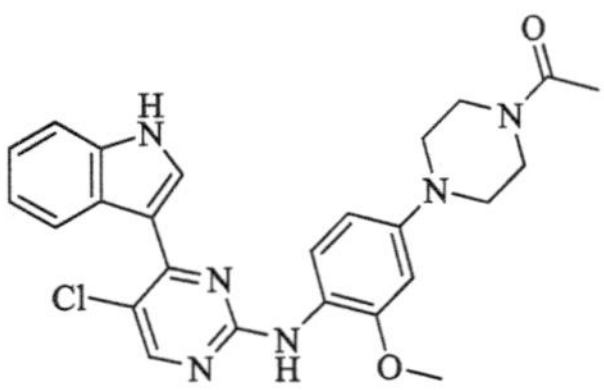

131
DM enzyme = 0.009 μM
DM cell = 0.77 μM
WT enzyme = 0.79 μM

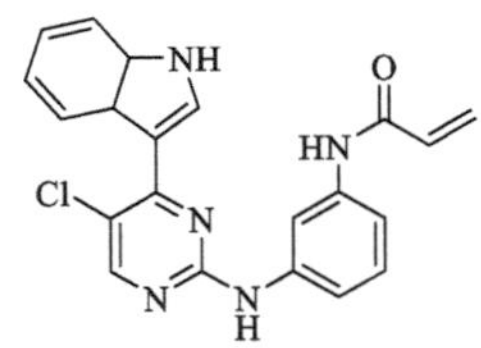

132
DM enzyme = 0.053 μM
AM cell = 0.27 μM
DM cell = 0.081 μM
WT cell = 3.5 μM

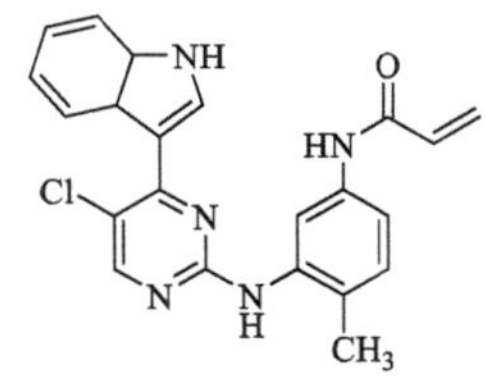

133
DM enzyme = 0.0063 μM
AM cell = 0.029 μM
DM cell = 0.022 μM
WT cell = 0.55 μM

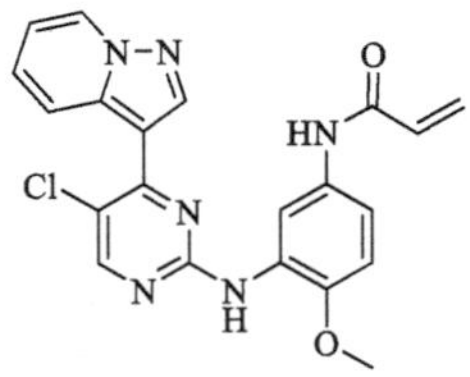

137b
AM cell = 0.15 μM
DM cell = 0.033 μM
WT cell = 12.8 μM

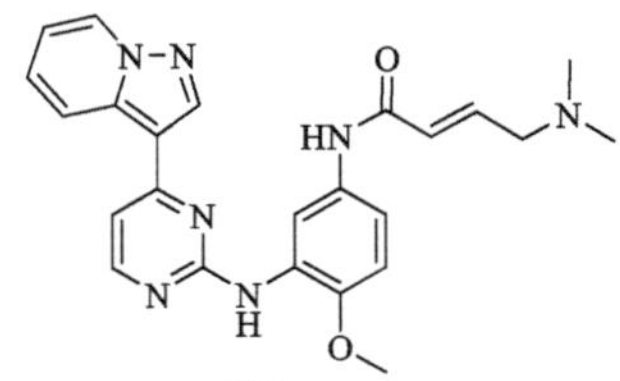

139b
AM cell = 0.39 μM
DM cell = 0.25 μM
WT cell = 20.0 μM

Chart 3.8 Antiproliferative and in vitro EGFR inhibitory activities.

cell assays (IC_{50} 0.033 and 0.15 μM), but with an improved margin to WT EGFR (390-fold) (Chart 3.8).

Although the DM activity was reduced for compound **139b**, compared to **139a** (0.25 vs 0.096 μM) the activity against the activating-mutant form was unchanged (0.39 vs 0.40 μM). The DM selectivity of **139b** was reduced to 80-fold from 237-fold for compound **139a**, potentially resulting from chlorine interacting more effectively with the methionine gatekeeper residue (Chart 3.8).

Xiao et al. reported the design and synthesis of the compounds **146a–j**, prepared by using commercially available 2,4-dichloro-5-(methylthio) pyrimidine **141** as the starting material (Scheme 3.18) [14]. Briefly, a direct nucleophilic coupling of tert-butyl 3-aminophenylcarbamate (**142**) with 2,4-dichloro-5-(methylthio)pyrimidine (**141**) produced tert-butyl (3-((2-chloro-5-(methylthio)pyrimidin-4-yl)oxy)phenyl)carbamate (**143**) as a white solid. With the catalysis of trifluoroacetic acid, nucleophilic coupling of a series of substituted aniline with compound (**143**) produced a variety of compounds (**144a–j**) with modest yield. Then the precursors (**144a–j**) were deprotected with trifluoroacetic acid and further acylated with acryloyl chloride to give the 5-(methylthio)pyrimidine derivatives (**146a–j**).

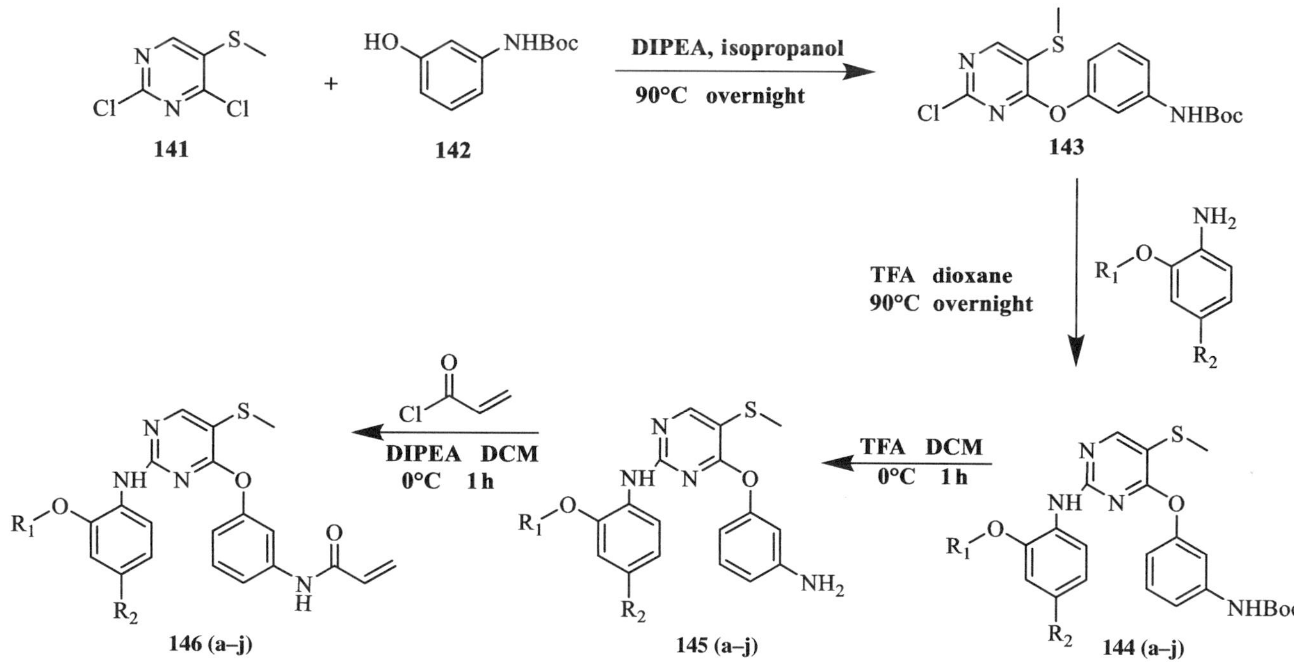

DIPEA = diisopropylethylamine, TFA = trifluoroacetic acid.

Scheme 3.18 Synthesis route of 5-(methylthio)pyrimidine derivatives (**146a–j**).

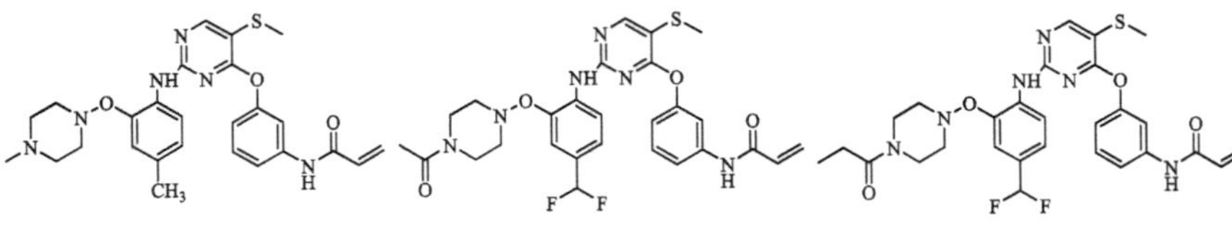

146a

H1975 (IC_{50}) = 0.4 μM
WT(IC_{50}) = 271.6 μM

146i

H1975 (IC_{50}) = 2.8 μM
WT(IC_{50}) = 38.8 μM

146j

H1975 (IC_{50}) = 3.7 μM
WT(IC_{50}) = 46.4 μM

Chart 3.9 Antiproliferative and in vitro EGFR inhibitory activities.

Similar to **146a**, compounds **146b–j** potently inhibited the enzymatic activities of EGFR L858R/T790M, with IC_{50} values of 0.7–15 nM, and the selectivity for the DM overwide type ranged from 13 to 216-fold (Chart 3.9). Most importantly, several compounds also displayed strong antiproliferative effects on Gefitinib-resistant H1975 cells bearing EGFR L858R/T790M with good selectivity over A431 human epithelial carcinoma cells with overexpressed WT EGFR. Compared to **146a**, compounds **146i** and **146j** not only showed 10-fold improved potency against H1975 cancer cell growth, but also 10-fold selectivity over A431 cells. These improvements might be attributed to the substitution of fluorine, which may play an important role in membranous permeability. Most importantly, compounds **146i** and **146j** harbored much more potency for H1975 cells and better selectivity over A431 cells than that of both WZ4002 and rociletinib (CO1686) (Chart 3.9).

Zhou et al. synthesized the pteridin-7(8*H*)-one analogs as shown in Scheme 3.19. Condensation of commercially available 2,4-dichloro-5-nitro-pyrimidine (**147**) with a primary arylamine in 1,4-dioxane containing *N,N*-diisopropylethylamine (DIPEA) provided intermediates with structure **148**. Compound **149** was prepared from intermediate **148** via nucleophilic substitution with another arylamine [15].

The nitro group in compound **149** was reduced, with Pd/C using hydrogenation in ethanol to obtain compound **150** with a nucleophilic amino group in good yields. The diaminopyrimidine compound **150** was cyclized with ethyl glyoxalate (or ethyl pyruvate) and acetic acid under refluxing with ethanol to form the pteridin-7(8*H*)-one core in compound **151**. Compound **151** was deprotected with trifluoroacetic acid in dichloromethane and further acylated by the corresponding acyl chloride to yield the pteridin-7(8*H*)-one derivatives **152e–f**, **153a–z**.

The addition of a dialkylaminomethyl group to the acrylamide fragment in clinical candidates such as EKB-569, BIBW2992, and HKI-272

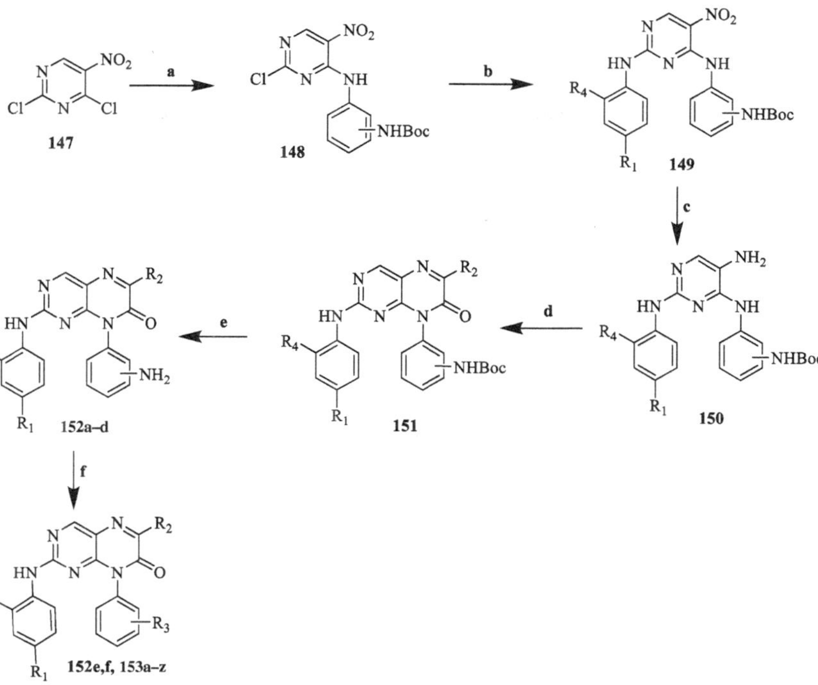

Scheme 3.19 Reagents and conditions: (a) $ArNH_2$, DIPEA, 1,4-dioxane, rt, 65%–82%; (b) $ArNH_2$, DIPEA, 1,4-dioxane, rt, 75%–85%; (c) Pd/C, H_2, EtOH, 83%–95%; (d) R_2COCO_2Et, HOAc, EtOH, reflux, 60%–80%; (e) trifluoroacetic acid, CH_2Cl_2, 0°C to rt, 85%–95%; and (f) acyl chloride, Et_3N, CH_2Cl_2, 0°C to rt or acyl chloride, 1-methyl-2-pyrrolidinone, CH_3CN, 0°C to rt, 30%–87%.

has been reported to greatly increase their potency, which the dialkylamino group can serve as an intramolecular catalyst for the Michael addition reaction. Inspired by this, Zhou et al. synthesized four compounds (**153f**, **153g**, **153h**, and **153i**) that possess a dialkylamino group attached to the acrylamide fragment (Chart 3.10). However, only **153f** displayed enhanced potency compared with the corresponding compounds (**153a**, **153b**, **153j**, and **153k**) that lacked the dialkylamino addition to the acrylamide moiety.

The R_1 methoxy group in **153a** and **153b** were replaced with 1-(4-methylpiperazinyl) (**153p** and **153q**) groups to evaluate the impact of the R_1 group.The results indicated that **153q** with a 1-(4-methylpiperazinyl) substituent exhibits the highest EGFR inhibitory potency, with IC_{50} values against WT EGFR, EGFR-L858R, and EGFR-T790M/L858R of 1.21, 1.32, and 0.677 nM, respectively (Chart 3.10).

Herein the combination of an R_4 methoxy group and a 1-(4-methylpiperazinyl) group leads to compound **153x**, which

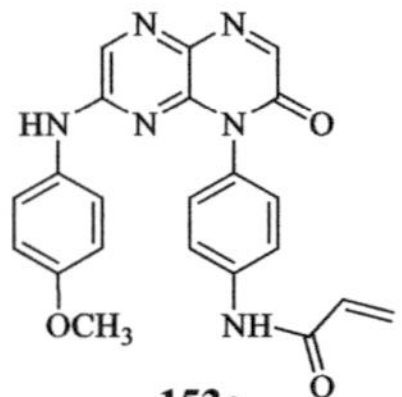

153a

WT(IC_{50}) = 51,292 nM
L858R(IC_{50}) = 3747 nM
T790M/L858R(IC_{50}) = 4802 nM

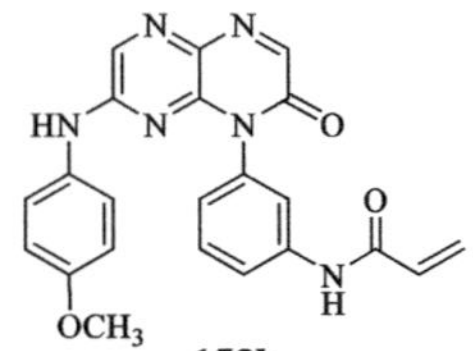

153b

WT(IC_{50}) = 13.8 nM
L858R(IC_{50}) = 14.8 nM
T790M/L858R(IC_{50}) = 5.6 nM

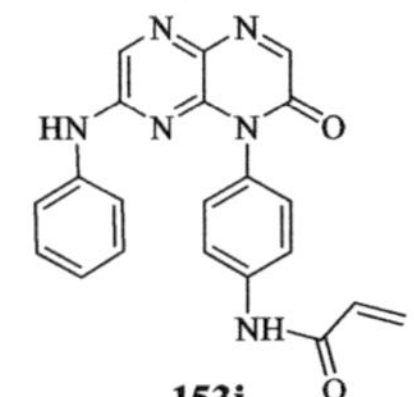

153j

WT(IC_{50}) = NA
L858R(IC_{50}) = NA
T790M/L858R(IC_{50}) = NA

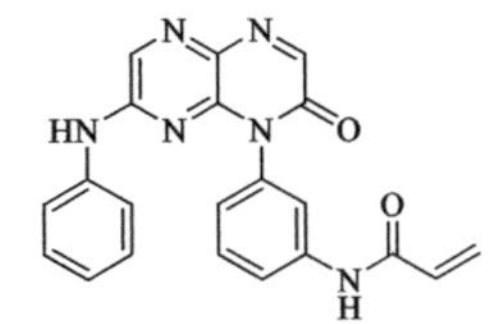

153k

WT(IC_{50}) = 12.80 nM
L858R(IC_{50}) = 11.10 nM
T790M/L858R(IC_{50}) = 4.94 nM

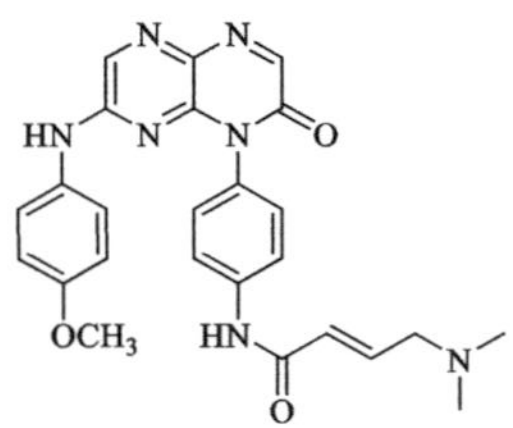

153f

WT(IC_{50}) = 59.1 nM
L858R(IC_{50}) = 76.5 nM
T790M/L858R(IC_{50}) = 31.3 nM

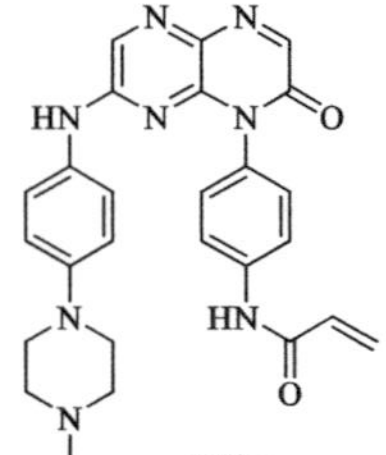

153p

WT(IC_{50}) = 1100 nM
L858R(IC_{50}) = 1510 nM
T790M/L858R(IC_{50}) = 691 nM

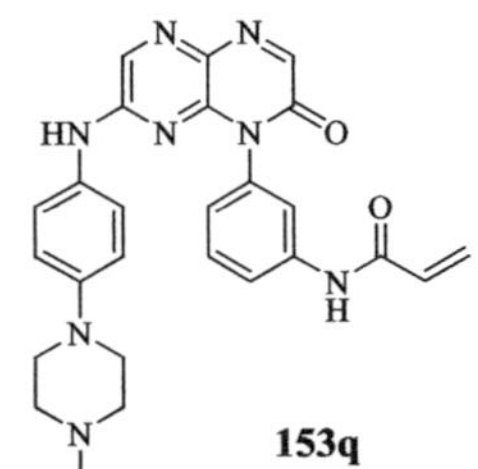

153q

WT(IC_{50}) = 1.21 nM
L858R(IC_{50}) = 1.32 nM
T790M/L858R(IC_{50}) = 0.677 nM

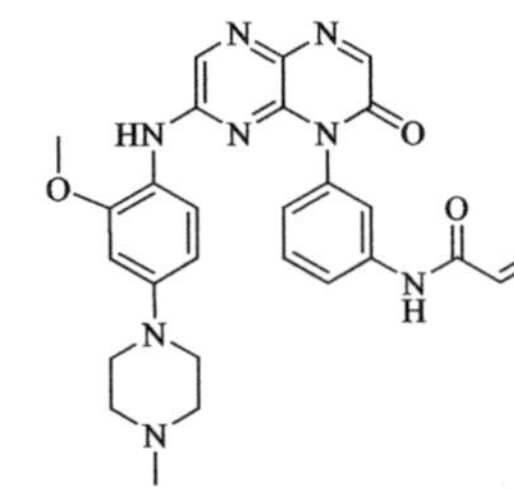

153x

WT(IC_{50}) = 3.82 nM
L858R(IC_{50}) = 2.56 nM
T790M/L858R(IC_{50}) = 1.07 nM

Chart 3.10 Antiproliferative and in vitro EGFR inhibitory activities.

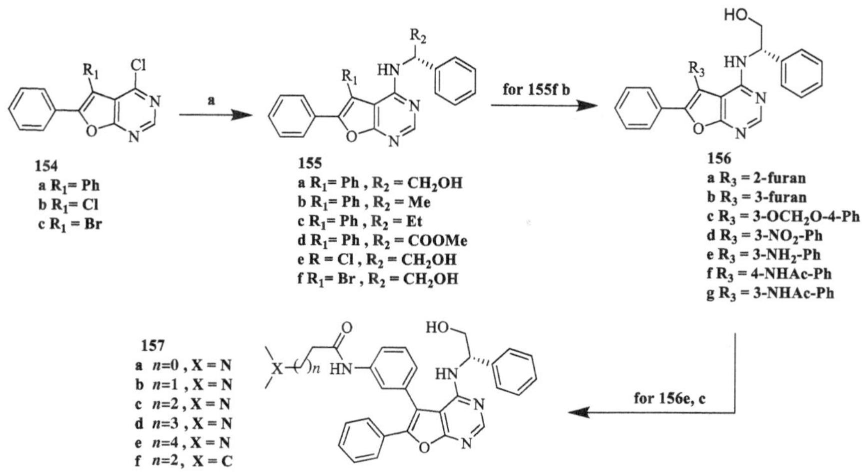

Scheme 3.20 Reagents and conditions: (a) amine, EtOH, Et_3N, reflux; (b) boronic acid, Pd(dppf)Cl_2, Na_2CO_3, dioxane, reflux; and (c) carboxylic acid, EDCI, DCM, rt.

demonstrated improved solubility while maintaining a strong inhibition against EGFR-WT, EGFR-L858R, and EGFR-T790M/L858R with IC_{50} values of 3.82, 2.56, and 1.07 nM, respectively.

Peng et al. synthesized 20 furanopyrimidine derivatives as shown in Scheme 3.20 [16]. Starting materials **154a–c** were prepared and reacted with various chiral amines via a S_NAr reaction to give chiral compounds **155a–f**. Suzuki coupling reactions were then employed to couple bromo compound **155f** with various substituted aryl boronic acids, yielding coupling products **156a–g**. Finally, reacting **156e** with the corresponding carboxylic acid in the presence of EDC yielded the desired compounds **157a–f**. **155a** inhibited WT EGFR and theL834R mutant, with IC_{50} values of 393 and 52 nM, respectively (Chart 3.11). Moreover, of the substituted phenyl derivatives **156d–e**, the amine-bearing substituent **156e** showed improved inhibitory activity compared with **155a** (Chart 3.11). In contrast the introduction of a nitro group in substituent **156d** decreased activity. As **156e** not only exhibited strong inhibition against *EGFR* but also possessed an extensible amino group, it was selected for further lead optimization.

To enhance the interactions of **156g** with the DFG motif, a series of metasubstituents bearing a *N,N*-dimethylamino group attached to different lengths of carbon chain were synthesized (Chart 3.11). Indeed the activity of amine-bearing **156g** derivatives (**157a–d**) toward both *WT* EGFR and L834R mutant was improved. Compound **157b**, containing a

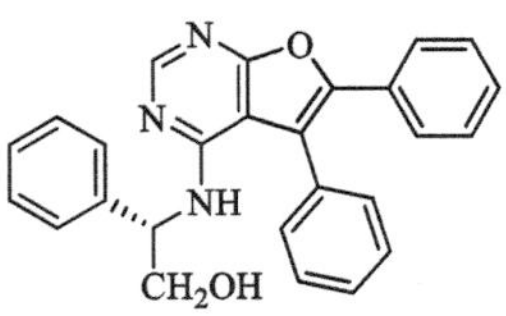

155a

WT(IC_{50}) = 393 nM

L858R(IC_{50}) = 52 nM

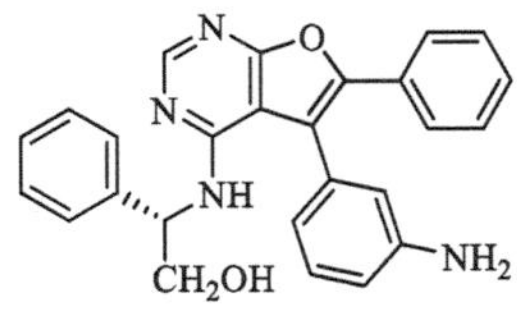

156e

WT(IC_{50}) = 202 nM

L858R(IC_{50}) = 22 nM

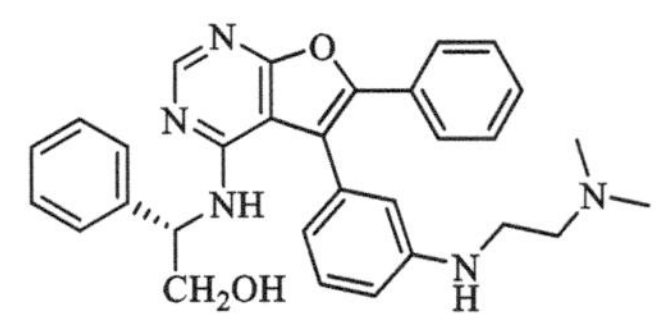

157b

WT(IC_{50}) = 8 nM

L858R(IC_{50}) = 2 nM

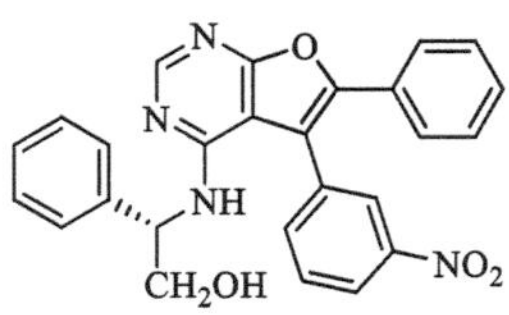

156d

WT(IC_{50}) = 820 nM

L858R(IC_{50}) = 49 nM

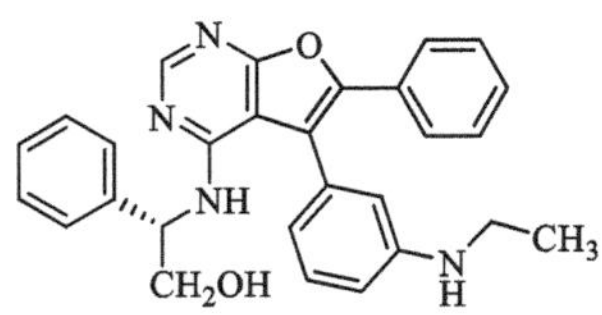

156g

WT(IC_{50}) = 218 nM

L858R(IC_{50}) = 29 nM

Chart 3.11 Antiproliferative and in vitro EGFR inhibitory activities.

two-carbon chain linker, was the most potent compound, with a 49-fold improvement in WT activity and a 26-fold improvement in L834R activity compared with **155a**. It also showed improved activity compared with gefitinib (fourfold and sevenfold improvement in WT and L834R mutant, respectively).

Concurrently with the aforementioned SAR study, **155a** was cocrystallized with the WT EGFR kinase domain and the structure complex was solved in order to elucidate the binding mode and obtain the structural insights necessary to design a more potent EGFR kinase inhibitor.

Zhou et al. synthesized the first class of purine irreversible compounds, as shown in Scheme 3.21 [17]. The aniline **162** was obtained in three steps, starting with acylation using propionyl chloride, methylation with methyl iodide, and hydrogenation of the 4-nitroaniline **158**. Tetrahydropyran (THP) protected 2,6-dichloropurine (**163**) was regioselectively aminated at the 6-position to afford intermediate **164**. Buchwald coupling with **162** in the presence of a carbene ligand provided **165**, which was converted into **166a** through acidic deprotection of THP, followed by acylation with acryloyl chloride. The purine analogs of **166b–e** were synthesized in an analogous manner using the appropriate amines and anilines.

A second sublibrary of pyrrolopyrimdine-derived acrylamides was synthesized, as depicted in Scheme 3.22. Regioselective iodination

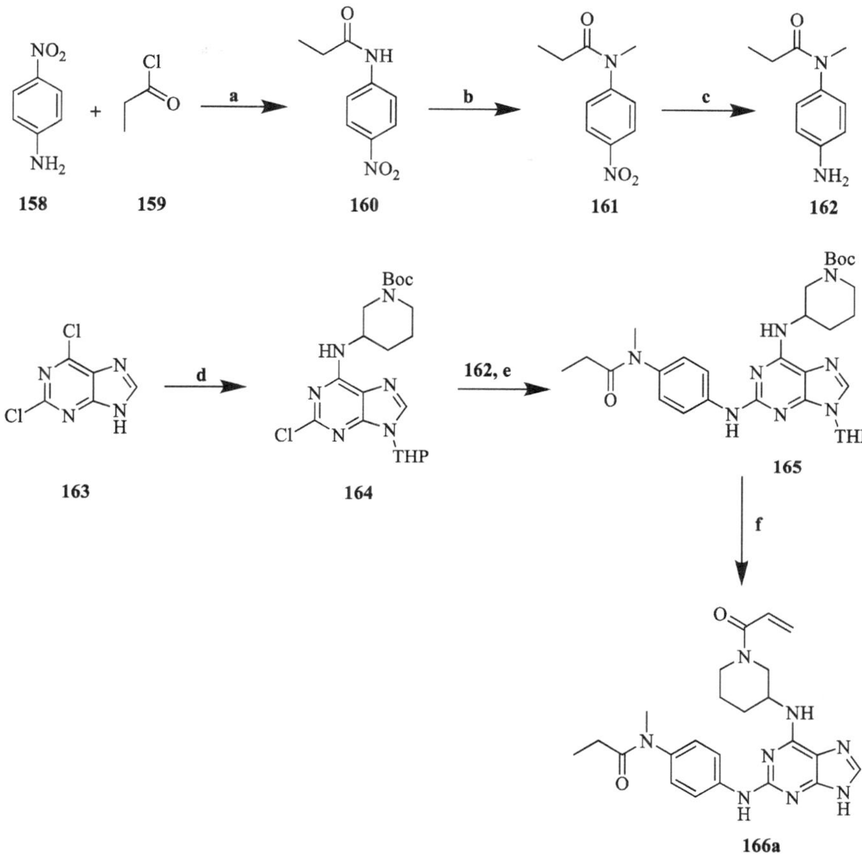

Scheme 3.21 Reagents and conditions: (a) pyridine, CH_2Cl_2; (b) NaH, MeI, DMF; (c) Pd/C, H_2, MeOH; (d) 3,4-2H-dihydropyran; 1-BOC-3-aminopiperidine, DIEA, BuOH, 60°C; (e) $Pd_2(dba)_3$, 1,3-bis(2,6-diisopropyphenyl) imidazonium chloride, NaOtBu, 1,4-dioxane, 95°C; and (f) (i) HCl, EtOH and (ii) acryloyl chloride, DIEA, CH_2Cl_2.

of 6-chlorodeazapurine (**167**) was achieved with *N*-iodosuccinimide. The resulting intermediate (**168**) was protected with SEM. Selective Suzuki coupling at the C5-position of **168** gave rise to **169**. Intermediate **170** was obtained by nucleophilic substitution at the C4-position. Compound **171** was obtained after acidic deprotection of **170**, followed by acylation with acryloyl chloride and the sequential removal of the hydroxymethyl group.

The synthetic scheme of one typical compound **177h** is shown in Scheme 3.23. 3-Nitrophenol was regioselectively introduced to the C4-position of 2,4,5-trichloropyrimidine to afford intermediate **174**. Amination of **174** with 4-(4-methylpiperazin-1-yl) aniline was accomplished in the presence of trifluoroacetic acid at 100°C. The nitro group in **175** was reduced by

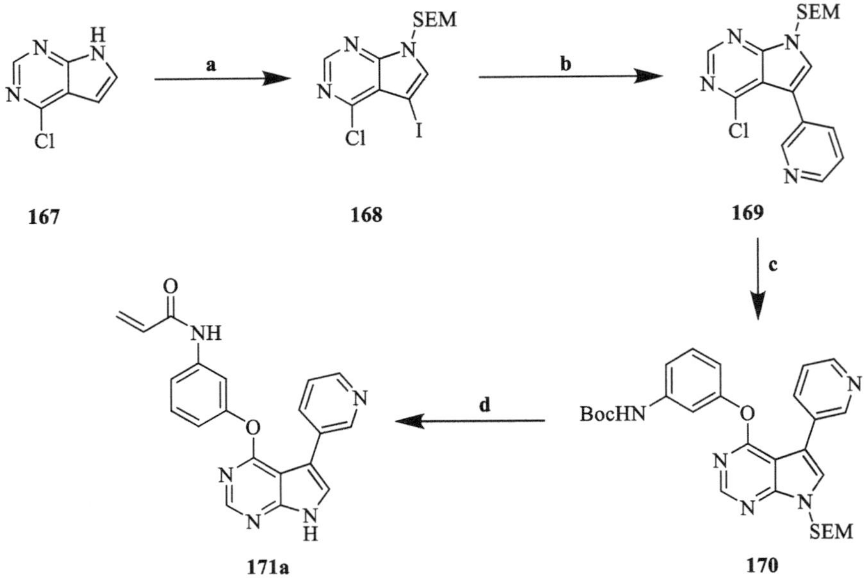

Scheme 3.22 Reagents and conditions: (a) (i) NIS, CH_2Cl_2; (ii) NaH, SEM–Cl, DMF; (b) pyridine-3-boronic acid, Pd(dppf)Cl_2, dppf, Na_2CO_3(aq.)/dioxane (1:2); (c) 3-BOC-aminophenol, K_2CO_3, DMSO, 100°C; and (d) (i) TFA, CH_2Cl_2; (ii) acryloyl chloride, DIEA, CH_2Cl_2; and (iii) $NaHCO_3$(aq.), THF.

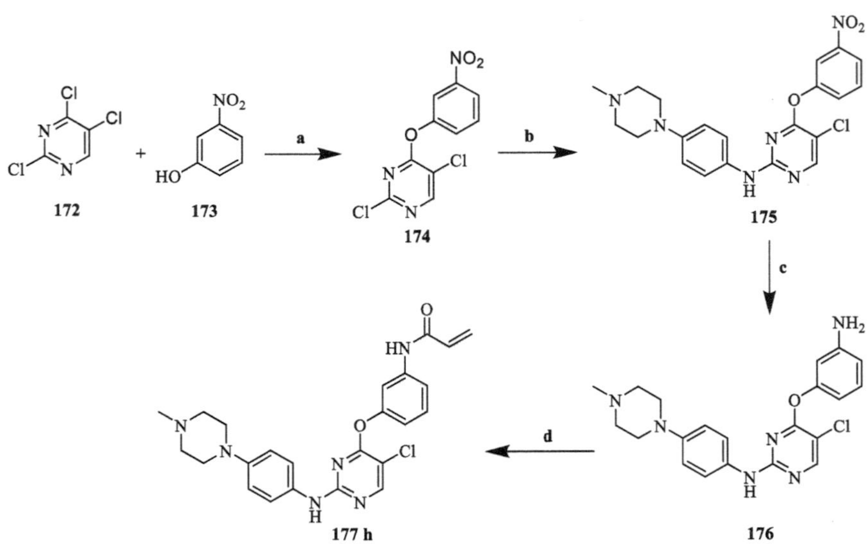

Scheme 3.23 Reagents and conditions: (a) K_2CO_3, DMF; (b) 4-(4-methylpiperazin-1-yl) aniline, TFA, 2-BuOH, 100°C; (c) PtO_2, H_2, MeOH; and (d) acryloyl chloride, DIEA, CH_2Cl_2.

PtO_2/H_2 without concominant reduction of the C5 chloride. Acylation of **176** by acryloyl chloride gave rise to **177h**. The pyrimidine analogs of **177** were obtained in a similar manner using the appropriate phenols and anilines.

Structure activity relationship explains that the potency of simple C_2-fluoroaniline compound **177a** may have been limited by its poor water solubility (Chart 3.12). The replacement of the fluoro group with functionality that would be expected to improve water solubility such as morpholine **177c**, 4-hydroxypiperidine **177d**, *N*-methylpiperidine **177e**, **177f**, **177g**, and piperazine **177i**, piperazine derivative **177j** resulted in 5–40-fold increases in potency. The removal of the methoxy group (**177h**) improved potency to the single digit nanomolar EC_{50} range. However, **177i** is more selective for *EGFR* than **177h** when evaluated in biochemical and cellular assays of other kinases. Next, they optimized the heteroatom linkage at the pyrimidine 4-position. The ether linkage was replaced with sulfur (**178a**, **178b**) and NH (**178c**) (Chart 3.12). Both replacements were tolerated as evidenced by **177i**, **178a**, and **178c**. The same trend was observed for analogs without substituents at the ortho position of C2-aniline (**177i** vs **178b**).

The analogs of **179** and **179a–c** with Cl replaced by H, Br, CF_3 were synthesized using the same route as depicted in Scheme 3.23. Among them, **179** hits the most potent (Chart 3.12) because it is likely that the chloro group engages in a dipole interaction with the thioether of the methionine at position 790 of *EGFR*.

Chan et al. reported the 2,4-diarylamino-pyrimidines as kinase inhibitors cotargeting *EGFR L858R/T790M* and IGF1R [18]. Compound **188a–x** and the derivatives were readily prepared by following procedures outlined in Scheme 3.24. In brief, a regioselective nucleophilic substitution of 2,4-dichloro pyrimidines (**180**) with commercially available chemicals **181** provided intermediates **182** or **185**.

Compounds **182** or **185** were then reacted with appropriately substituted nitroanilines to produce **183** or **186** either by a p-toluenesulfonic acid (p-TSA) catalyzed S_NAr reaction or under Buchwald-Hartwig coupling conditions (for compounds where X = O). Subsequent iron-mediated nitro reduction of **183** or **186** then yielded key intermediates **184** or **187**. Compounds **184** or **187** were coupled with (*E*)-4-(dimethylamino)-but-2-enoic acid to produce the final compounds **188a–x**.

Among the synthesized compounds, **188g** potently inhibited both of the kinases with low nanomolar IC_{50} values, but was significantly less potent at inhibiting WT *EGFR*. This compound also displayed a good to moderate kinase selectivity profile against a panel of 468 kinases. Moreover,

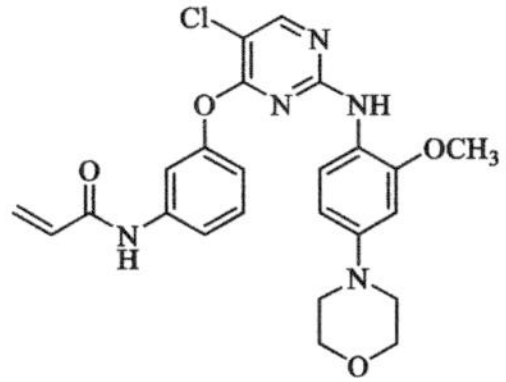

177c
T790M(IC_{50}) = 0.15 μM
EGFR del(IC_{50}) = ND

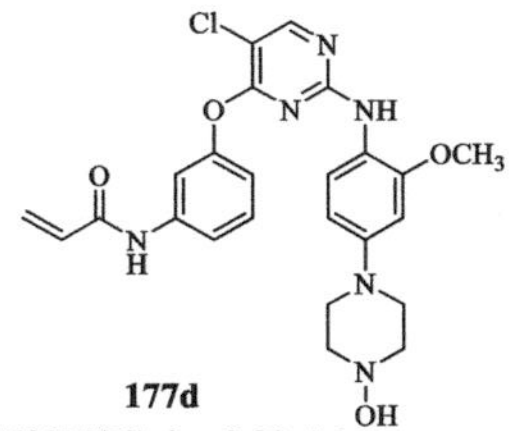

177d
T790M(IC_{50}) = 0.80 μM
EGFR del(IC_{50}) = 0.20 μM

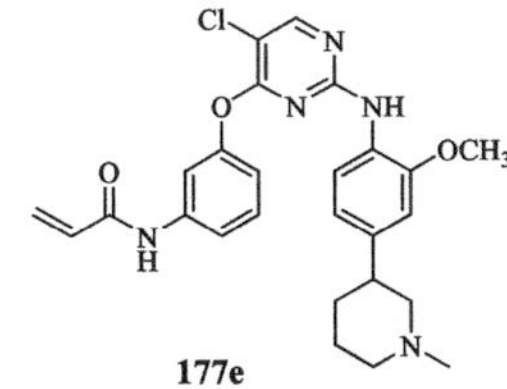

177e
T790M(IC_{50}) = 0.20 μM
EGFR del(IC_{50}) = 0.10 μM

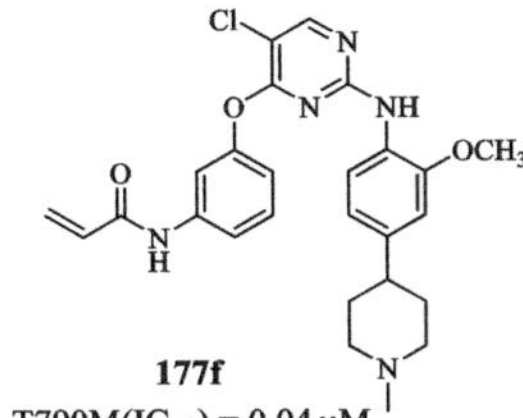

177f
T790M(IC_{50}) = 0.04 μM
EGFR del(IC_{50}) = 0.10 μM

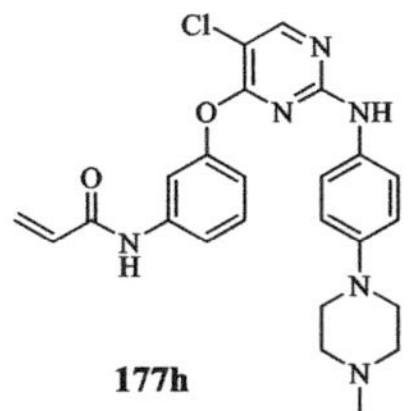

177h
T790M(IC_{50}) = 0.003 μM
EGFR del(IC_{50}) = 0.015 μM

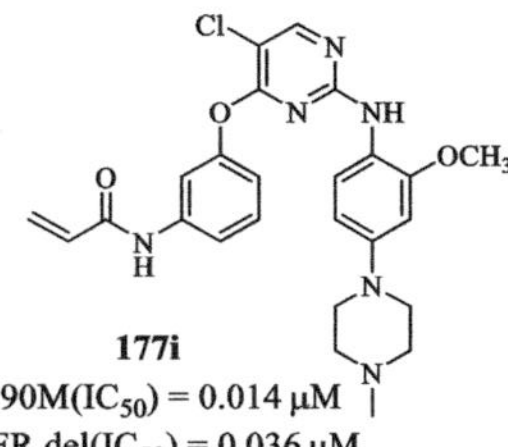

177i
T790M(IC_{50}) = 0.014 μM
EGFR del(IC_{50}) = 0.036 μM

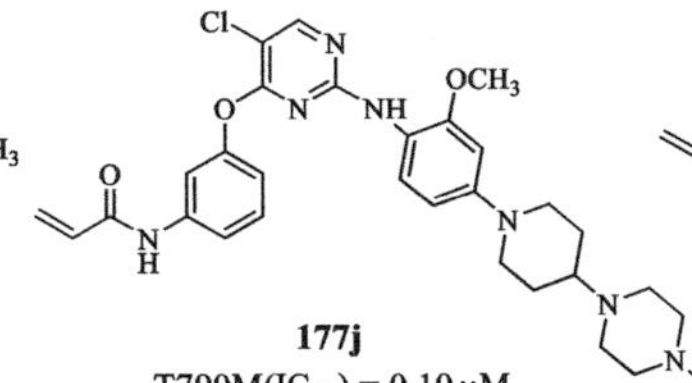

177j
T790M(IC_{50}) = 0.10 μM
EGFR del(IC_{50}) = 0.080 μM

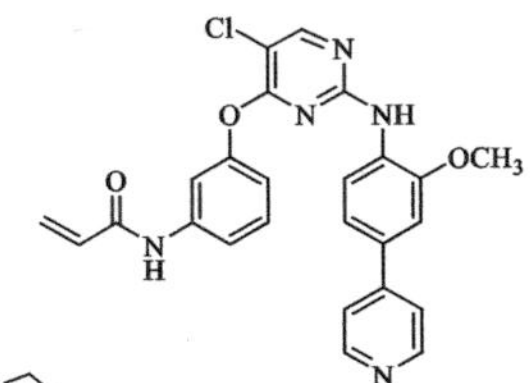

177a
T790M(IC_{50}) = 0.80 μM
EGFR del(IC_{50}) = ND

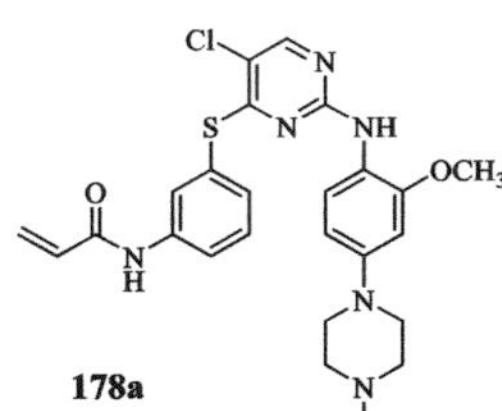

178a
T790M(IC_{50}) = 0.014 μM
EGFR del(IC_{50}) = 0.020 μM

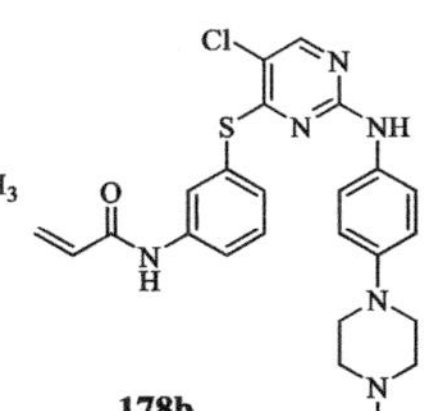

178b
T790M(IC_{50}) = 0.008 μM
EGFR del(IC_{50}) = 0.006 μM

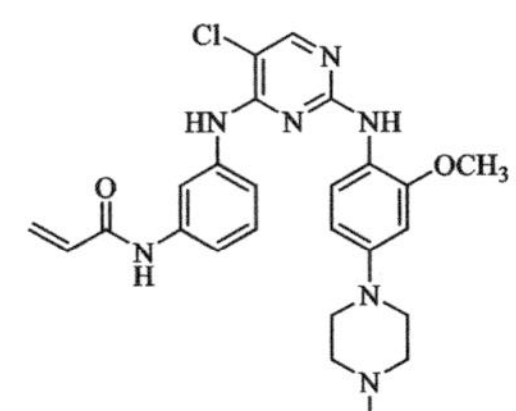

178c
T790M(IC_{50}) = 0.020 μM
EGFR del(IC_{50}) = 0.040 μM

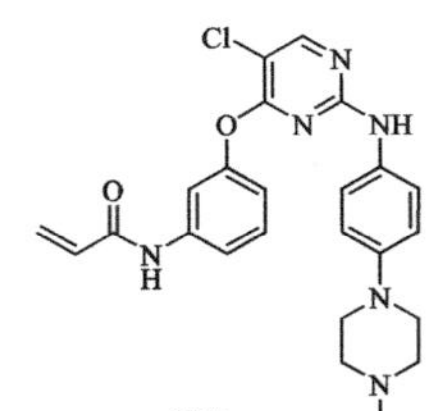

179
T790M(IC_{50}) = 0.003 μM
EGFR del(IC_{50}) = 0.015 μM

Chart 3.12 Antiproliferative and in vitro EGFR inhibitory activities.

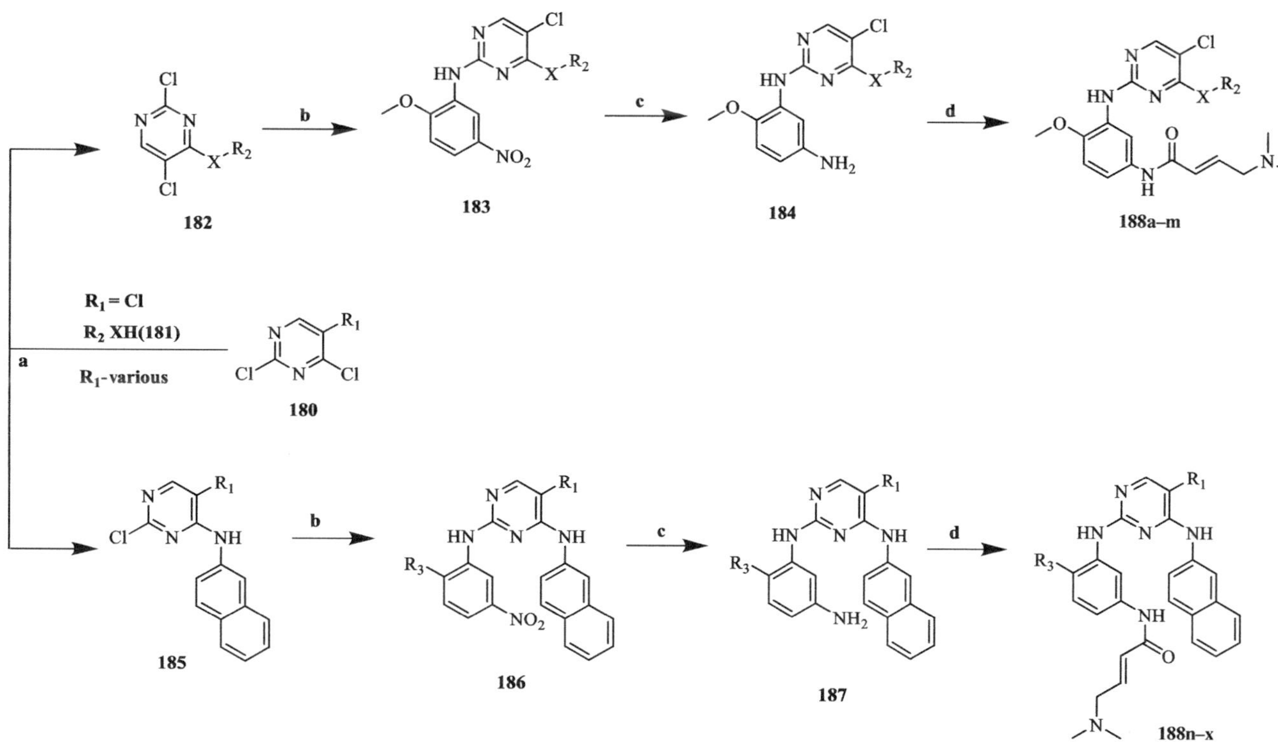

Scheme 3.24 Reagents and conditions: (a) Na_2CO_3, EtOH, rt or *n*-BuLi, THF, −40°C, 68%–90%; (b) p-TSA, 2-pentanol, reflux or $Pd(OAc)_2$, Xantphos, Cs_2CO_3, 1,4-dioxane, reflux, 83%–95%; (c) Fe, NH_4Cl, EtOH, water, 60°C, 72%–90%; and (d) (*E*)-4-(dimethylamino) but-2-enoic acid, 1-[bis(dimethylamino) methylene]-1*H*-1,2,3-triazolo[4,5-*b*]pyridinium 3-oxid hexafluorophosphate (HATU), *N*,*N*-diisopropylethylamine (DIEA), THF, rt, 60%–88%.

188g strongly suppresses the proliferation of rociletinib (CO1686)-resistant H1975-IGF1R cancer cells, suggesting its promising potential as a new lead compound for drug discovery overcoming acquired resistance to the second- and third-generation *EGFR* inhibitors. The α-naphthylamino compound **188f** was designed and synthesized to potently inhibit EGFR L858R/T790M and IGF1R with IC_{50} values of 0.018 and 0.015 μM, respectively. This compound also spared WT EGFR with a ratio of 14-fold [IC_{50} (WT EGFR)/IC_{50} (EGFR-L858R/T790M)], as shown in Chart 3.13.

Wurz et al. designed a synthesis of the 7-oxopyrido[2,3-*d*]pyrimidinyl core that would allow for versatile analoging (Scheme 3.25) [19]. Commercially available ethyl-4-chloro-2-(methylthio)-pyrimidine-5-carboxylate (**189**) underwent an S_NAr-type displacement with mono-BOC-protected 1,3-phenylenediamine to produce diaryl amine **190** in a high yield. The reduction of the ethyl ester with lithium aluminum hydride afforded the corresponding alcohol, which was subsequently oxidized with MnO_2 to aldehyde **191**. Treatment of the aldehyde with the appropriate Grignard reagent followed by the oxidation of the resulting secondary alcohol to ketone **192** provided a substrate for an aldol-like cyclocondensation reaction with ester enolates to afford the 7-oxopyridopyrimidinyl scaffold (**193**). In this manner, analogs with the desired substitution patterns at the 5- and 6-positions of the pyridone ring could be prepared. The BOC protecting group was removed with TFA, and the acrylamide warhead was introduced using acryloyl chloride to provide intermediate **194**. Oxidation of the methyl(thio) pyrimidinyl motif to the sulfoxide with m-CPBA furnished sulfoxide (**195**), which underwent S_NAr displacement with the desired aniline to afford the final product **196**.

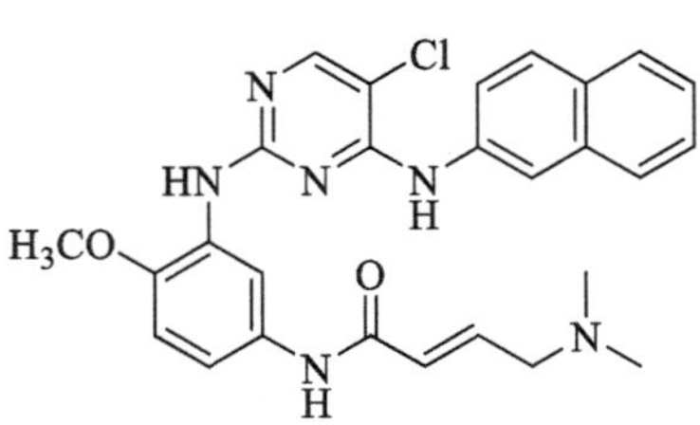

188f

EGFR (WT) (IC_{50}) = 0.260 μM
EGFR (DM) (IC_{50}) = 0.018 μM
IGFIR (IC_{50}) = 0.015 μM

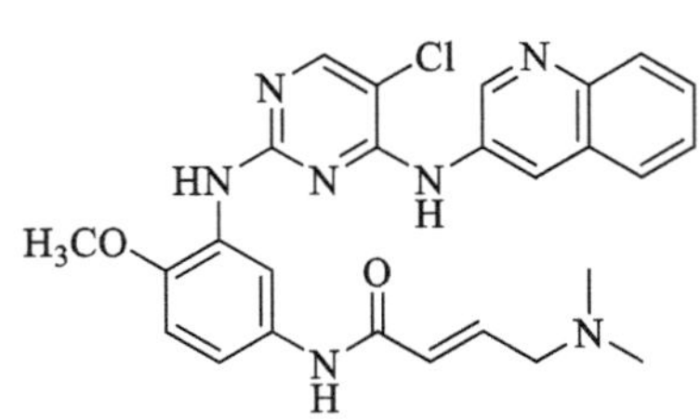

188g

EGFR (WT) (IC_{50}) = 0.848 μM
EGFR (DM) (IC_{50}) = 0.038 μM
IGFIR (IC_{50}) = 0.013 μM

Chart 3.13 Antiproliferative and in vitro EGFR inhibitory activities.

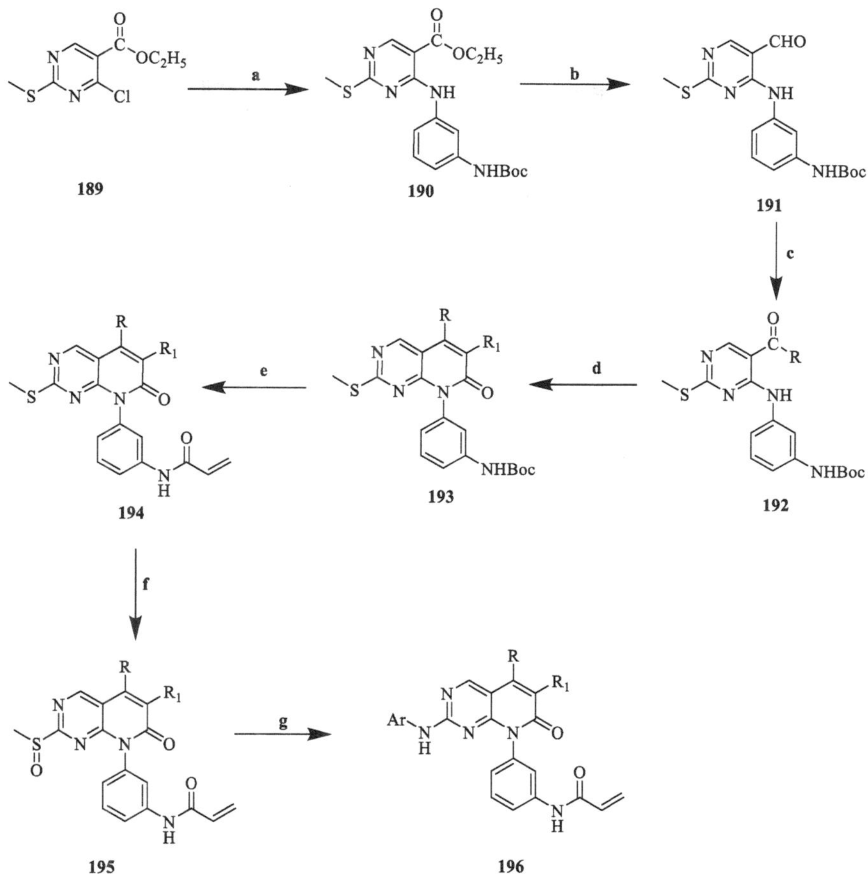

Scheme 3.25 Reagents and conditions: (a) K_2CO_3 (2 equiv.), *m*-NH_2-C_6H_4-NHBoc, DMF, 80°C, 86%; (b) (i) $LiAlH_4$ (2.5 equiv.), THF, 0°C and (ii) MnO_2, $CHCl_3$, RT, 66% (2 steps); (c) (i) RMgBr (3.2 equiv.), THF, 0°C rt, 1 h and (ii) MnO_2, $CHCl_3$, 60°C, 16 h; (d) $R_1CH_2CO_2Et$, LiHMDS, THF, −78°C rt; (e) (i) TFA, CH_2Cl_2, RT 30 min and (ii) acryloyl chloride, Hunig's base, CH_2Cl_2, 0°C rt, 1 h; (f) *m*-CPBA (1.1 equiv.), CH_2Cl_2, 0°C, 1 h; and (g) DMAC, Ar-NH_2 (2.2 equiv.), 95°C, 6 h.

As demand for larger quantities of material grew to facilitate additional studies, a more efficient synthetic route was developed (Scheme 3.26).

This alternative route used the commercially available 5-bromo-2,4-dichloropyrimidine (**197**) and a S_NAr displacement with the mono-BOC protected 1,3-phenylenediamine to furnish **198**. Heck coupling with crotonic acid followed by intramolecular dehydration of the resulting adduct with Ac_2O afforded the 7-oxopyrido[2,3-*d*]pyrimidine core **199** in modest yields. Subsequent BOC deprotection using HCl, followed by neutralization with NaOH and a treatment with acryloyl chloride, yielded the

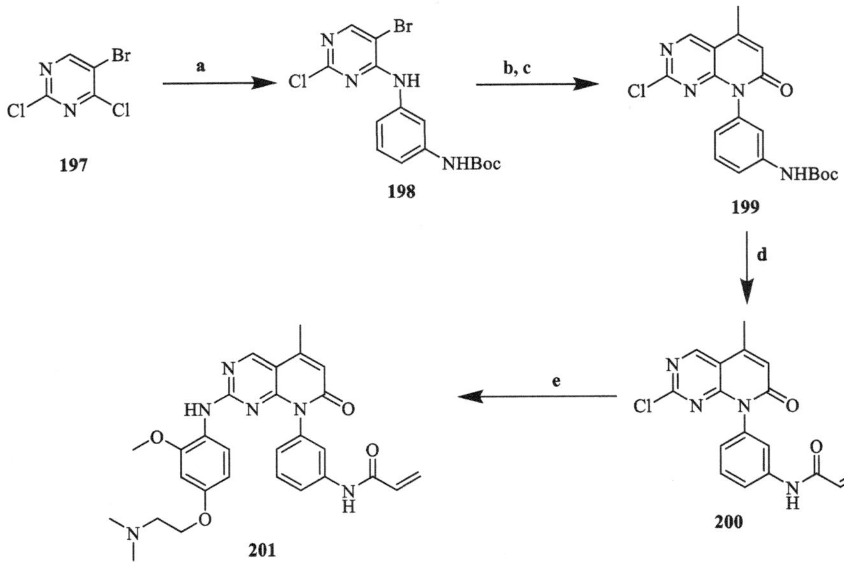

Scheme 3.26 Reagents and conditions: (a) K_2CO_3 (1.3 equiv.), *m*-NH_2–C_6H_4-NHBoc, DMF, RT, 16 h, 98%; (b) $PdCl_2(PhCN)_2$, $P(otolyl)_3$, crotonic acid, DIEA, PhMe, 60°C; (c) Ac_2O, 60°C, 1 h, 60% (two steps); (d) (i) 4N HCl, 1,4-dioxane, 50°C and (ii) 10N NaOH, acryloyl chloride, 92% (two steps); and (e) Ar-NH_2·HCl (1.1 equiv.), 2,2,2-trifluoroethanol, reflux, 16 h, 51%.

compound **200**. Anilines could be introduced via the S_NAr displacement of the chloride in refluxing 2,2,2-trifluoroethanol to achieve **201**.

A novel class of T790M mutant-selective EGFR inhibitors was designed and synthesized, as shown in Chart 3.14. Compound **201g** was highly selective against a panel of 100 kinases and had excellent cellular potency in proliferation assays. Compound **201g** was orally efficacious in an H1975 plug pharmacodynamic assay in mice resulting in a >50% inhibition of p-EGFR for >12 h. Its selectivity was also confirmed in the MSD assay with

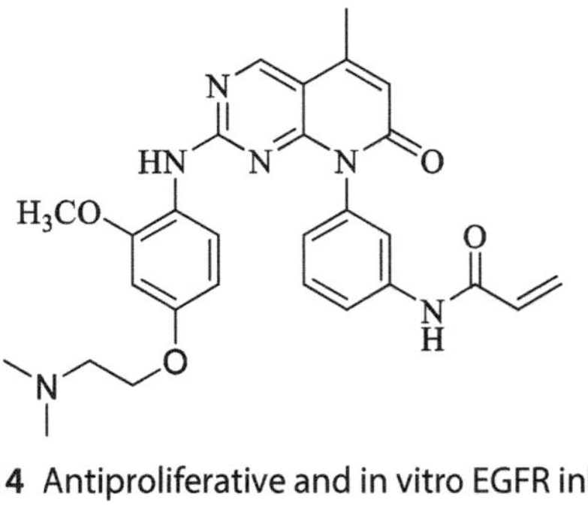

201g
H1975 (IC_{50}) = 4 nM
HCC 827 (IC_{50}) = 28 nM
A431 (IC_{50}) = 2400 nM

Chart 3.14 Antiproliferative and in vitro EGFR inhibitory activities.

>100-fold selectivity over WT EGFR in an H1975 xenograft model, oral dosing of compound **201g** at 30 mg/kg daily for 2 weeks led to significant tumor growth inhibition with no observed weight loss. In an exploratory 4-day rat toxicology study, no hepatobiliary toxicity was observed.

Chang et al. prepared the designed compounds **212a–v** and **213a–b** by using ethyl-4-chloro-2-(methylthio)-pyrimidine-5-carboxylate (202) as the starting material (Scheme 3.27) [20]. Briefly a direct nucleophilic coupling of compound **202** with tert-butyl-3-aminophenyl carbamate (**203**) almost quantitatively produced ethyl-4-(3-(tertbutoxycarbonyl)phenylamino)-2-(methylthio)pyrimidine-5-carboxylate (**204**) as a white solid. The reduction of compound **204** with lithium aluminum hydride yielded tert-butyl 3-(5-(hydroxymethyl)-2-(methylthio)pyrimidin-4-ylamino)-phenylcarbamate (**205**), which was followed by selective oxidation to produce the aldehyde **206** in a good yield. A variety of diamine intermediates **207** could be prepared by a reductive amination of aldehyde **206** with different amines. With diamine compounds **207** in hand the key intermediates **209** were readily synthesized by direct cyclization with triphosgene and followed by oxidation. Compound **209** coupled with different substituted anilines to produce precursor compounds **210**, which were further deprotected **211** with trifluoroacetic acid and reacted with acryloyl chloride to obtain the conformation-constrained EGFR inhibitors **212a–v**, **213a–b**.

All the synthesized compounds significantly inhibited the enzymatic activities of different mutants of EGFR as well as the WT EGFR kinase. The most potent compounds **212q** and **212s** displayed good potencies, with IC_{50} values in the high picomolar range against all types of EGFR kinase, including the clinical resistance-related EGFR-T790M mutant as shown in Chart 3.15. A further kinase profiling study also suggested that the compounds displayed remarkable selectivity on EGFR.

Ji et al. reported the design, synthesis, and biological evaluation of novel 6-alkenylamides substituted of 4-anilinothieno[2,3-*d*]pyrimidines as irreversible EGFR inhibitors compounds **221a–o** was synthesized according to Scheme 3.28 [21]. The thieno[2,3-*d*]pyrimidine core **216** was obtained from commercially available compound **214** via cyclization in condition of microwave heating. The key intermediate **220** was generated from compound **216** via nitrification, chlorination, amination, and reduction. Compounds **221a–f** and **221l–n** were obtained from compound **220** via condensation with different groups substituted acetyl chlorides.

Compound **221o** was obtained from compound **220** via condensation with chloroacetyl chloride. Compounds **221g–k** was obtained from

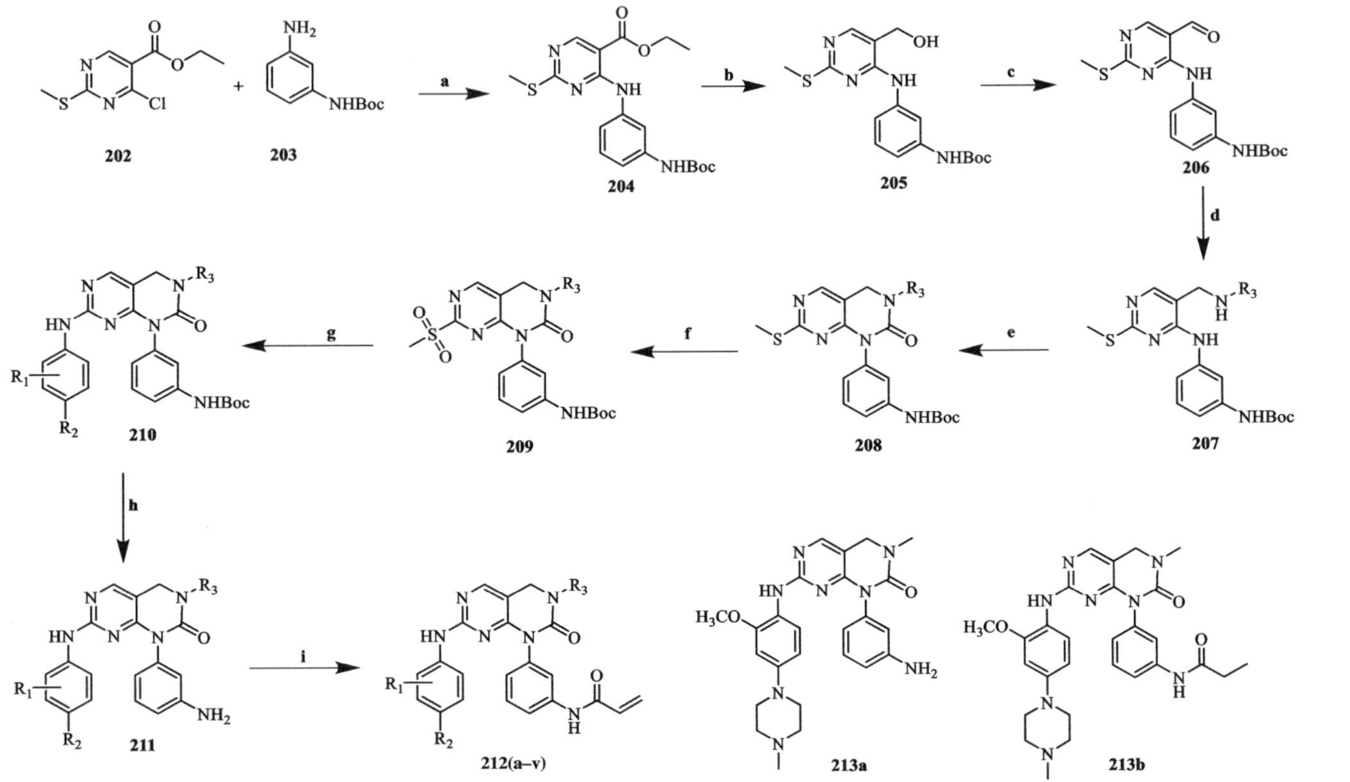

Scheme 3.27 Reagent and conditions: (a) K_2CO_3 (2.0 equiv.), dimethylformamide (DMF), 60°C, overnight, 97%; (b) lithium aluminum hydride (LAH, 2.0 equiv.), THF, −40°C to rt, 57%; (c) MnO_2, dichloromethane, rt, overnight, 85%; (d) R_2NH_2, AcOH, $NaBH_4$, MeOH, 0°C to rt, overnight; (e) triphosgene (0.34 equiv.), Et_3N (4.0 equiv.), 0°C to rt, 2.0 h; (f) 3-chloroperbenzoic acid (m-CPBA, 3.0 equiv.), dichloromethane, rt, 3.0 h; (g) R_1NH_2 (1.0 equiv.), trifluoroacetic acid (1.0 equiv.), 2-BuOH, 110°C, sealed tube, 18.0 h, 30%–60%; (h) trifluoroacetic acide, dichloromethane, rt; and (i) acryloyl chloride (1.0 equiv.), *N,N*-diisopropylethylamine (DIPEA, 1.0 equiv.), 0°C to rt.

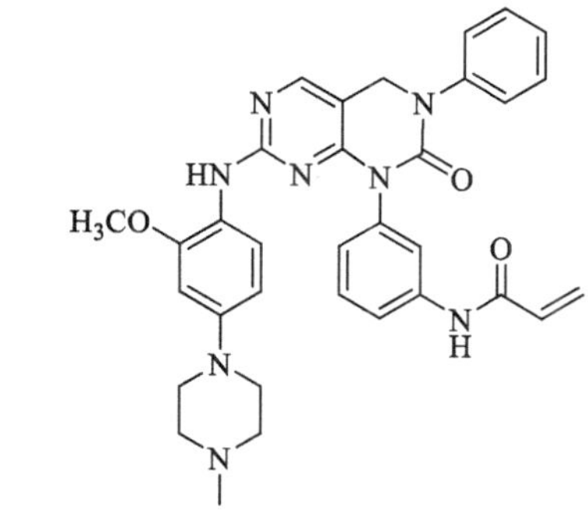

212q
WT(IC_{50}) = 0.30 nM
T790M(IC_{50}) = 0.51 nM
L858R(IC_{50}) = 0.34 nM
L861Q(IC_{50}) = 0.93 nM
L858R/T790M(IC_{50}) = 0.98 nM

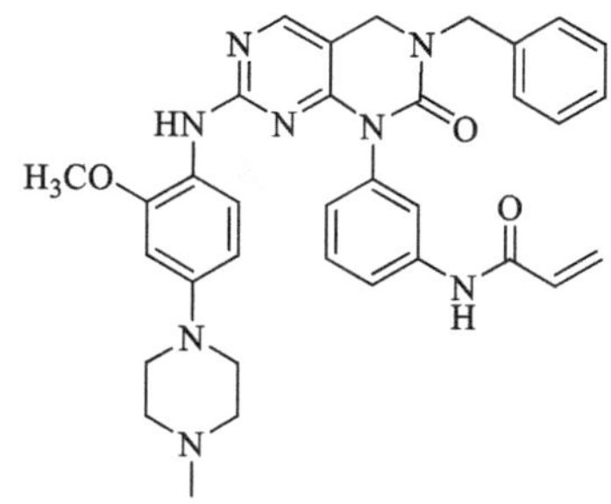

212s
WT(IC_{50}) = 0.29 nM
T790M(IC_{50}) = 0.67 nM
L858R(IC_{50}) = 0.38 nM
L861Q(IC_{50}) = 0.72 nM
L858R/T790M(IC_{50}) = 0.93 nM

Chart 3.15 Antiproliferative and in vitro EGFR inhibitory activities.

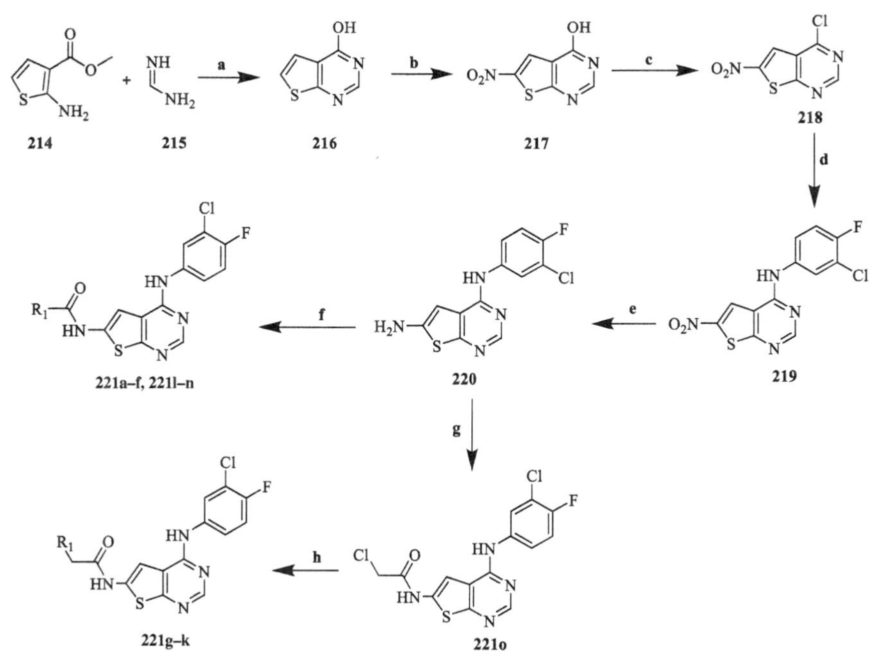

Scheme 3.28 Reagents and conditions: (a) macrowave, 4–5 min; (b) H_2SO_4/HNO_3, 90°C; (c) $POCl_3$, 110°C or $POCl_3$, TEA, CH_3CN, 70°C; (d) 3-chloro-4-fluoroaniline, i-PrOH, 80°C; (e) Fe, NH_4Cl, EtOH, 45°C; (f) different groups substituted acetyl chloride, TEA, THF; (g) chloroacetyl chloride, pyridine, THF, 60°C; and (h) different amine, THF, 60°C.

compound **221o**, which underwent a nucleophilic reaction with different amines. The synthesis of compounds **226a–z** was similar with the synthesis of compounds **221a–o** (Scheme 3.29). The difference was that compound **225** was obtained from compound **220** via condensation with compound **224**. Compounds **226a–z** were afforded from compound **225** underwent a nucleophilic reaction with different amines. The intermediate **224** was synthesized from crotonic acid, which was similar with synthesis of neratinib.

All the synthesized compounds (**221a–o** and **226a–z**) were evaluated for the in vitro inhibitory activities on WT EGFR and EGFR T790M/L858R mutant, as well as the antiproliferative effects on human epithelial carcinoma A431 cells and gefitinib-resistant NCI-H1975 lung adenocarcinoma cells, which expressed high levels of EGFR-WT or EGFR T790M/L858R, respectively (Chart 3.16).

As shown in Chart 3.16, compounds **221a–o** showed good and moderate inhibitory activities against *EGFR*-WT (IC_{50} = 0.016–10 μM) and

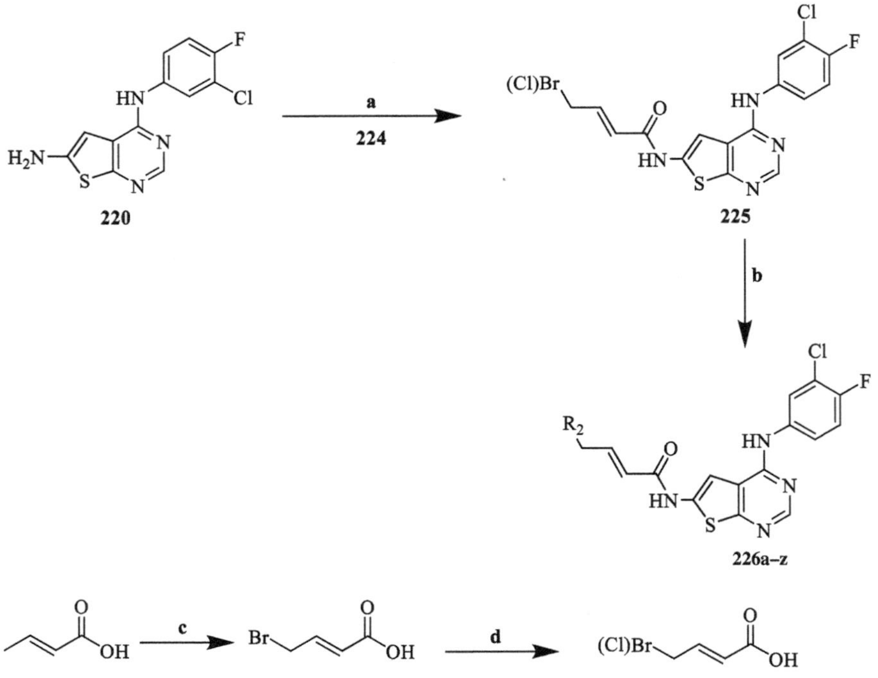

Scheme 3.29 Reagents and conditions: (a) (*E*)-4-bromobut-2-enoyl chloride, pyridine, THF, 60°C; (b) different amine, NaI, DMF, overnight; (c) NBS, $(PhCO)_2O$, CCl_4, reflux; and (d) $SOCl_2$, DCM, rt.

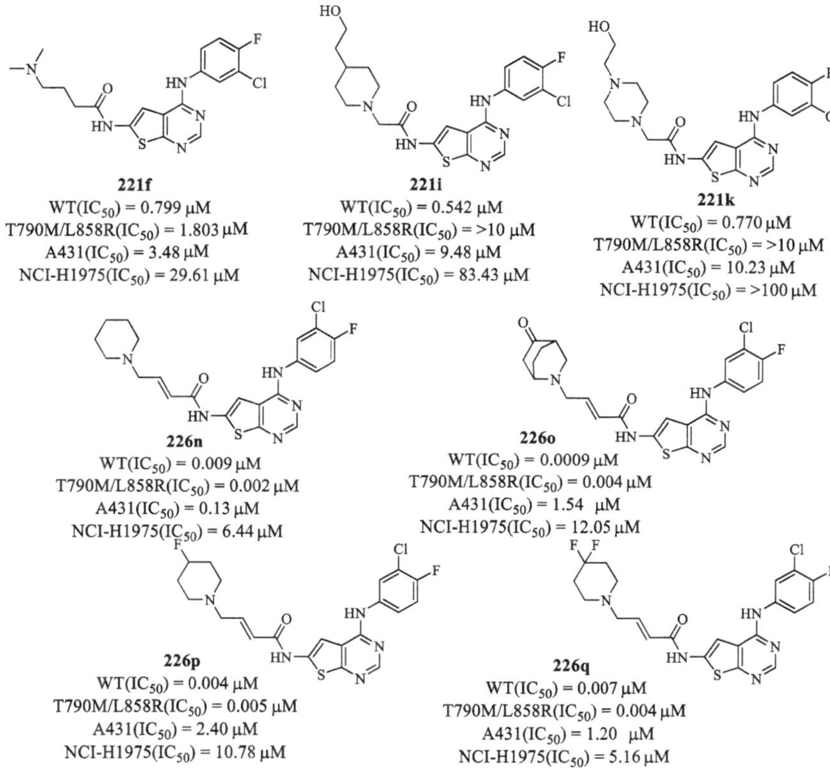

Chart 3.16 Antiproliferative and in vitro EGFR inhibitory activities.

EGFR T790M/L858R (IC_{50} = 0.024–10 μM). When R_1 was aliphatic alkane groups (compounds **221a–c**) the inhibitory activities against *WT EGFR* and *EGFR T790M/L858R* were very weak (IC_{50} > 10.000 μM). Introduction of O and N atoms in aliphatic alkane groups (compounds **221d–k**), led to an improvement in inhibitory activities against *EGFR*: compounds **221f** (*WT EGFR*, IC_{50} = 0.799 μM), **221i** (*WT EGFR*, IC_{50} = 0.542 μM), and **221k** (*WT EGFR*, IC_{50} = 0.770 μM).

Compounds **226a–z** were designed synthesized and evaluated as shown in Chart 3.16. Compounds **226n** (WT EGFR, IC_{50} = 0.009 μM; EGFR T790M/L858R, IC_{50} = 0.002 μM), **226o** (WT EGFR, IC_{50} = 0.0009 μM; EGFR T790M/L858R, IC_{50} = 0.004 μM), **226p** (WT EGFR, IC_{50} = 0.004 μM; EGFR T790M/L858R, IC50 = 0.005 μM), and **226q** (WT EGFR, IC_{50} = 0.007 μM; EGFR T790M/L858R, IC_{50} = 0.004 μM) showed excellent inhibitory activities against WT EGFR and EGFR T790M/L858R [21].

Wu et al. designed and synthesized tetrahydropyridothieno-[2,3-*d*]-pyrimidine scaffold-based *EGFR* kinase inhibitors [22]. Compounds **231**, **231a–d**, **232**, and **232a–d** were synthesized as shown in Scheme 3.30. Appropriate cyclic ketones **227a–c** were condensed with ethyl cyanoacetate under basic conditions and then cyclized in the presence of sulfur to construct the thiophene cores **228a–c** through Gewald reaction. The thieno[2,3-*d*]pyrimidine ring system was then constructed using a modified Niementowski quinazoline synthesis by condensation of **228a–c** with formamide to afford **229a–c**. Finally, chlorination of pyrimidone **229a–c** with phosphrous oxychloride gave the intermediate **230a–c**, which underwent nucleophilic reaction with appropriate amines to give the desired compounds **231**, **231a–d**, **232**, and **232a–d**.

For the preparation of thieonpyrimidines **235–251** with the nitrogen bridge in the alicyclic ring, the key intermediate **230c** was constructed starting from 4-*oxo*-piperidine-1-carboxylic acid tert-butyl ester (**227c**) using a similar sequence of reaction as that for obtaining **230a–c** (Scheme 3.30). Nucleophilic reaction of **244c** with appropriate primary amines gave **233a–f**, which were then subjected to BOC deprotection using TFA resulting in the intermediates **234a–f**. The desired compounds **235–237** and **239–242** were obtained by amide coupling reaction of appropriate secondary amine (**234**) and the acid (RCO_2H) using EDCI as a coupling agent. For the preparation of compounds **244–251** a two-step reaction sequence was utilized to get

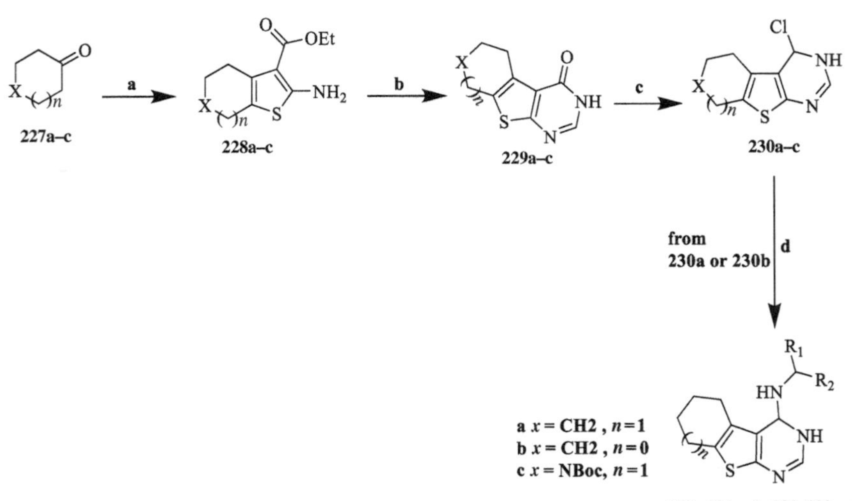

Scheme 3.30 Reagents and conditions: (a) $NCCH_2CO_2Et$, S8, EtOH, Et_3N, reflux or rt, 16 h; (b) formamide, reflux, 2 h or formamidine acetate, DMF, 100°C, 16 h; (c) $POCl_3$, 0–60°C, 3 h; and (d) primary amine ($R_1R_2CHNH_2$), *n*-BuOH, reflux, 8 h.

the final products; an initial amide coupling of 4-bromo crotonoic acid with appropriate intermediate **233**, followed by an S_N2 reaction with different secondary amines (R_2NH), was carried out. For the sulfonamide derivative **238**, acylation of **234a** with sulfonyl chloride was carried out (Scheme 3.31).

Designing in an unusual *S*-chiral 2-phenyl-2-aminoethanol side chain identified from a counter-screening approach, followed by the introduction of a Michael acceptor group into the tricyclic core through a diversity point "NH" in HTS hit **231**, led to the development of **235** and **243**. Both **235** and **243** showed >300-fold enhanced WT EGFR kinase inhibition and >3000-fold enhanced HCC827 antiproliferative activity compared to HTS hit **231** and also inhibited gefitinib-resistant DM (T790M/L858R) EGFR kinase at nanomolar concentration. The presence of *S*-chiral 2-phenyl-2-aminoethanol side chain along with the acrylamide-type Michael acceptor group imparts maximal activity toward gefitinib-resistant DM EGFR kinase. Six of the compounds (**235**, **243–245**, **248**, **249**) possess excellent antiproliferative activity against a HCC827 lung cancer cell line in a single-digit nanomolar range, with compound **243** showing in vivo efficacy in nude mice bearing HCC827 tumor xenografts (Chart 3.17).

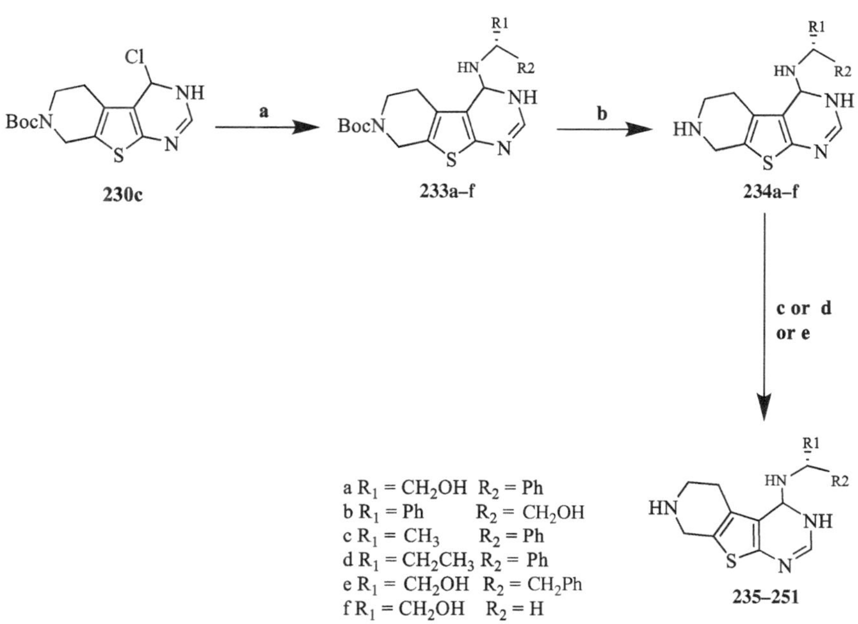

Scheme 3.31 Reagents and conditions: (a) amine ($R_1R_2CHNH_2$), EtOH, reflux, 8 h; (b) TFA, CH_2Cl_2, 0°C rt, 2 h; (c) acid (RCO_2H), EDCI, CH_2Cl_2, rt, 2 h; (d) $Cl(CH_2)_2SO_2Cl$, CH_2Cl_2, rt, 4 h; and (e) (i) (*E*)-$BrCH_2CHdCHCO_2H$, EDCI, CH_2Cl_2, rt, 2 h and (ii) secondary amine (R_2NH), THF, 0°C, 2 h.

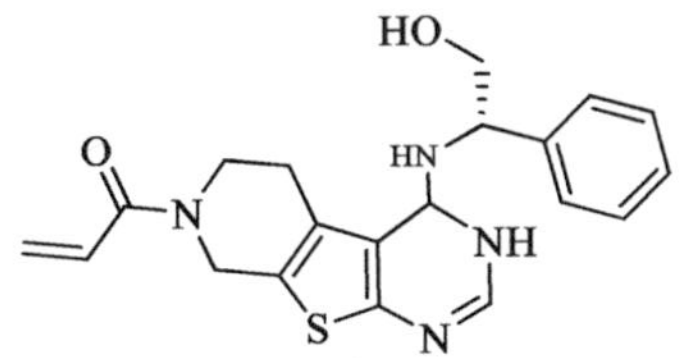

235
WT (IC_{50}) = 0.008 μM
DM (IC_{50}) = 0.100 μM
HCC827 (IC_{50}) = 0.003 μM

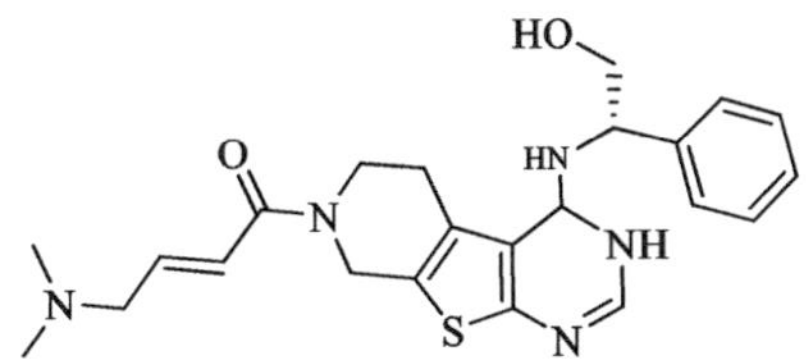

243
WT (IC_{50}) = 0.007 μM
DM (IC_{50}) = 0.305 μM
HCC827 (IC_{50}) = 0.002 μM

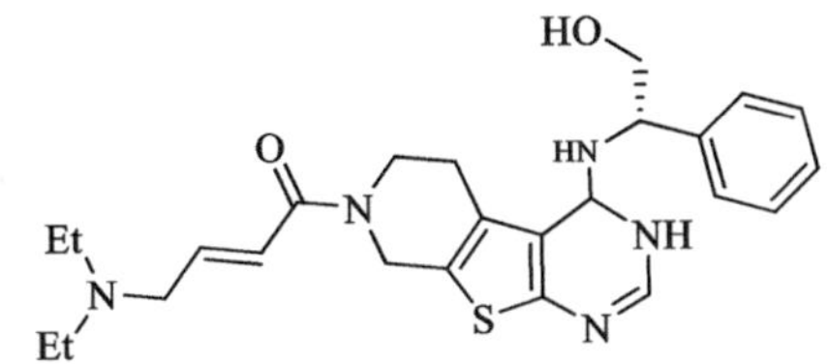

244
WT (IC_{50}) = 0.009 μM
DM (IC_{50}) = 0.930 μM
HCC827 (IC_{50}) = 0.008 μM

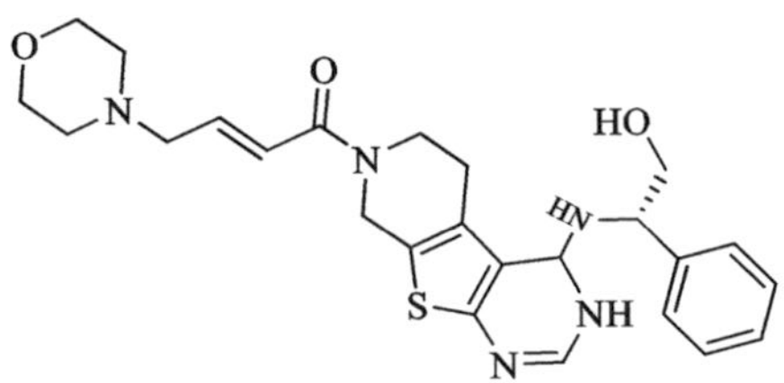

245
WT (IC_{50}) = 0.021 μM
DM (IC_{50}) = 1.399 μM
HCC827 (IC_{50}) = 0.004 μM

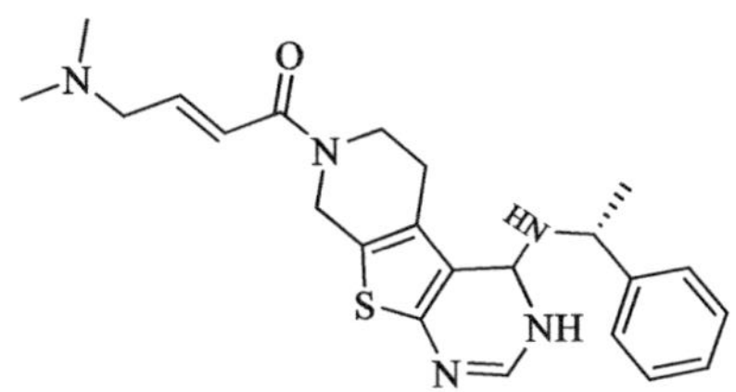

248
WT (IC_{50}) = 0.012 μM
DM (IC_{50}) = 0.596 μM
HCC827 (IC_{50}) = 0.002 μM

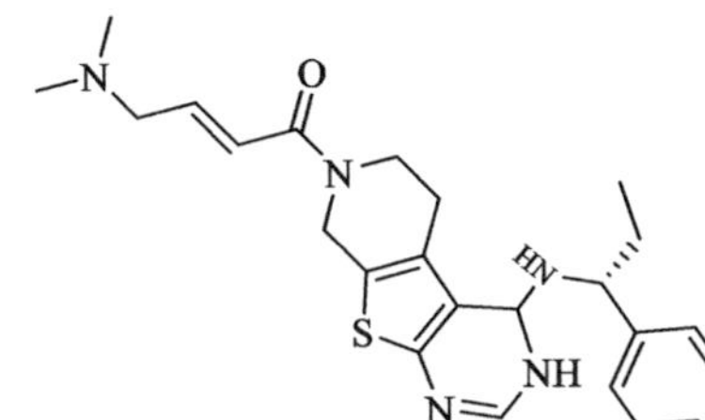

249
WT (IC_{50}) = 0.012 μM
DM (IC_{50}) = 5.750 μM
HCC827 (IC_{50}) = 0.007 μM

Chart 3.17 Antiproliferative and in vitro EGFR inhibitory activity.

Hennessy et al. prepared the potent compound **257** as shown in Scheme 3.32, in which the acrylamide and the benzimidazole ring are linked through a meta-substituted arene [23]. The synthesis of this compound commenced with the preparation of aniline **253**, which was coupled to aminopyrazine carboxylate **254** followed by cyclization in refluxing acetic acid to give benzimidazole **255**. Suzuki coupling with an *N*-protected pyrazolopiperidine boronate ester followed by Boc deprotection and reductive amination afforded the *N*-methylpiperazine intermediate **256**. Hydrolysis of the amide group followed by acylation with acryloyl chloride yielded the target compound **257**.

Significantly, this compound was found to be a very potent inhibitor of EGFR T790M/L858R enzyme and also effectively inhibited the autophosphorylation of this DM EGFR in NCI-H1975 cells.

Among the synthesized compounds as shown in Chart 3.18, compound **257** was found to be potent compound in NCI-H1975 cells. Hennessy et al. next surveyed the effects of this compound in other cell lines containing EGFR of different mutation status. In PC9 cells, which have an activating mutation

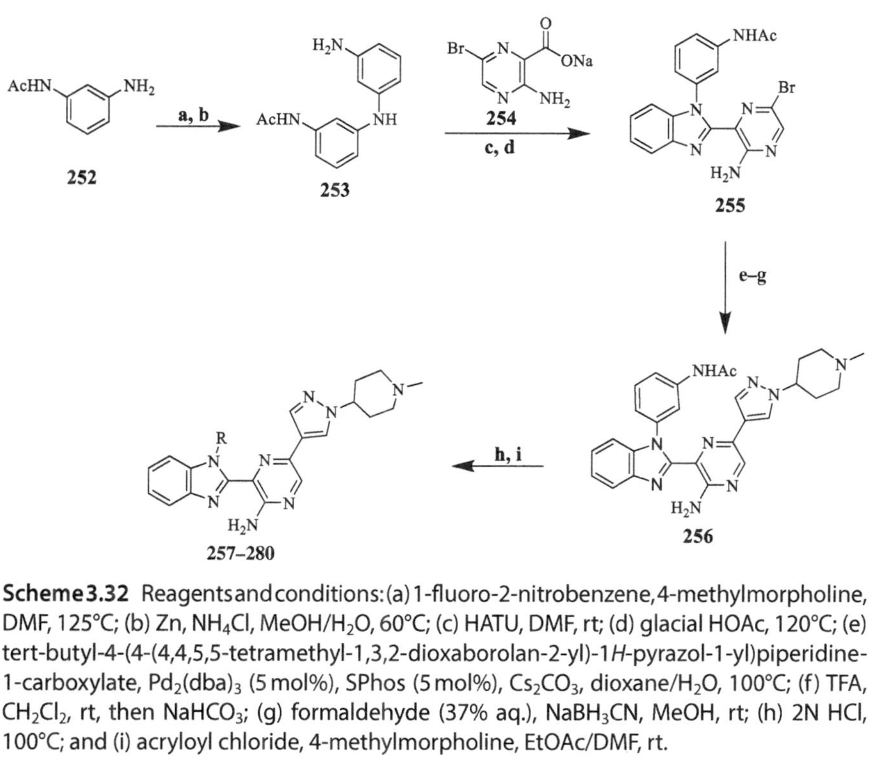

Scheme 3.32 Reagents and conditions: (a) 1-fluoro-2-nitrobenzene, 4-methylmorpholine, DMF, 125°C; (b) Zn, NH_4Cl, MeOH/H_2O, 60°C; (c) HATU, DMF, rt; (d) glacial HOAc, 120°C; (e) tert-butyl-4-(4-(4,4,5,5-tetramethyl-1,3,2-dioxaborolan-2-yl)-1*H*-pyrazol-1-yl)piperidine-1-carboxylate, $Pd_2(dba)_3$ (5 mol%), SPhos (5 mol%), Cs_2CO_3, dioxane/H_2O, 100°C; (f) TFA, CH_2Cl_2, rt, then $NaHCO_3$; (g) formaldehyde (37% aq.), $NaBH_3CN$, MeOH, rt; (h) 2N HCl, 100°C; and (i) acryloyl chloride, 4-methylmorpholine, EtOAc/DMF, rt.

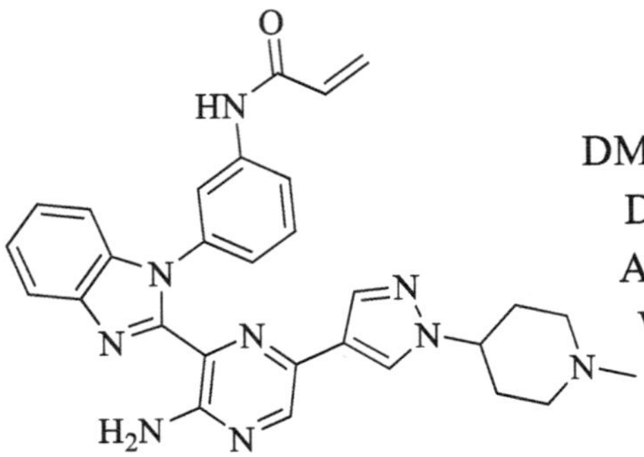

257
DM enzyme (IC_{50}) = <0.001 μM
DM cell (IC_{50}) = 0.014 μM
AM cell (IC_{50}) = 0.009 μM
WT cell (IC_{50}) = 0.98 μM

Chart 3.18 Antiproliferative and in vitro EGFR inhibitory activities.

(exon 19 deletion) in EGFR that would predict sensitivity to established EGFR inhibitors such as gefitinib and erlotinib, this compound is similarly a very effective inhibitor of EGFR autophosphorylation. Moreover, these levels of cellular potency translate to pronounced antiproliferative effects in both cell lines in a 72-h assay (GI_{50} for **257** in NCI-H1975 = 25 nM; PC9 GI_{50} = 16 nM). Interestingly, in cell lines with WT EGFR, **257** is a much poorer inhibitor of EGFR autophosphorylation (EC_{50} in LoVo = 0.98 μM) and cell proliferation (GI_{50} in Calu6 = 1.7 μM).

Cheng et al. reported the discovery of 1-{(3*R*,4*R*)-3-[5-chloro-2-(1-methyl-1*H*-pyrazol-4-ylamino)-7*H*-pyrrolo[2,3-*d*]pyrimidin-4-yloxymethyl]-4methoxypyrrolidin-1-yl}propenone (PF-06459988), a potent, WT-sparing, irreversible inhibitor of T790M-containing EGFR mutants [24]. The synthesis of compounds **292–294** began with the incorporation of the trimethylsilylethoxymethyl or SEM protecting group to 2,4-dichloro-7*H*-pyrrolo[2,3-*d*]pyrimidine **281** using SEM-Cl and LiHMDS to give **282** (Scheme 3.33). The phenoxy linker in intermediate **283** was then introduced using tert-butyl(3-hydroxyphenyl)carbamate and potassium carbonate. The attachment of the key hinge binding heteroaromatic amino side chains was achieved using Buchwald-Hartwig aminations under microwave conditions to yield the amines **284–286**. A selective deprotection of the BOC group in compound **284** and **285** followed by acryloylation gave the acrylamides **287** and **288**. A two-step deprotection of the SEM protecting group was done using trifluoroacetic acid in the first step, which led to partial cleavage of the SEM protecting group, giving intermediates **289** and **290** containing an *N*-hydroxymethyl moiety. The deprotection was completed by treating these intermediates with potassium carbonate to yield the final acrylamides **292** and **293**.

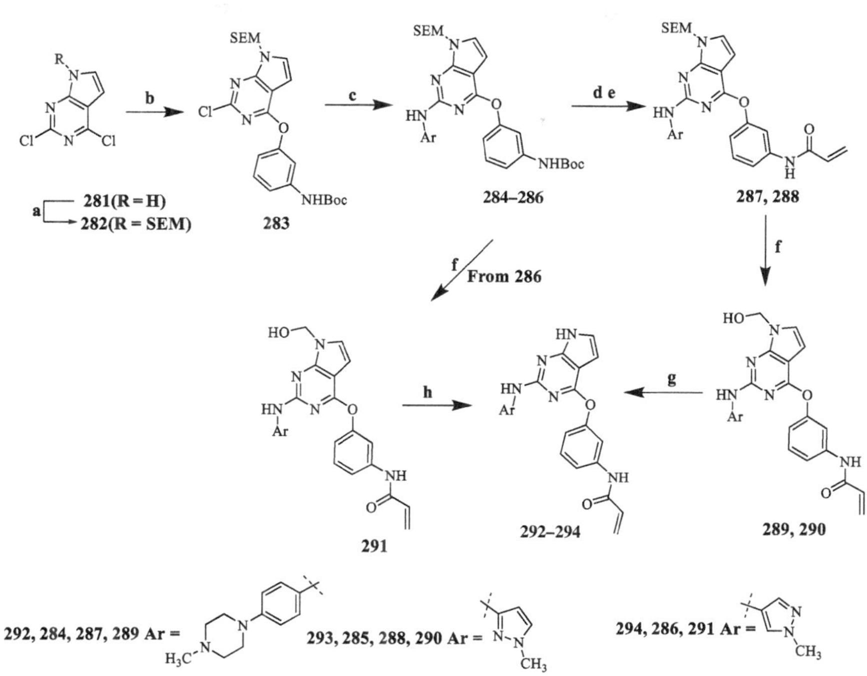

Scheme 3.33 Reagents and conditions: (a) $TMSCH_2CH_2OCH_2Cl$, LiHMDS, THF, 79%; (b) tert-butyl(3-hydroxyphenyl)carbamate, K_2CO_3, MeCN, 90%; (c) $ArNH_2$, $Pd_2(dba)_3$, Cs_2CO_3, Xantphos, 1,4-dioxane, μW, 43%–87%; (d) HCl, 1,4-dioxane; (e) $CH_2{=}CHCOCl$, DCM, or THF 65%–70% (steps d and e); (f) TFA, DCM; (g) K_2CO_3, MeOH or EtOH, 46%–77% (steps f and g); and (h) (i) $CH_2{=}CHCOCl$, DCM and (ii) K_2CO_3, EtOH, 47% (steps f and h).

Alternatively, acrylamide **294** was obtained by deprotection of the BOC group and partial deprotection of the SEM protecting group in compound **286** using TFA; acryloylation followed by final deprotection. Compounds **302** and **303** were synthesized in a similar fashion starting from pyrrolo[2,3-*d*]pyrimidine **282** and reacting it with the alcohols **295** and **296** under basic conditions (Scheme 3.34).

The installation of the aminopyrazole was done using Buchwald-Hartwig amination conditions; this was followed by the aforementioned two-step deprotection protocol to remove the BOC and SEM protecting groups. Acryloylation of the corresponding pyrrolidines to the acrylamides **302a–d** was carried out under carefully monitored conditions using DIPEA and acryloyl chloride to minimize bis-acryloylated side products. Route optimization efforts were initiated to overcome the harsh conditions required for the Buchwald-Hartwig amination step and to avoid the

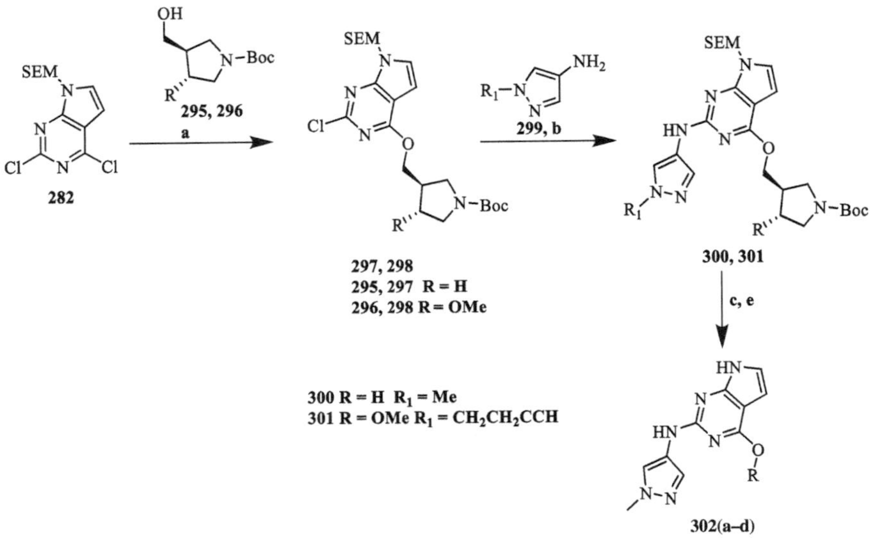

Scheme 3.34 Reagents and conditions: (a) KHMDS, 1,4-dioxane; (b) 1-alkyl-1H-pyrazol-4-amine, $Pd_2(dba)_3$, Cs_2CO_3, Xantphos, 1,4-dioxane, µW; (c) TFA, DCM; (d) K_2CO_3, MeOH, H_2O or NH_4OH, 1,4-dioxane; and (e) $CH_2{=}CHCOCl$, DIPEA, DCM or $CH_2{=}CHCOCl$, THF.

need for the capricious *N*-SEM protecting group. To this end, a 196-well screen using two solvents; 1,4-dioxane and toluene, two bases; cesium carbonate and potassium tert-pentoxide and 48 catalyst-ligand combinations was utilized to review the amination of an unprotected model substrate, 4-(benzyloxy)-2-chloro-7*H*-pyrrolo[2,3-*d*]pyrimidine **304**, with 1-methyl-1*H*-pyrazol-4-amine (Scheme 3.35).

Further refinement of the sequence led to the discovery of an efficient one-pot protocol (Scheme 3.36), wherein excess potassium tert-pentoxide (4 equiv.) was used as the base to append the alkoxy linker (**296**) to the unprotected 2-chloro-pyrrolo[2,3-*d*]pyrimidine core (**306**) using an appropriate alcohol and 1,4-dioxane as solvent in the first step to obtain **307**.

After 30 min of aging, tert-BuXPhos palladacycle and the appropriate amine were added to the same pot, and the reaction mixture was warmed to 90°C to complete the incorporation of the heteroaromatic amino side chain (**308**). After purification the BOC protecting group was cleaved using trifluoroacetic acid to give **309**. Further optimization of the final acryloylation step indicated that Schotten-Baumann conditions using aqueous sodium bicarbonate and acryloyl chloride in ethyl acetate were most optimal for a robust acryloylation (**310, 316–320**). The synthesis of acrylamide **310, 316–320** (Scheme 3.36) illustrates the effectiveness of this atom-economic sequence.

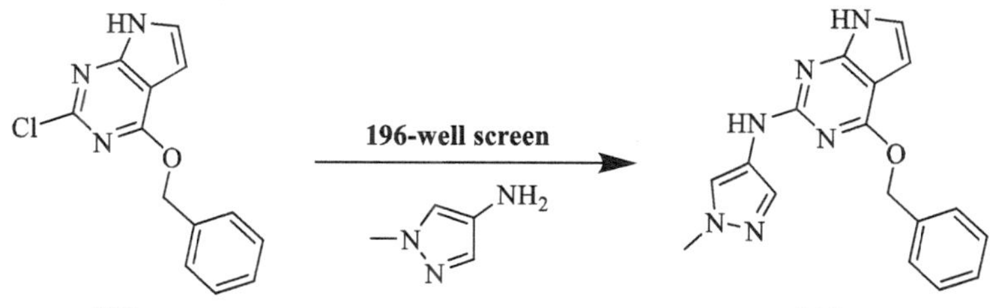

Scheme 3.35 Reagents and condition: 196-well screen using two solvents; 1,4-dioxane and toluene, two bases; cesium carbonate and potassium *tert*-pentoxide.

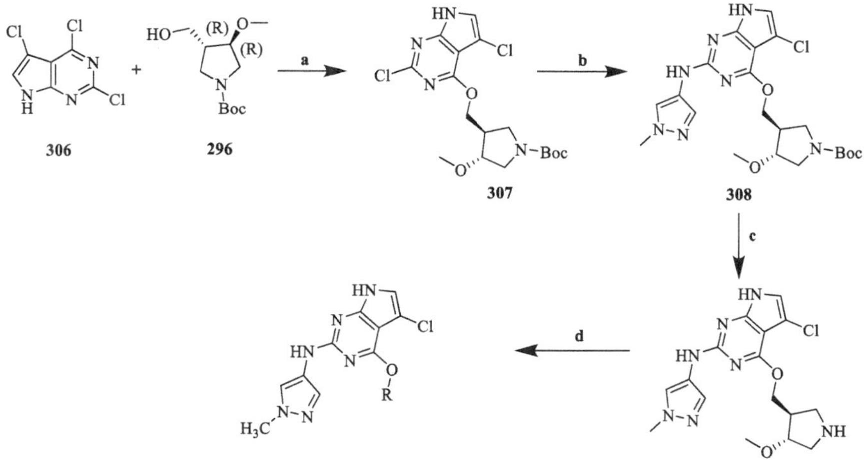

Scheme 3.36 (a) Potassium tert-pentoxide, 1,4-dioxane; (b) 1-methyl-1*H*-pyrazol-4-amine, tert-BuXPhos palladacycle, 79% over two steps; (c) TFA, DCM, 99%; and (d) $CH_2{=}CHCOCl$, aq. $NaHCO_3$, EtOAc, 77%.

The synthesis of the chiral alcohol **296** was achieved as outlined in Scheme 3.37. A [3+2]− cycloaddition reaction of methyl(*E*)-3-methoxyacrylate **311** with an azomethine ylide generated in situ from *N*-benzyl-1-methoxy-*N*-((trimethylsilyl)methyl)methanamine **312** provided the racemic pyrrolidine ester **313**. Chiral separation of this racemic ester using chiral supercritical fluid chromatography (SFC) yielded the requisite enantiomer **314**.The benzyl protecting group in **314** was then removed under hydrogenolysis conditions and replaced with a BOC protecting group using Boc_2O.The final step involved the reduction of the ester using lithium borohydride which yielded (3*R*,4*R*)-**296**.

Activity data of all the synthesized compounds are shown in Chart 3.19. Compound **292** was an attractive lead, potently inhibiting

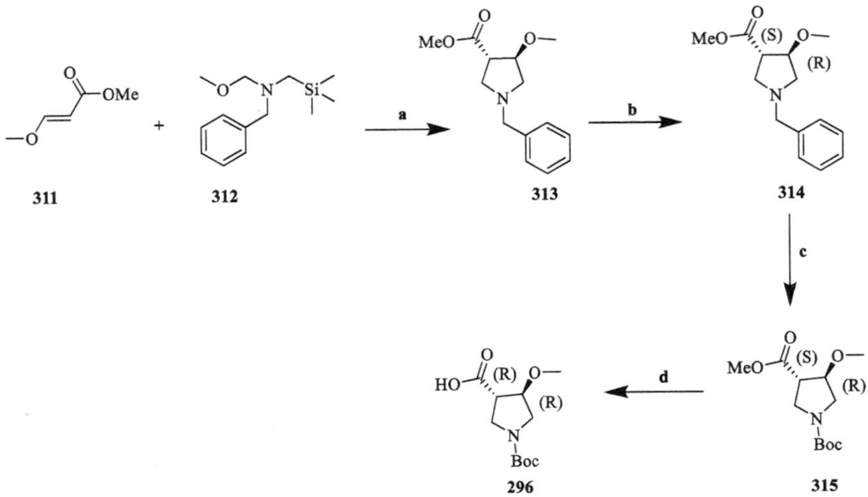

Scheme 3.37 (a) 312, TFA, 73%; (b) chiral-SFC (Chiralpak AD-H) 32%; (c) (i) H_2 (1 atm), 10% $Pd(OH)_2$ (6%), EtOH and (ii) Boc_2O, 98%; and (d) $LiBH_4$, THF, 92%.

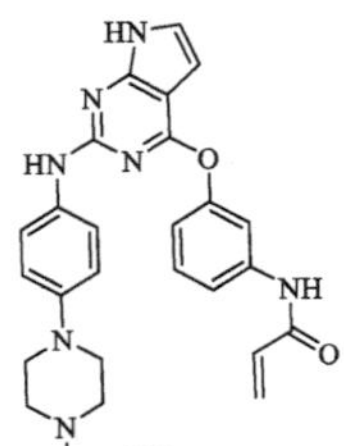

292
H1975 IC_{50} = 4.0 nM
WT IC_{50} = 0.79 nM
WT/DM IC_{50} = 198 nM

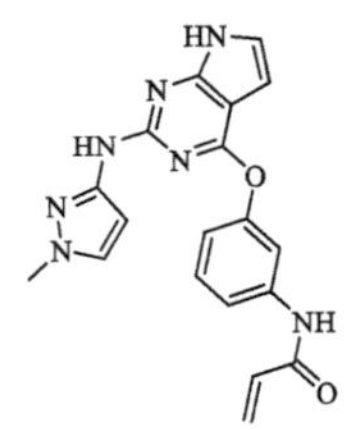

293
H1975 IC_{50} = 27 nM
WT IC_{50} = 7.6 nM
WT/DM IC_{50} = 281 nM

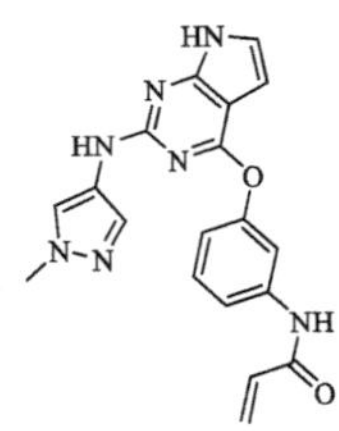

294
H1975 IC_{50} = 13 nM
WT IC_{50} = 1.6 nM
WT/DM IC_{50} = 123 nM

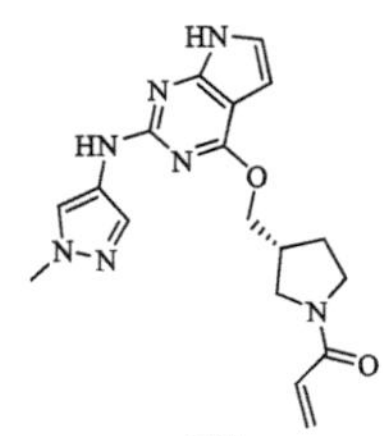

302a
H1975 IC_{50} = ND
WT IC_{50} = ND
WT/DM IC_{50} = ND

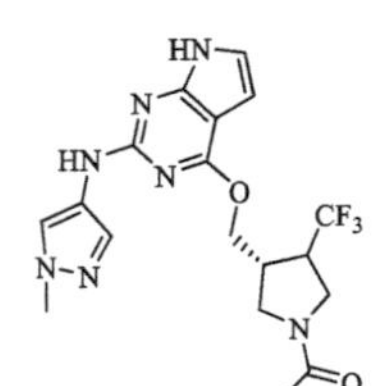

302b
H1975 IC_{50} = 47 nM
WT IC_{50} = >5 nM
WT/DM IC_{50} = >106 nM

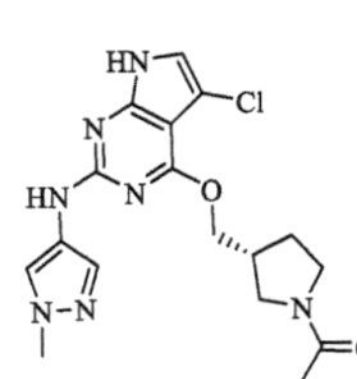

302c
H1975 IC_{50} = 45 nM
WT IC_{50} = >3.3 nM
WT/DM IC_{50} = >73 nM

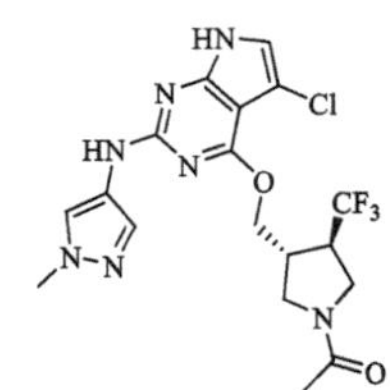

302d
H1975 IC_{50} = 8 nM
WT IC_{50} = 3.5 nM
WT/DM IC_{50} = 438 nM

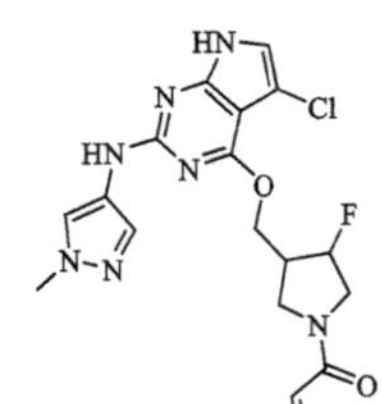

316
H1975 IC_{50} = 9 nM
WT IC_{50} = 4.2 nM
WT/DM IC_{50} = 467 nM

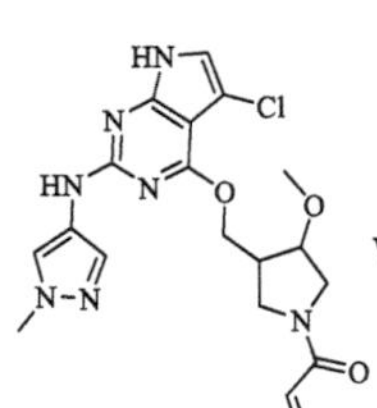

310
H1975 IC_{50} = 13 nM
WT IC_{50} = 5.1 nM
WT/DM IC_{50} = 392 nM

Chart 3.19 Antiproliferative and in vitro EGFR inhibitory activities.

EGFR L858R/T790M autophosphorylation (IC_{50} = 4 nM) in the H1975 cell line but only modestly inhibiting *WT EGFR* autophosphorylation (IC_{50} = 0.79 μM) in the A549 cell line. Relative to the benchmark compound canertinib in a glutathione conjugation assay, compound **292** exhibits a sixfold reduction in intrinsic chemical reactivity with thiol nucleophiles (Chart 3.19).

Furthermore the lower lipophilicity of aminopyrazole functionality might enable the removal of the basic amine found in the piperazinyl-aniline, resulting in a more ligand-efficient lead. To investigate this possibility, the *N*1-methyl-3-amino- and *N*1-methyl-4-amino-pyrazoles **293** and **294** were prepared (Chart 3.19). Both compounds demonstrated an increased permeability and decreased dofetilide-binding activity relative to **292** while maintaining good EGFR L858R/T790M potency and selectivity over EGFR-WT (>100-fold). Ligand efficiency (LE, calculated from Kiest) for **293** and **294** are 0.41 and 0.45, respectively, which are significantly higher than LE of 0.37 for **292**.

The 4-amino-pyrazole **294** better matched the EGFR T790M binding affinity, inactivation rate (kobs/I), cellular potency, and intrinsic chemical reactivity to GSH of **292** but showed increased microsomal clearance relative to **292** and **293**.The modeling of **302a** suggested that a small lipophilic group at the pyrrolidine 4-position would be suited, thus a CF_3 group was installed (chosen for metabolic stability benefits over methyl or ethyl), resulting in a compound **302b** (Chart 3.19).This molecule provided a 17-fold increase in potency (Kiest 35 nM) and an increase in irreversible activation (kobs/I of 5410 $M^{-1} s^{-1}$ at 1 μM) relative to **302a**.

Compound **302d** had high permeability (RRCK 13×10^{-6} cm/s) with significant cellular activity (p*EGFR* H1975 IC_{50} 45 nM) and a high selectivity over *WT EGFR* (>73×). However, the high metabolic clearance needed to be addressed in order to achieve a desirable human PK profile. The cocrystal structure of **302c** reveals that the CF_3 binds very tightly in the G-loop hydrophobic pocket, with the CF_3 directly interacting with the Phe723 side chain phenyl ring. As half of the CF_3 moiety was exposed to solvent, it presented an opportunity to modulate metabolic stability by replacing the CF3 with small polar groups. Based on the outcome from modeling studies and in silico ADME predictions –F, –OMe, –OH, –CH_2OH, and –CH_2OMe replacements for –CF_3 were explored (Chart 3.19).

Pyrrolidine derivatives **310** (PF-06459988) demonstrates potent affinity and high inactivation rates with an improvement in metabolic stability (Chart 3.19). Compound **310** emerged as providing the best balance of overall properties including potency, WT selectivity, intrinsic chemical reactivity, and ADME properties. This suggests the optimum alignment of the acrylamide warhead with Cys797 is critical for the observed high potency for **310** (Chart 3.19).

Han et al. reported the synthesis and evaluation of 2-anilinopyrimidines bearing 3-aminopropamides as potential *EGFR* inhibitors. Compounds **332a–i** were synthesized as depicted in Scheme 3.38 [25]. Starting from compound **321**, a methylation reaction using methyl iodide and a following amination at 5-position by *N*-methylpiperazine furnished methylphenyl ether **323**, which underwent a nitro reduction to produce aniline **324**. 3-Nitrophenol **326** was regioselectivity coupled to the 4-position of 2,4,5-trichloropyrimidine **325** to afford aryl ether **327** at the room temperature. The coupling reaction of **324** with **327** was accomplished in the presence of trifluoroacetic acid under reflux, offering aniline pyrimidine **328**. The nitro group in **328** was reduced by iron powder/NH_4Cl to produce arylamine **329**. Acylation of anilino group of **329** using acryloyl chloride gave compound **330**. The compounds **332a–i** were obtained through the Michael addition reaction of **330** with various secondary amines (**331a–i**) in THF/MeOH under reflux.

Novel compounds **332a–i** were synthesized and biologically evaluated as shown in Chart 3.20. Several exhibited stronger inhibitory activity than gefitinib against *EGFR L858R/T790M* and antiproliferative effects on H1975 and HCC827 cells. The 3-aminopropamide in compounds like **332h** could be converted to the active acrylamide in the presence of arginine. Importantly, **332h** showed improved stability relative to compound **330** whose structure is same to **332h** excepting an acrylamide moiety. Interestingly, **332i**, a NO donating compound of **332h**, showed more potent and selective inhibition than **332h** on H1975 cells.

Significantly, **332i** produced high levels of NO in H1975 cells but not in the nontumorous 16HBE cells, and its inhibition was diminished by NO scavenger. Furthermore, **332i** produced dose-dependent inhibitory effects on EGFR downstream signaling in H1975 cells.

Cha et al. synthesized and evaluated pyrimidine-based dual inhibitors of human epidermal growth factor receptor 1 (HER-1) and HER-2 tyrosine kinases [26]. As shown in Scheme 3.39, the synthesis of analogs **337a–c** was initiated using 4(3*H*)-pyrimidinone **333** treated with *N*-iodosuccinimide at 100°C in acetic acid to yield 5-iodopyrimidin-4(3*H*)-one **334**. The resulting compound **334** was chlorinated with phosphorus oxychloride under reflux conditions and subjected to substitution with aniline **335** in refluxing

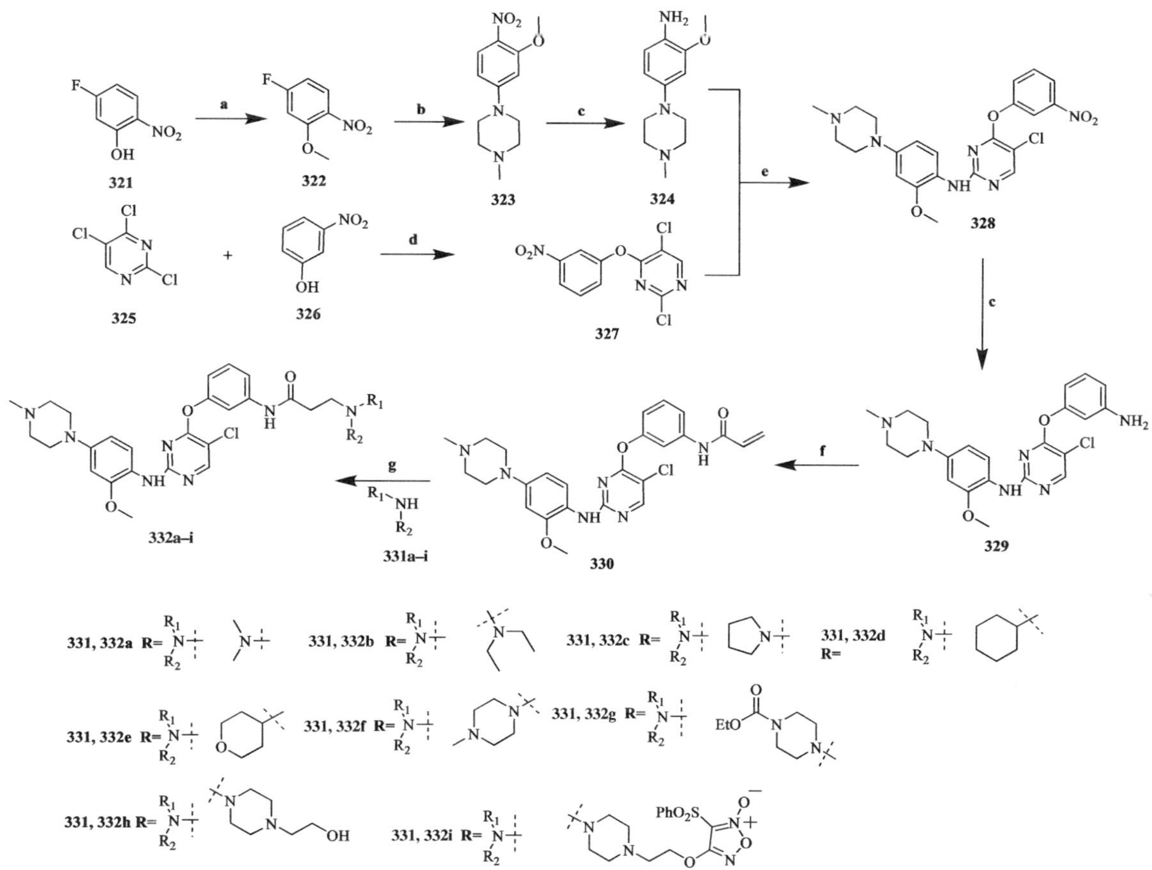

Scheme 3.38 Reagents and conditions: (a) CH_3I, K_2CO_3, acetone, reflux, 4 h; (b) *N*-methyl-piperazine, K_2CO_3, DMSO, rt, overnight (c) iron powder, NH_4Cl, THF/H_2O (1/1, v/v), reflux, 3 h; (d) K_2CO_3, DMF, rt, 2 h; (e) 5, TFA, 2-BuOH, reflux, 3 h; (f) acryloyl chloride, Et_3N, anhydrous CH_2Cl_2, 0°C, 1 h; and (g) **331a–i**, THF/MeOH, reflux, 4 h.

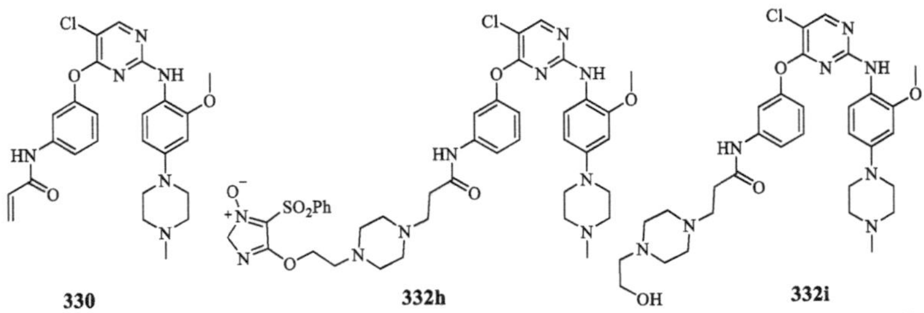

330	332h	332i
WT $IC_{50} = 0.0044\ \mu M$	WT $IC_{50} = 2.388\ \mu M$	WT $IC_{50} = 0.201\ \mu M$
L858R $IC_{50} = 0.0048\ \mu M$	L858R $IC_{50} = 5.866\ \mu M$	L858R $IC_{50} = 0.262\ \mu M$
L858R/T790M $IC_{50} = 0.0007\ \mu M$	L858R/T790M $IC_{50} = 2.419\ \mu M$	L858R/T790M $IC_{50} = 0.263\ \mu M$
H1975 $IC_{50} = 0.046\ \mu M$	H1975 $IC_{50} = 0.931\ \mu M$	H1975 $IC_{50} = 0.108\ \mu M$
HCC827 $IC_{50} = 0.008\ \mu M$	HCC827 $IC_{50} = 0.799\ \mu M$	HCC827 $IC_{50} = 0.130\ \mu M$
A431 $IC_{50} = 1.108\ \mu M$	A431 $IC_{50} = 0.632\ \mu M$	A431 $IC_{50} = 4.469\ \mu M$
A549 $IC_{50} = 2.102\ \mu M$	A549 $IC_{50} = 2.816\ \mu M$	A549 $IC_{50} = 8.251\ \mu M$
BEAS-2b $IC_{50} = 1.158\ \mu M$	BEAS-2b $IC_{50} = 1.657\ \mu M$	BEAS-2b $IC_{50} = 3.368\ \mu M$
16HBE $IC_{50} = 1.355\ \mu M$	16HBE $IC_{50} = 2.056\ \mu M$	16HBE $IC_{50} = 2.876\ \mu M$

Chart 3.20 Antiproliferative and in vitro EGFR inhibitory activities.

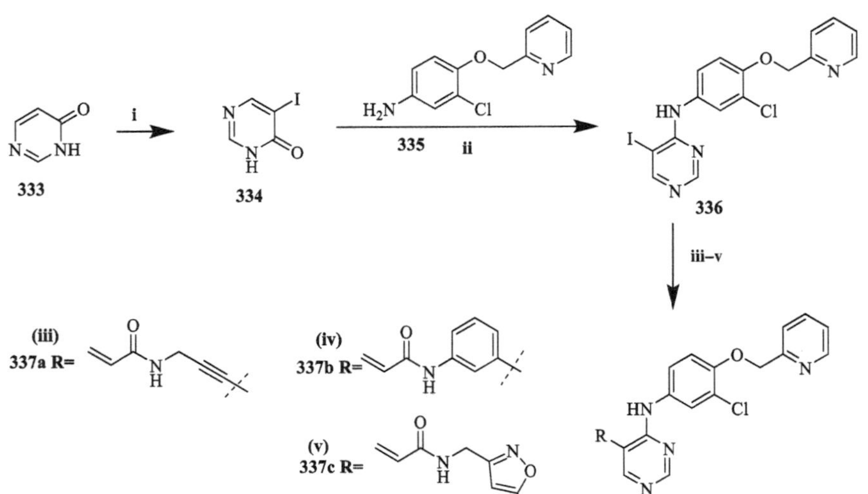

Scheme 3.39 Reagents and conditions: (i) *N*-iodosuccinimide, AcOH, 100°C; (ii) (a) $POCl_3$, reflux and (b) 8,2-propanol, reflux; (iii) (a) *N*-bocpropargylamine, $PdCl_2(PPh_3)_2$, CuI, TEA, THF, rt; (b) TFA, DCM, rt; and (c) acryloyl chloride, $NaHCO_3$, aq. THF (THF/H_2O = 5:1), 0°C; (iv) (a) 3-aminophenylboronic acid, $Pd(OAc)_2$, K_2CO_3, MeOH, 70°C and (b) acryloyl chloride, $NaHCO_3$, aq. THF (THF/H_2O = 5:1), 0°C; and (v) (a) trimethylsilyl acetylene, $PdCl_2(PPh_3)_2$, CuI, TEA, THF, rt; (b) TBAF, THF, 0°C; (c) 2-(1,3-dioxoisoindolin-2-yl)-*N*-hydroxyacetimidoyl chloride, TEA, THF, reflux; (d) hydrazine monohydrate, EtOH, 65°C; and (e) acryloyl chloride, $NaHCO_3$, aq. THF (THF/H_2O = 5:1), 0°C.

2-propanol to yield *N*-(3-chloro-4-(pyridin-2-ylmethoxy)-phenyl)-5-iodopyrimidin-4-amine **336** as a key intermediate. This intermediate **336** was reacted with BOC-propargylamine under Sonogashira coupling conditions in the presence of bis(triphenylphosphine)palladium(II) dichloride and copper(I) iodide. Subsequently, the BOC group was removed upon treatment with 30% trifluoroacetic acid in dichloromethane, followed by amide coupling with acryloyl chloride in the presence of sodium bicarbonate at 0°C to yield propargyl analog **337a**. Phenyl analog **337b** was synthesized through Suzuki coupling reaction of the key intermediate **336** with 3-aminophenylboronic acid in the presence of palladium (II) acetate, followed by *N*-acryloylation. For the preparation of isoxazolylmethyl analog **337c** the key intermediate **336** was treated with trimethylsilyl (TMS) acetylene under the identical Sonogashira coupling conditions used above.

The desired analog **337c** was synthesized through a series of reactions including TMS deprotection using tetrabutylammonium fluoride (TBAF), 1,3-dipolar cycloaddition with 2-(1,3-dioxoisoindolin-2-yl)-*N*-hydroxyacetimidoyl chloride, phthalimide deprotection using hydrazine, and subsequent N-acryloylation. For the synthesis of oxazolylmethyl analog **337d**, ethyl-4-((3-chloro-4-(pyridin-2-ylmethoxy)phenyl)amino) pyrimidine-5-carboxylate (**339**) was prepared by condensation of diethyl ethoxymethylenemalonate (**338**) with formamidine, chlorination using thionyl chloride and phosphorus oxychloride in the presence of diisopropylethylamine, and substitution with aniline **335** in the refluxing 2-propanol (Scheme 3.40).

Then the resulting ethyl carboxylate **339** was hydrolyzed in 2*N* sodium hydroxide aqueous solution and coupled with DL-serine methyl ester using EDCI and HOBt to provide β-hydroxyamide **340**, which was converted to oxazole **341** via the mild cyclodehydration reaction using diethylaminosulfur trifluoride (DAST) and oxidation using bromotrichloromethane and 1,8-diazabicyclo[5.4.0]undec-7-ene (DBU). Finally, oxazolylmethyl analog **337d** was prepared by the reduction of methyl ester of oxazole **341**, substitution of hydroxyl group to azide, Staudinger reduction of azide to amine, and subsequent N-acryloylation.

The synthesis of the amine-containing analog of **346** is outlined in Scheme 3.41. 4,6-Dichloropyrimidine-5-carbaldehyde **342** was chlorinated under radical conditions, followed by amide coupling with DL-serine methyl ester to give β-hydroxyamide **343**. The C_6-amine was introduced to **343** using ammonia gas in 50% ethanolic toluene in a sealed tube, followed by oxazoline formation using DAST. The C_4-chloride of oxazoline

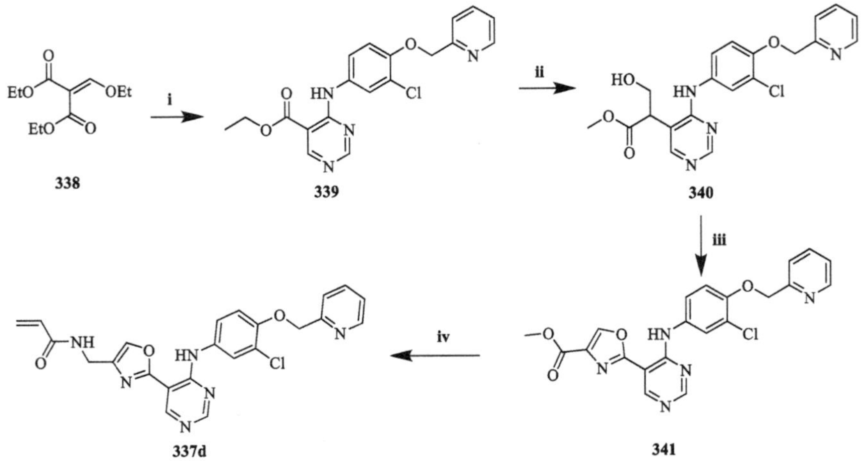

Scheme 3.40 Reagents and conditions: (i) (a) formamidine hydrochloride, 160°C; (b) $POCl_3$, $SOCl_2$, DIPEA, reflux; and (c) 351, 2-propanol, reflux; (ii) (a) 2N NaOH aq. solution, EtOH, 80°C and (b) DL-serine methyl ester hydrochloride, EDCI, HOBt, DIPEA, DMF, rt; (iii) (a) DAST, DCM, −78°C to 0°C and (b) DBU, $CBrCl_3$, DCM, −40°C to rt; and (iv) (a) LAH, THF, 0°C; (b) NaN_3, PPh_3, DMF/CCl_4 (4:1), 90°C; (c) PPh_3, H_2O, THF, 60°C; and (d) acryloyl chloride, $NaHCO_3$, aq. THF (THF/H_2O = 5:1), 0°C.

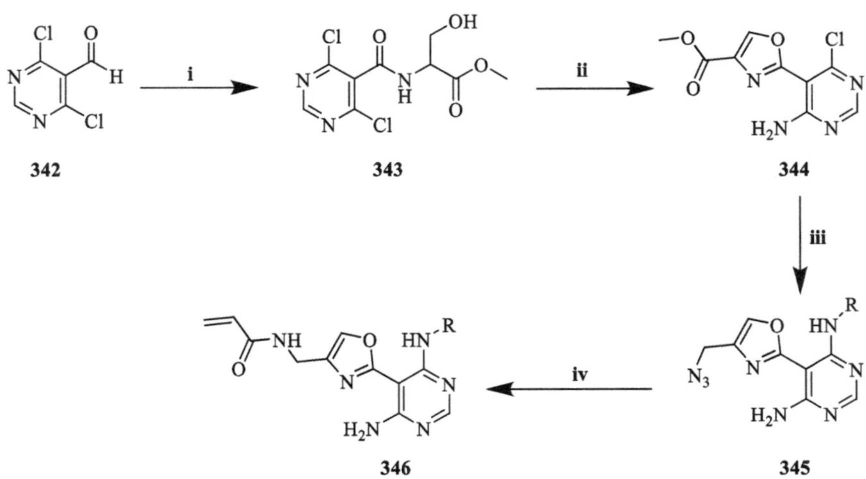

Scheme 3.41 Reagents and conditions: (i) (a) $SOCl_2$, AIBN, CCl_4, 80°C and (b) DL-serine methyl ester hydrochloride, DIPEA, THF, 0°C to rt; (ii) (a) NH_3 (gas), toluene/EtOH (1:1), sealed tube, 60°C and (b) DAST, DCM, −78°C to 0°C; (iii) (a) R-NH_2, 4N HCl in 1,4-dioxane, *t*-BuOH, 60°C; (b) DBU, $CBrCl_3$, DCM, −40°C to rt; (c) LAH, THF, 0°C; and (d) NaN_3, PPh_3, DMF/CCl_4 (4:1), 90°C; (iv) (a) PPh_3, H_2O, THF, 60°C and (b) acryloyl chloride, $NaHCO_3$, aq. THF (THF/H_2O = 5:1), 0°C.

344 was substituted with various anilines (R-NH_2) under acidic conditions and oxidized to oxazole using bromotrichloromethane and DBU. After the reduction of methyl ester in **344** and the substitution of alcohol to azide, the resulting azide moiety in azidomethyloxazole **345** was reduced to amine in the presence of triphenylphosphine subsequently N-acryloylated to obtain the desired analog **346**.

HER-2 is overexpressed in 20%–25% of metastatic breast or gastric cancers and is associated with an aggressive tumor type and reduced survival rate. Therefore HER-2-targeted therapy such as trastuzumab or lapatinib is beneficial for cancer patients with a high HER-2 expression level. The cellular and enzymatic data of the synthesized compounds are given in Chart 3.21. However, most cancers that initially respond to HER-2-targeted therapy begins to progress again within 1 year. The selected compound (**346a**) showed excellent HER-1 and HER-2 inhibition activity and also showed excellent selectivity in an enzyme-based inhibition assay with a panel of 20 kinases. As shown in Chart 3.21 the HER-1/HER-2 cellular inhibitory activity was decreased by the replacement of 3-chloro (**346a**, IC_{50} = 11 nM for A431 and IC_{50} = 0.7 nM for SK-Br3) with 3-fluoro (**346b**, IC_{50} = 63 nM for A431 and IC_{50} = 6.1 nM for SKBr3) at the R_2-position or the replacement of 4-(2-pyridine) (**346a**) with 4-(3-fluorophenyl) (**346c**, IC_{50} = 63 nM for A431 and IC_{50} = 6.1 nM for SK-Br3) at the R_1-position, which shares the same structure-activity relationship pattern of phenyl substituents in the case of Lapatinib. Further, they confirmed that **346a** has an irreversible binding mode to HER-1 and HER-2 confirmed by a cell washout test in A431 and SK-Br3 cell lines. In contrast to the currently approved HER-2-targeted agent, their selected compound (**346a**) is an irreversible small-molecule inhibitor of HER-1 and HER-2 tyrosine kinases with the potential to overcome clinically relevant, mutation-induced drug resistance. **346a** showed excellent inhibitory activity toward cellular growth of lapatinib-sensitive or resistant cancer cells. The excellent pharmacokinetic profiles of **346a** in rats and mice clearly demonstrate its merit for further investigation as a novel

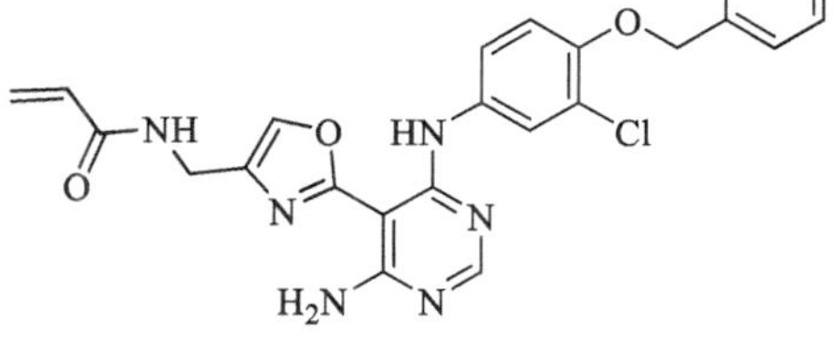

346a
WT IC_{50} = 3.3 nM
T790M IC_{50} = 3.2 nM
HER-2 IC_{50} = 22.4 nM
HER-4 IC_{50} = 4.0 nM
A431 IC_{50} = 11.0 nM

Chart 3.21 Antiproliferative and in vitro EGFR inhibitory activities.

therapeutic agent for HER-2-targeting treatment of solid tumors, especially HER-2-positive breast or gastric cancer.

Hao et al. reported the discovery and structural optimization of *N*5-substituted 6,7-dioxo-6,7-dihydropteridines as potent and selective *EGFR* inhibitors against *L858R/T790M* resistance mutation [27]. As outlined in Scheme 3.42, compounds **355a–e** were prepared starting from commercially available 1,3-benzendiamine (**348**), followed by treatment with di-tert-butyl dicarbonate to achieve BOC-protected amine (**349**). Reacting compound **349** with 2,4-dichloro-5-nitropyrimidine yielded tert-butyl(3-((2-chloro-5-nitropyrimidin-4-yl)amino)phenyl) carbamate (**350**) in a high

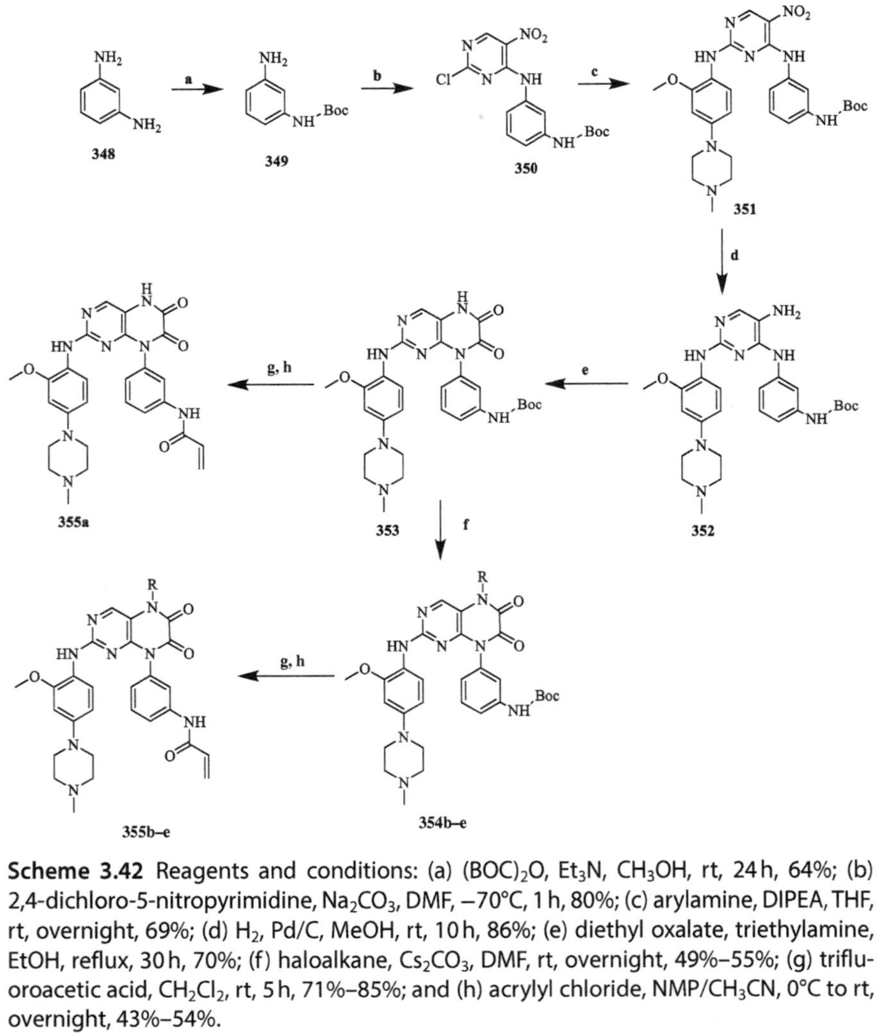

Scheme 3.42 Reagents and conditions: (a) $(BOC)_2O$, Et_3N, CH_3OH, rt, 24 h, 64%; (b) 2,4-dichloro-5-nitropyrimidine, Na_2CO_3, DMF, −70°C, 1 h, 80%; (c) arylamine, DIPEA, THF, rt, overnight, 69%; (d) H_2, Pd/C, MeOH, rt, 10 h, 86%; (e) diethyl oxalate, triethylamine, EtOH, reflux, 30 h, 70%; (f) haloalkane, Cs_2CO_3, DMF, rt, overnight, 49%–55%; (g) trifluoroacetic acid, CH_2Cl_2, rt, 5 h, 71%–85%; and (h) acrylyl chloride, NMP/CH_3CN, 0°C to rt, overnight, 43%–54%.

yield (80%). Then, compound **350** was converted to the intermediate (**351**) by the displacement of 2-chloro in the pyrimidine with appropriate arylamines in a good yield (69%), a subsequent reduction of the nitro group, and a reaction with diethyl oxalate to obtain the ring-closure compound (**353**) in an acceptable 60% overall yield, which was deprotected with trifluoroacetic acid and reacted with acryloyl chloride to produce the compound **355a** with a moderate 41% overall yield of the two-step procedure. Meanwhile, compound **353** was alkylated according to the reported procedure with appropriate alkyl halides to afford the N5-alkylated derivatives **354b–e** in moderate 49%–55% yields.

Finally, compounds **354b–e** were deprotected with trifluoroacetic acid in dichloromethane and reacted with acryloyl chloride to produce the final target compounds **354b–e** with 31%–46% overall yields. Compounds **355f–j** were prepared by a method similar to that for compound **355b** in similar overall yields.

The introduction of bulkier substitutions at the 5-position resulted in the further enhancement of inhibitory activities against EGFR L858R/T790M as observed for the 5-isopropyl (**355d**) and 5-propyl (**355e**) derivatives. Both of these compounds strongly inhibited EGFR L858R/T790M with IC_{50} values below 1 nM. However, their WT EGFR inhibitory potency (2.0 and 2.2 nM, respectively) was also improved relative to compound **355b**, resulting in a slight decrease of selectivity between EGFR L858R/T790M and WT EGFR. The effects of all three compounds were further investigated upon the activation of H1975 and A431 cells (Chart 3.22). It was shown that compounds **355c** and **355e** exhibited moderate potency in suppressing the growth of H1975 cells with IC_{50} values of 126 and 119 nM, respectively, whereas their corresponding values for A431 cells were about 18–70-fold greater (Chart 3.22). Additionally the most enzymatically potent compound **355d** also exhibited potent growth-inhibitory activities in the H1975 cells with an IC_{50} value of 18 nM, which was comparable to that of compound rociletinib (CO1686).

On the basis of compound **355a**, they introduced alkyl groups, such as methyl and ethyl at the N-5 moiety, to improve selectivity over WT EGFR kinase, and the resulting compounds **355b** and **355c** indeed showed 101- and 67-fold selectivity between EGFR L858R/T790M and WT EGFR. These results confirmed that hydrophobic optimization at 5-position to strengthen nonpolar contacts is extremely important to inhibitors of L858R/T790M EGFR mutants with selectivity over WT EGFR. Furthermore, substituting groups at the 5-position are actually size constrained; namely, introduction

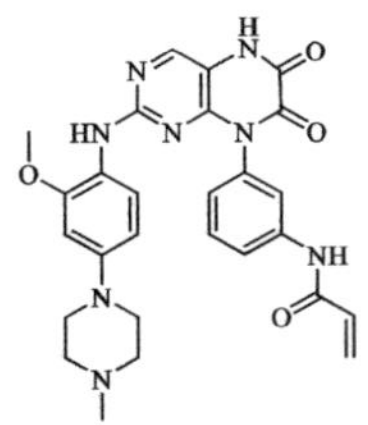

355a

WT IC_{50} = 2.8 nM
L858R/T790M IC_{50} = 2.5 nM
A431 IC_{50} = >10,000 nM
H1975 IC_{50} = >10,000 nM

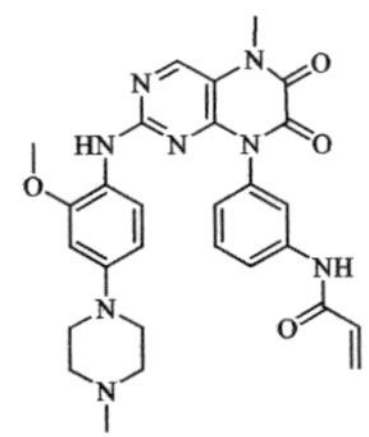

355b

WT IC_{50} = 474 nM
L858R/T790M IC_{50} = 4.7 nM
A431 IC_{50} = >10,000 nM
H1975 IC_{50} = 1772 nM

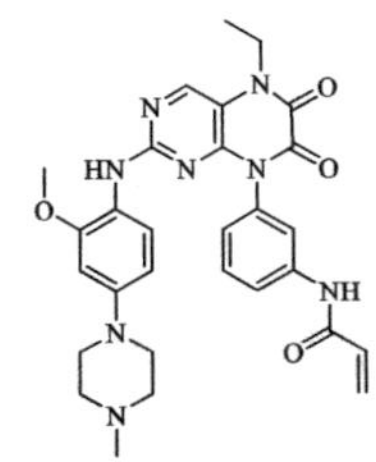

355c

WT IC_{50} = 81 nM
L858R/T790M IC_{50} = 1.2 nM
A431 IC_{50} = 8815 nM
H1975 IC_{50} = 126 nM

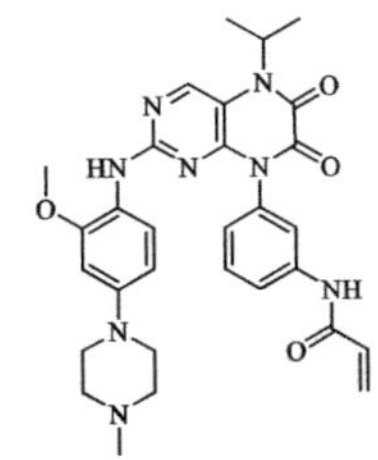

355d

WT IC_{50} = 2.0 nM
L858R/T790M IC_{50} = 0.3 nM
A431 IC_{50} = 7773 nM
H1975 IC_{50} = 18 nM

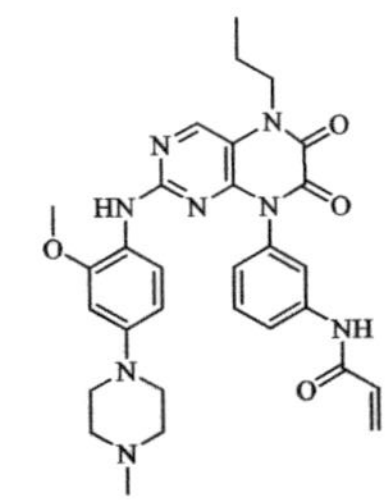

355e

WT IC_{50} = 2.2 nM
L858R/T790M IC_{50} = 0.6 nM
A431 IC_{50} = 2143 nM
H1975 IC_{50} = 119 nM

Chart 3.22 Antiproliferative and in vitro EGFR inhibitory activities.

of bulkier substitutions such as isopropyl (**355d**) and propyl (**355e**) at the 5-position resulted slight decrease of selectivity over WT EGFR kinase, suggesting that bulkier substituents may improve the inhibitory activities against EGFR-WT through the hydrophobic interactions within the gatekeeper pocket, therefore resulting in a decrease of selectivity between EGFRL858R/T790M and EGFRWT (Chart 3.22).

Kawakita et al. designed and synthesized pyrrolo[3,2-*d*]pyrimidine as a human epidermal growth factor receptor 2 (HER2)/*EGFR* as a dual inhibitors [28]. The *N*-tert-butylcarboxamide derivative **362** was synthesized as shown in Scheme 3.43. The condensation of commercially available 2-chloro-1-fluoro-4-nitrobenzene (**356**) with phenol **357** was carried out in the presence of potassium carbonate (K_2CO_3) in dimethylformamide (DMF), followed by a reduction of the nitro group with platinum on carbon (Pt/C) under a hydrogen atmosphere to afford substituted aniline **359** in two steps (95% yield).

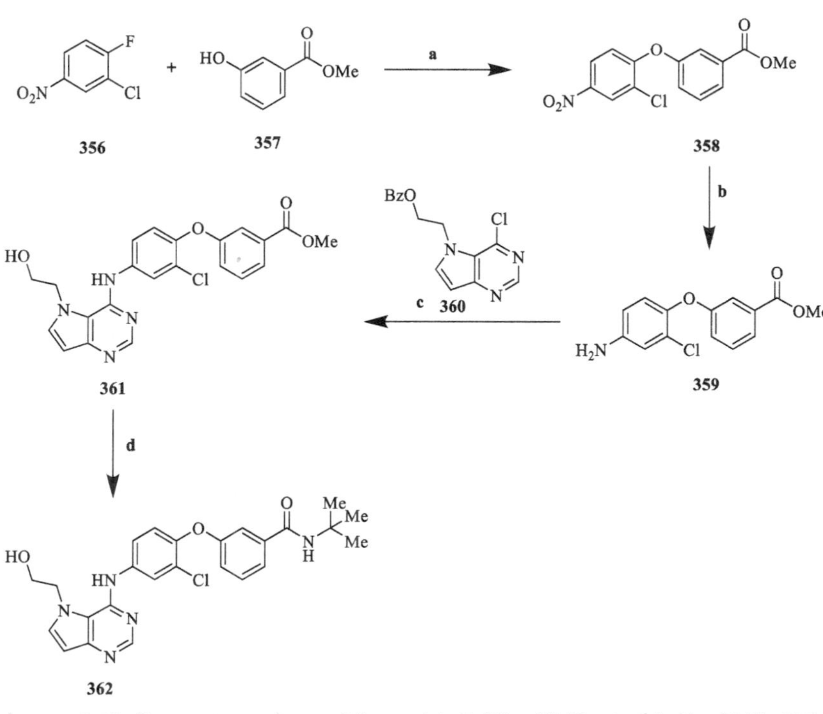

Scheme 3.43 Reagents and conditions: (a) K_2CO_3, DMF, rt; (b) H_2, Pt/C, EtOAc, rt; (c) (i) 2-propanol, 80°C and (ii) 1N NaOH, MeOH, rt; and (d) EDC, HOBt, Et_3N, 2-methylpropan-2-amine, DMF, rt.

After the reaction of the obtained aniline **359** with 2-(4-chloro-5*H*-pyrrolo[3,2-*d*]-pyrimidin-5-yl)ethyl benzoate (**360**), the terminal benzoyl and methyl esters were hydrolyzed with 1N aqueous sodium hydroxide (NaOH), resulting in an 88% yield of the corresponding carboxylic acid (**361**). Amidation of the carboxylic acid **361** with 2-methylpropan-2-amine in the presence of triethylamine, 1-ethyl-3-(3-dimethylaminopropyl)carbodiimide hydrochloride (EDC), and 1-hydroxybenzotriazole monohydrate (HOBt) provided the corresponding desired carboxamide **362** in a 66% yield. Anilines with a 1,3-dihydro-2*H*-indol-2-one ring (**367c**, **d**, **g**, and **h**) were prepared as shown in Scheme 3.44. Reaction of **356** with the commercially available 1*H*-indol derivatives **363c**, **d**, **g**, and **h** gave the corresponding nitrobenzenes **364c**, **d**, **g**, and **h**, respectively.

Bromination of compounds **364c**, **d**, **g**, and **h**with *N*-bromosuccinimide (NBS) yielded the intermediate compounds **365c**, **d**, **g**, and **h**, which on hydrolysis using 6N HCl provided 1,3-dihydro-2*H*-indol-2-one derivatives **366c**,**d**, **g**, and **h**. Finally, the nitro groups of the compounds **366c**, **d**, **g**, and **h** were reduced with Pt/C under a hydrogen atmosphere to afford desired aniline compounds **367c**, **d**, **g**, and **h** in yields of 61%–96%. Synthesis of the aniline compounds **367a** and **e** bearing the corresponding 2,3-dihydro-1*H*-isoindol-1-one ring, is shown in Scheme 3.45. The reaction of the hydroxy methyl benzoates **368a** and **368e** with **356** in the presence of K_2CO_3 resulted in 92%–96% yields of the condensed compounds **369a** and **369e**, which were reacted with NBS to provide 37% and

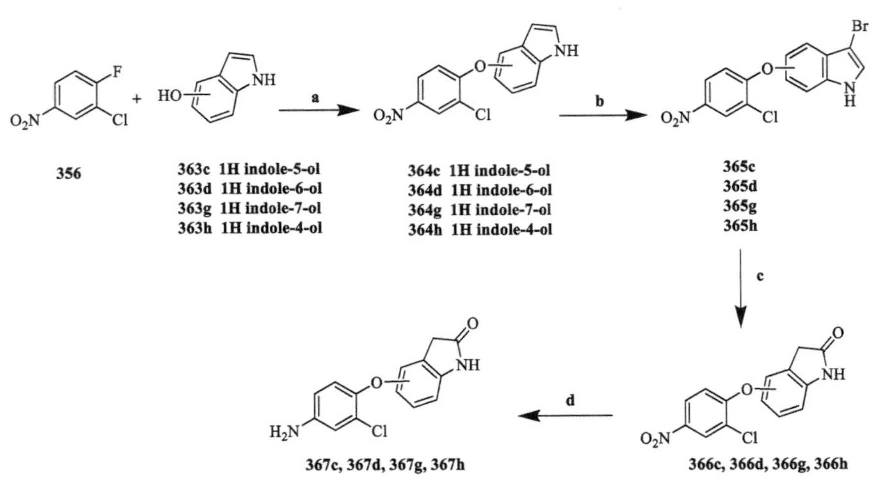

Scheme 3.44 Reagents: (a) K_2CO_3, DMF, rt; (b) NBS, tert-BuOH, THF, rt; (c) 1N HCl, THF, 70°C; and (d) H_2, Pt/C, MeOH, THF, rt.

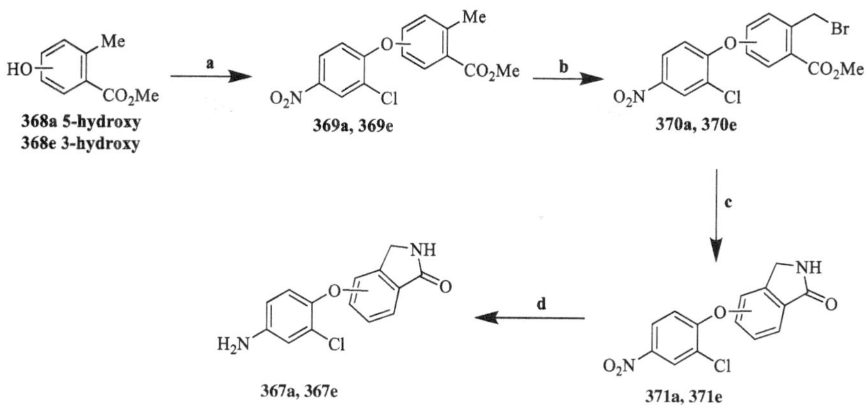

Scheme 3.45 Reagents: (a) K_2CO_3, DMF, rt; (b) NBS, AIBN, CF_3–Ph, 100°C; (c) 28% aq. NH_3, THF, rt; and (d) H_2, Pt/C, MeOH, THF, rt.

70% yields, respectively, of the methyl 2-(bromomethyl)benzoate derivatives **370a** and **370e**. Amination of the benzylbromides **370a** and **370e** with 28% aqueous ammonia solution was carried out in tetrahydrofuran (THF), followed by the subsequent cyclization of the amine intermediates in situ to afford 2,3-dihydro-1*H*-isoindol-1-one derivatives **371a** and **371e** in yields of 82% and 96%, respectively. The obtained products **371a** and **371e** were converted to the desired aniline compounds **367a** and **367e** in 50%–70% yields under the usual hydrogenation conditions.

The 2,3-dihydro-1*H*-isoindol-1-one derivative **367b** was synthesized as shown in Scheme 3.46. The bromination of methyl-4-methoxy-2-methylbenzoate (**372**) with NBS in the presence of AIBN in trifluoromethylbenzene, followed by a treatment with 28% aqueous ammonia solution, resulted in a 39% yield of cyclic compound **374**. The deprotection of the methyl group of **374** with aqueous hydrogen bromide yielded phenol compound **375**, which on further coupling with nitrobenzene gave **376** in 74% yield. Standard hydrogenation of **376** using Pt/C led to the desired aniline compound **367b** in quantitative yield.

Synthesis of the aniline derivatives **367f** is given in Scheme 3.47. Bromination of ethyl 2-methoxy-6-methylbenzoate (**377**) provided dibrominated benzoate **378** followed by a conversion with K_2CO_3 provided a coupled product **382**, and subsequent reduction of the nitro group of **382** afforded the desired aniline **367f** in quantitative yield. As shown in Scheme 3.48, the preparation of the aniline compound **367i** bearing an 1*H*-indole ring was achieved by the reduction of the nitro group of **364h** using reduced iron and calcium chloride in EtOH (73% yield).

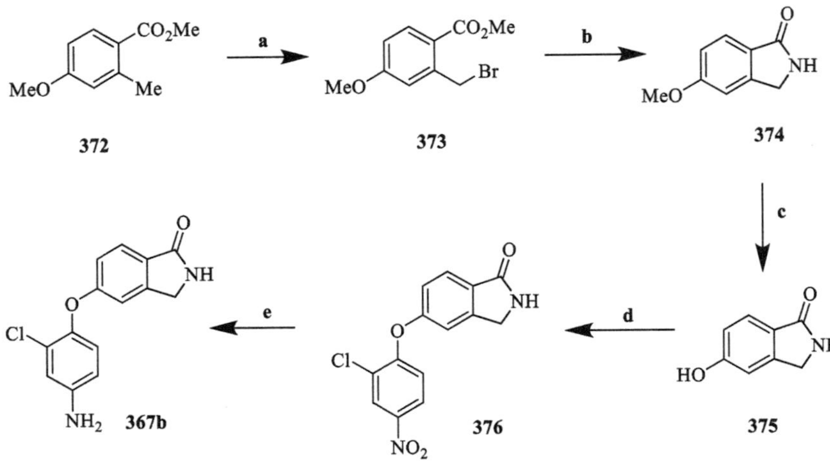

Scheme 3.46 Reagents: (a) NBS, AIBN, CF_3–Ph, 100°C; (b) 28% aq. NH_3, THF, rt; (c) 48% aq. HBr, 100°C; (d) 4, K_2CO_3, DMF, rt; and (e) H_2, Pt/C, MeOH, THF, rt.

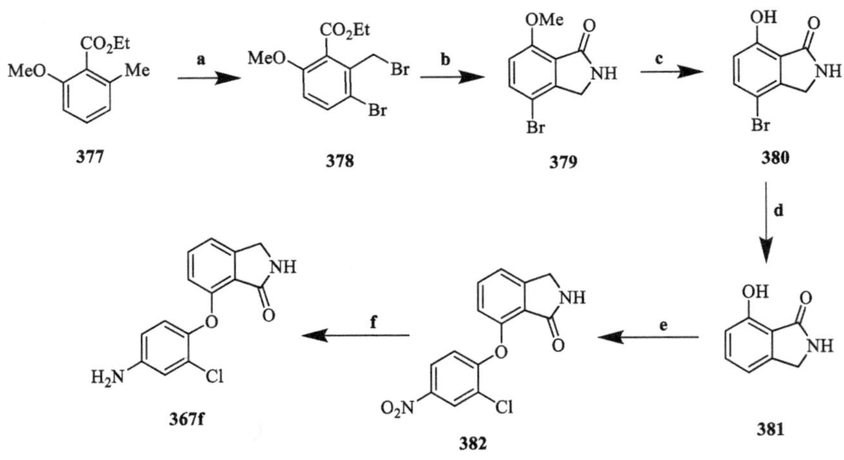

Scheme 3.47 Reagents and conditions: (a) NBS, AIBN, CF_3–Ph, 100°C; (b) 28% aq. NH_3, THF, rt; (c) 48% aq. HBr, 100°C; (d) H_2, Pt/C, MeOH, rt; (e) 376, K_2CO_3, DMF, rt; and (f) H_2, Pt/C, MeOH, THF, rt.

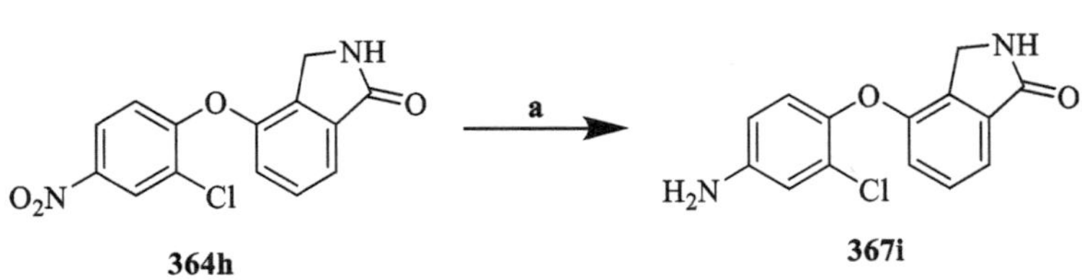

Scheme 3.48 Reagents and conditions: (a) Fe, $CaCl_2$, aq. EtOH, 100°C.

The aniline compound **367j** bearing the 1-benzothiophene ring was synthesized as shown in Scheme 3.49. Aromatization of 6,7-dihydro-1-benzothiophen-4(5*H*)-one (**383**) was carried out with phenyltrimethylammonium tribromide in THF, followed by a treatment with lithium carbonate and lithium bromide in DMF to provide 1-benzothiophene-4-ol 31 in a 91% yield. The obtained phenol compound **384** was coupled with **356** in the presence of K_2CO_3 to provide **385** in 67% yield. The reduction of the nitro group of **385** with reduced iron powder gave the desired aniline compound **367j** in a 80% yield.

The synthesis of the aniline compounds **367k** and **367l** bearing 1*H*-indazole ring is shown in Scheme 3.50. Conversion of commercially available 1*H*-indazol-4-amine **386** into 1*H*-indazol-4-ol **387k** using sodium nitrite, followed by the reaction of **387k** with **356**, afforded a coupled nitrobenzene compound **388k**. The nitro group of **388k** was reduced with Pt/C under hydrogen atmosphere to provide the desired aniline compound **367k**. Another methylated aniline derivative **367l** was synthesized in 59% yield from commercially available 1-methyl-1*H*-indazol-4-ol (**387l**) in a manner similar to the synthesis of **367k**.

The aniline compound **367m** bearing an 1,2-benzisothiazole ring was synthesized as shown in Scheme 3.51. The coupling of the commercially available 2-fluoro-6-methoxybenzaldehyde (**389**) with phenylmethanethiol was carried out in the presence of sodium tert-butoxide in DMF (91% yield), followed by the cyclization of the obtained compound **390** with hydroxylamine-*o*-sulfonic acid in the presence of thioanisole in aqueous acetonitrile to afford 1,2-benzisothiazole **391** in a 85% yield. Subsequently

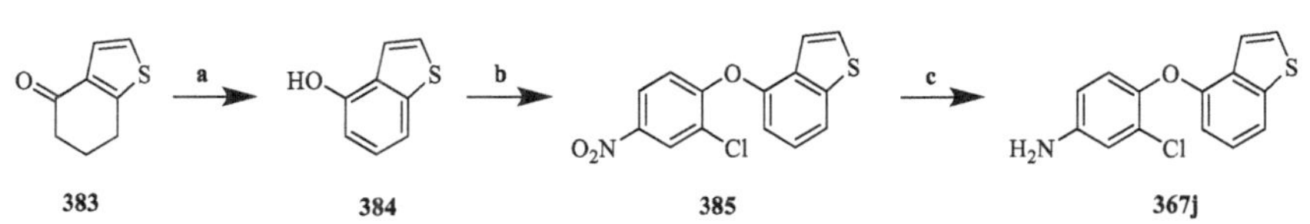

383 384 385 367j

Scheme 3.49 Reagents and conditions: (a) (i) $PhMe_3NBr_3$, THF, rt and (ii) Li_2CO_3, LiBr, DMF, 150°C; (b) 376, K_2CO_3, DMF, rt; and (c) Fe, $CaCl_2$, aq. EtOH, 100°C.

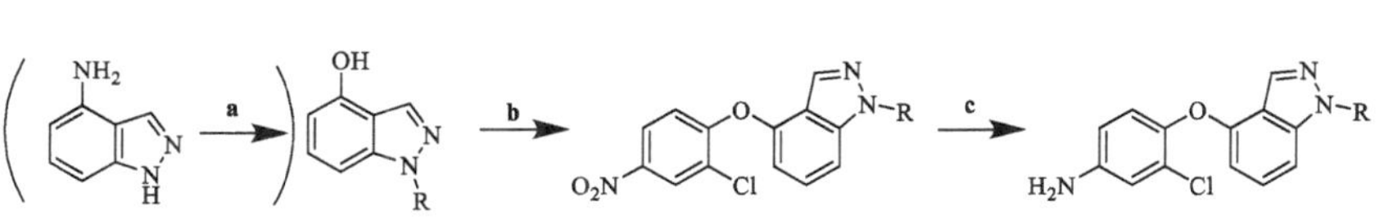

386 387k R = H 387l R = CH3 388k, 388l 367k, 367l

Scheme 3.50 Reagents and conditions: (a) $NaNO_2$, H_2SO_4, 100°C; (b) 376, K_2CO_3, DMF, rt; and (c) H_2, Pt/C, AcOEt, rt.

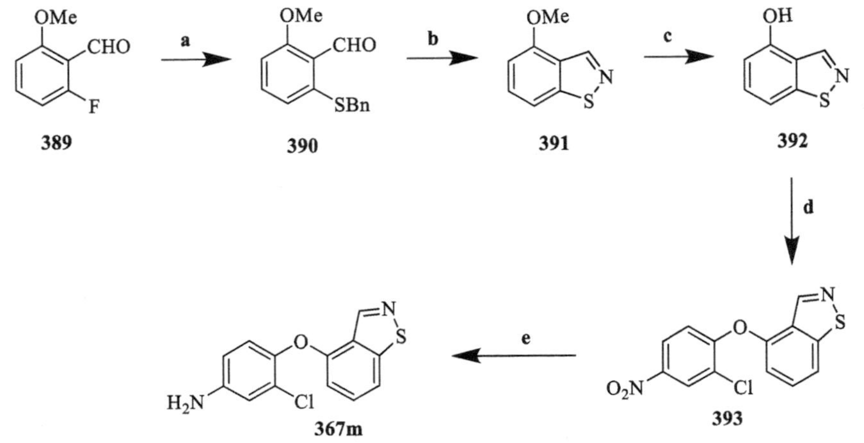

Scheme 3.51 Reagents and conditions: (a) BnSH, tert-BuONa, DMF, rt; (b) H_2NOSO_3H, thioanisole, CH_3CN, H_2O, rt; (c) PyHCl, 195°C; (d) 4, K_2CO_3, DMF, rt; and (e) Fe, 1N HCl, EtOH, 80°C.

the removal of the methyl group from **391** was achieved using pyridine hydrochloride with heating at 195°C for 2h in a 78% yield. Attempts at the demethylation of **391** and **392** using HBr resulted in decomposition to form complex mixtures. The obtained phenol derivative **392** was coupled with **356** in DMF to give nitrobenzene derivative **393** in a 96% yield. Finally the nitro group of **393** was reduced with reduced iron powder under acidic conditions to afford the desired aniline derivative **367m** in 98% yield.

The synthesis of the 4,5-substituted pyrrolo[3,2-*d*]-pyrimidines **395(a–m)** was conducted as shown in Scheme 3.52. The aniline derivatives **367a–m** were reacted with pyrrolo[3,2-d]pyrimidine **360** in 2-propanol at 80°C, followed by the hydrolysis of the benzoyl group to afford the desired compounds **395a–m** in 27%–98% yield over two steps.

The preparation of the *N*5 carboxamide derivatives **398–404m** is summarized in Scheme 3.53. The condensation of **396** with the aniline **367m** was carried out in 2-propanol, followed by the removal of the BOC protecting group with 4N HCl in EtOAc to afford the key intermediate **397** in a 80% yield over two steps. The amidation of **397** with a variety of carboxylic acids in the presence of EDC and HOBt provided the corresponding carboxamide derivatives **(398–401)m** in a 41%–80% yield. The synthesis of the carboxamide derivative **402m** was achieved using 2,2-dimethylpropanoic acid anhydride, and the product isolated as a salt of methanesulfonic acid in 63% yield.

The cellular and enzymatic data of the synthesized compounds are shown in Chart 3.23. To develop novel *HER2/EGFR* kinase inhibitors, Kawakita et al. explored pyrrolo[3,2-*d*]pyrimidine derivatives bearing novel back-

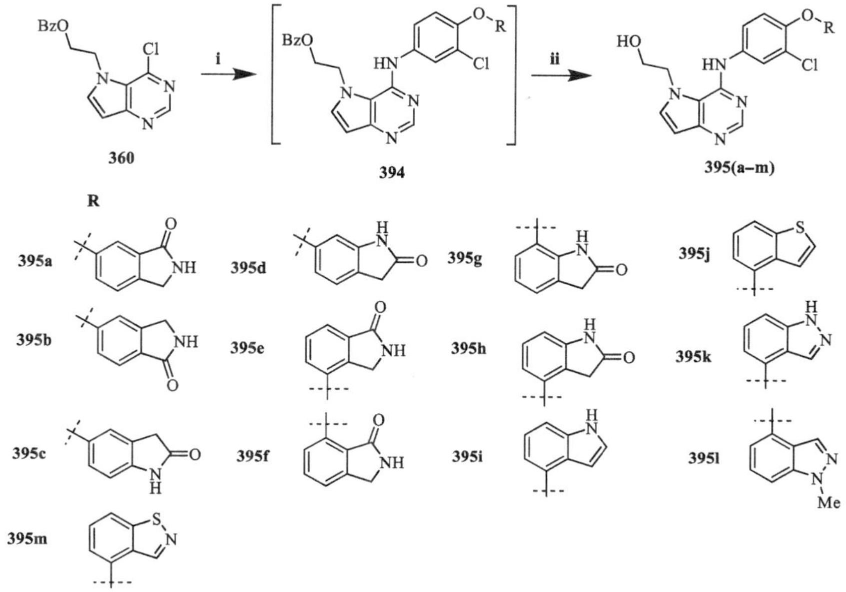

Scheme 3.52 Reagents and conditions: (i) aniline **387a–m**, 2-propanol, 80°C and (ii) 1N NaOH, MeOH, rt.

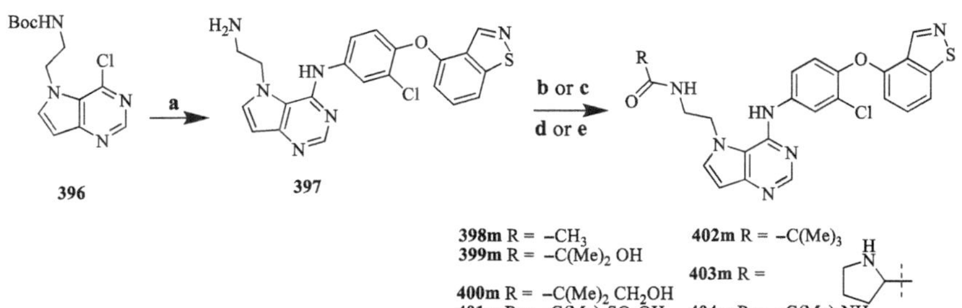

Scheme 3.53 Reagents and conditions: (a) (i) 367m, 2-propanol, 80°C and (ii) 4N HCl in EtOAc, EtOH, 80°C; (b) RCO_2H or 1-(tert-butoxycarbonyl)proline or *N*-(tertbutoxycarbonyl)-2-methylalanine, EDC, HOBt, Et_3N, DMF, rt; (c) 2,2-dimethylpropanoic acid anhydride, Et_3N, DMF, rt; (d) 4N HCl in EtOAc, EtOH, 80°C or 1N HCl, EtOH, 80°C; and (e) MsOH, EtOAc, or DMSO, rt.

pocket binders. Among the prepared derivatives, compound **404m** showed both potent *HER2/EGFR* inhibitory activities (IC_{50}, 0.98/2.5 nM), as well as breast cancer cell BT-474 GI (GI_{50}, 2.0 nM) activities.

Besides HER2 and EGFR, compound **404m** exhibited HER4 inhibitory activity with an IC_{50} value of 79 nM. Reflecting these strong in vitro activities, compound **404m** exhibited potent tumor regressive efficacy against both HER2- and EGFR-overexpressing tumor xenograft models in

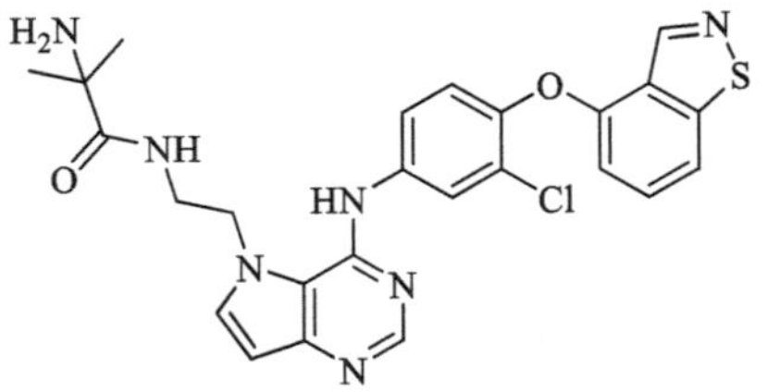

404m
HER2 IC_{50} = 0.98 nM
EGFR IC_{50} = 3.0 nM
HER4 IC_{50} = 79 nM
HER2 GI_{50} = 13 nM

Chart 3.23 Antiproliferative and in vitro EGFR inhibitory activities.

mice at doses of 50 and 100 mg/kg. These preclinical studies demonstrated that **404m** has the potential for use as a HER2/EGFR dual kinase inhibitor to cure- and EGFR-overexpressing cancer.

Lelais et al. reported the discovery of *N*-(7-chloro-1-(1-[4-(dimethylamino)but-2-enoyl]azepan-3-yl)-1*H*-benzo[*d*]imidazol-2-yl)-2-methylisonicotinamide (EGF816) **478**, a novel, potent, and WT-sparing covalent inhibitor of oncogenic (*L858R*, ex19del) and resistant (*T790M*) *EGFR* mutants for the treatment of *EGFR* mutant nonsmall-cell lung cancers [29]. Compounds were prepared from the commercially available 1-fluoro-2-nitroarenes according to the general procedures described in Schemes 3.54–3.56. SNAr reaction with primary amines or appropriately

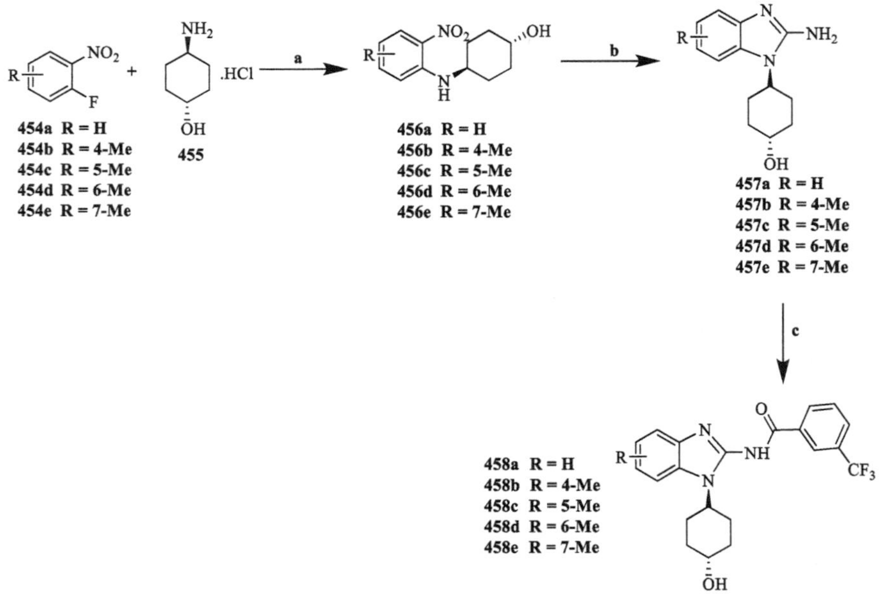

Scheme 3.54 Reagents and conditions: (a) DMF, DIPEA, 120°C; (b) Pd/C, H_2, MeOH; (c) CNBr, MeCN/H_2O/MeOH 55°C; and (d) 3-(trifluoromethyl)benzoic acid, HATU, DIPEA, DMF or 3-(trifluoromethyl)benzoic acid, HOBt, EDCI, Et_3N, DMF.

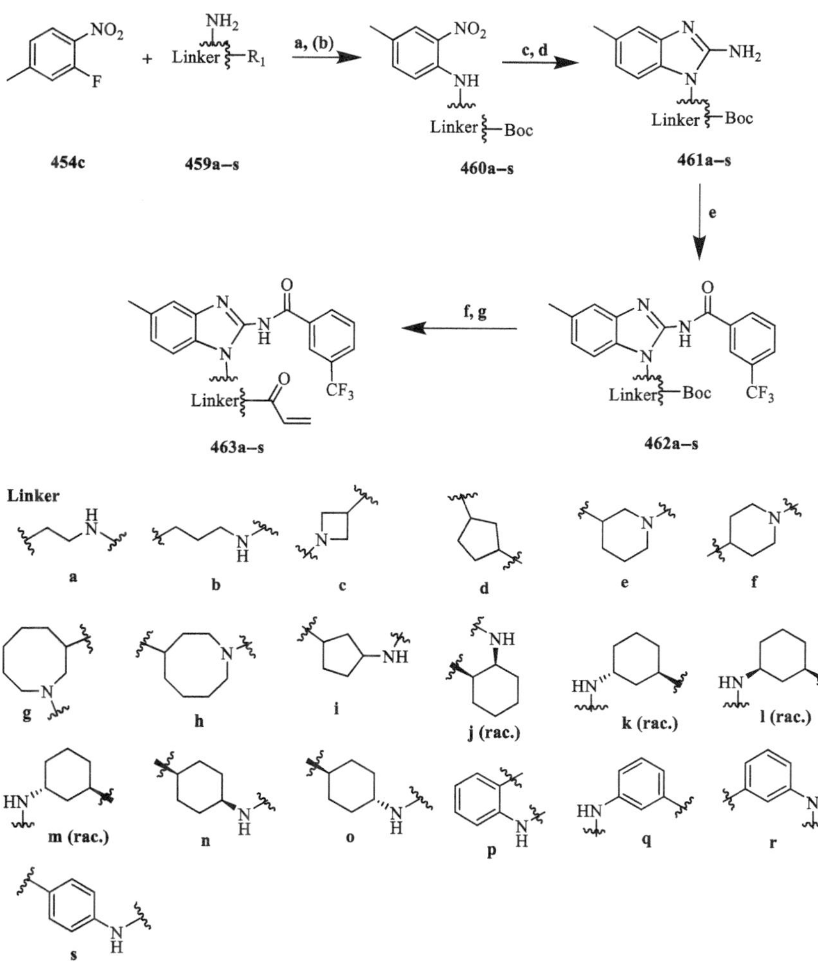

Scheme 3.55 Reagents and conditions: (a) R_1 = BOC: DMF, base (DIPEA or K_2CO_3), 110–130°C; R_1 = H: neat or DMF, 130–150°C; (b) for p, q, s, R_1 = H: Boc_2O, Et_3N, dioxane; (c) Pd/C, H_2, MeOH; (d) CNBr, MeCN/H_2O/MeOH 45–50°C; (e) 3-(trifluoromethyl)benzoic acid, HATU, DIPEA, DMF; (f) TFA, CH_2Cl_2; and (g) acryloyl chloride, with or without Et_3N, CH_2Cl_2, 0–20°C.

protected diamines followed by nitro reduction and cyclization with cyanogen bromide afforded 2-aminobenzimidazole intermediates **457a–e**, **461a–s**, **468**, or **469** in moderate yields (14%–85% over several steps). Amide coupling of **457a–e**, **461a–s**, or **468** with 3-(trifluoromethyl)benzoic acid provided **458a–e** (Scheme 3.54) or intermediates **462a–s** and **470**, respectively.

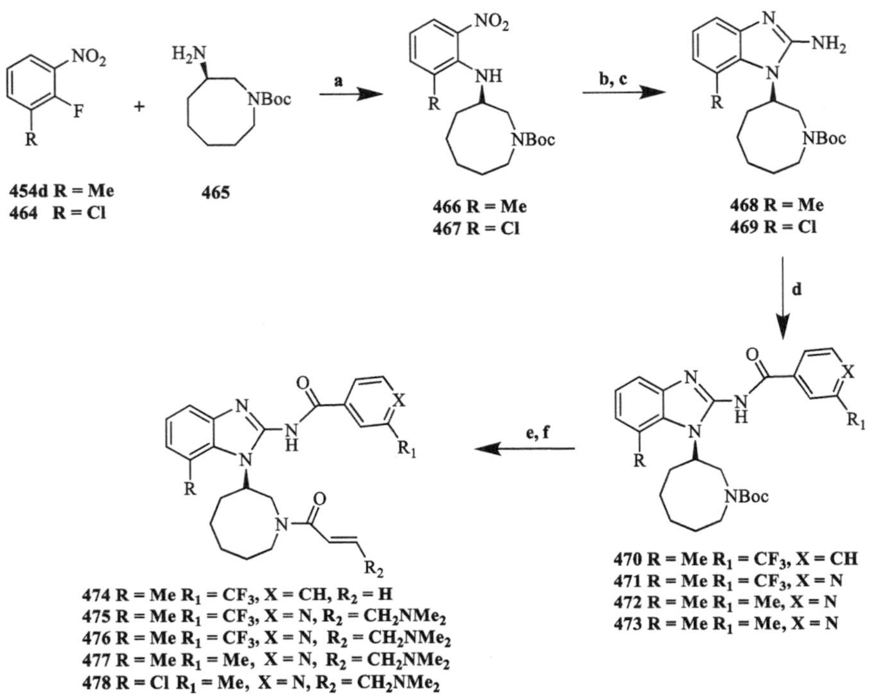

Scheme 3.56 Reagents and conditions: (a) neat, DMA or DMF, DIPEA, 110–140°C; (b) for **466**: Pd/C, H_2, MeOH. For **467**: Zn, AcOH, rt; (c) CNBr, $MeCN/H_2O/MeOH$ 55°C; (d) acid, HATU, Et_3N, CH_2Cl_2 or acid, HATU, DIPEA, DMF; (e) TFA, CH_2Cl_2 or HCl, dioxane/MeOH; and (f) For **474**: acryloyl chloride, CH_2Cl_2, 0–20°C. For **475–478**: (*E*)-4-(dimethylamino) but-2-enoic acid hydrochloride, HOBt, EDCI, Et3N, DMF or (*E*)-4-(dimethylamino)but-2-enoic acid, HATU, Et_3N, CH_2Cl_2.

The latter were deprotected and transformed to the desired products **463a–s** (Scheme 3.55) or **474** and **475** (Scheme 3.56) in acceptable yields (17%–71%). Amide coupling of **468** and **469** with 2-substituted isonicotinic acids afforded **471–473** that were deprotected and further coupled with (*E*)-4-(dimethylamino)-but-2-enoic acid to provide **476–478** in moderate yields (35%–61%, Scheme 3.56).

Biological evaluation data of the synthesized compounds are shown in Chart 3.24. Structure-activity relationship demonstrated that the unsubstituted derivative **458a** was well tolerated, demonstrating twofold improved biochemical activity on the L858R-T790M construct. Comparable potency was observed with the 6-methyl analog (**458d**), while slight potency improvements were noted with compounds that had a methyl substitution at positions 5 and 7 (**458c** and **458d**). Conversely, as predicted by modeling, compound **458b** was less potent, likely due to the methyl group pointing toward the hinge binding domain of EGFR. Importantly, in all cases the

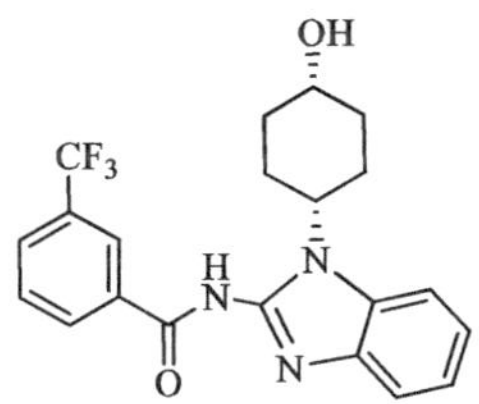

458a

L858R/T790M IC_{50} = 0.26 nM
L858R IC_{50} = 2.0 nM
WT IC_{50} = >50 nM

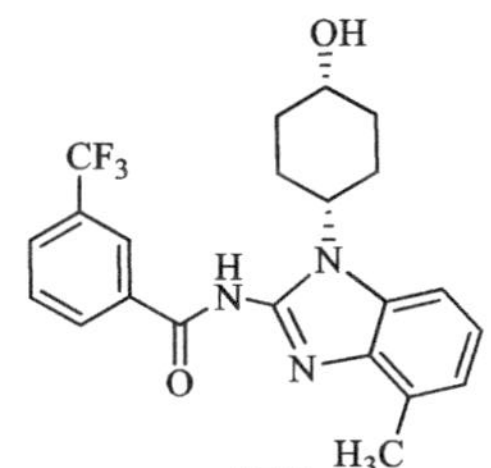

458b

L858R/T790M IC_{50} = 1.37 nM
L858R IC_{50} = 11.75 nM
WT IC_{50} = >50 nM

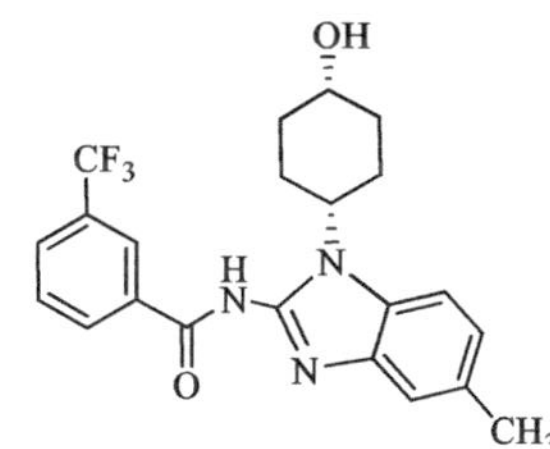

458c

L858R/T790M IC_{50} = 0.14 nM
L858R IC_{50} = >8.48 nM
WT IC_{50} = >11.78 nM

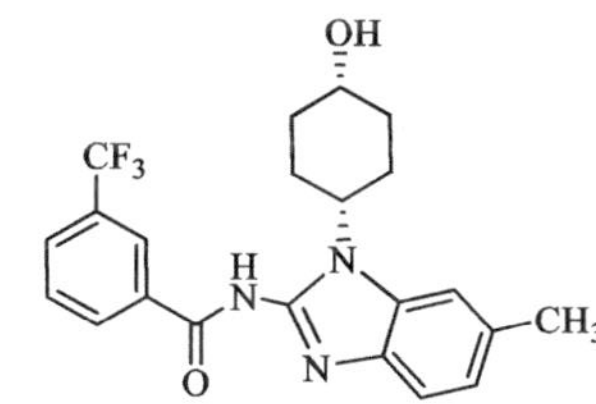

458d

L858R/T790M IC_{50} = 0.28 nM
L858R IC_{50} = >16 nM
WT IC_{50} = >50 nM

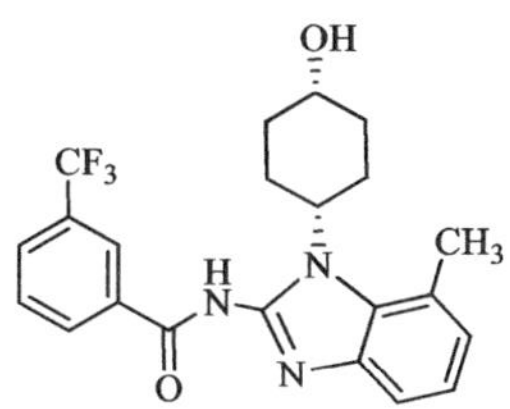

458e

L858R/T790M IC_{50} = 0.09 nM
L858R IC_{50} = 0.68 nM
WT IC_{50} = 5.27nM

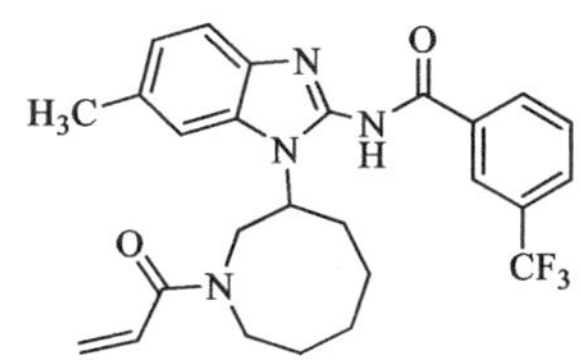

463g

H1975 IC_{50} = 6.1 nM
H3255 IC_{50} = 74.9 nM
HCC827 IC_{50} = 15.9 nM
HaCaT(WT) IC_{50} = 4330 nM

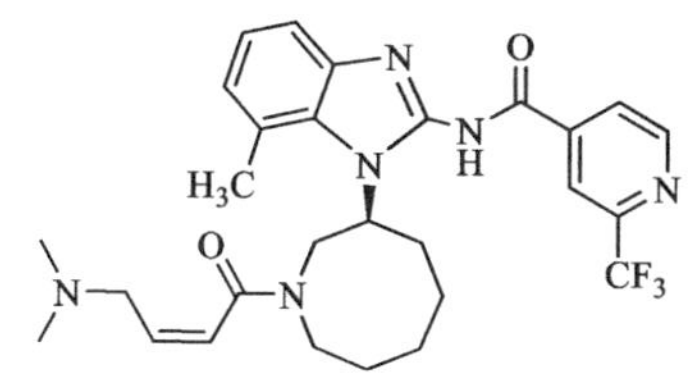

478

H1975 IC_{50} = 4.18 nM
H3255 IC_{50} = 6.11 nM
HCC827 IC_{50} = 0.015 nM
HaCaT(WT) IC_{50} = 4.330 nM

Chart 3.24 Antiproliferative and in vitro EGFR inhibitory activities.

selectivity against WT EGFR was maintained, and hints of activity on the oncogenic L858R mutant were observed with **458a** and **458e**.

The cellular potency on H1975 was in good agreement with the biochemical inhibitory concentration on the L858R-T790M construct, with **463g** demonstrating superior activity (IC_{50} = 0.006 μM), followed by **463e** (IC_{50} = 0.02 μM; R-enantiomer). In addition, compound **463g** was highly potent across all mutant EGFR lines and displayed acceptable WT EGFR selectivity (≥50-fold).

Significantly, **478** had single-digit activity on the mutant cell lines (4, 6, 2 nM IC_{50} in H1975, H3255, and HCC827, respectively) with less than a twofold loss of potency on H3255. At the same time, it displayed a reduced activity on HaCaT (0.18 μM), providing similar WT EGFR selectivity (~36-fold). Importantly, **478** demonstrated improved ADME and PK properties as compared to other compounds. Compound **478** is currently being tested in numerous clinical trials, and it has already demonstrated encouraging efficacy in NSCLC patients with T790M mutations (Chart 3.24).

Jeff Bruce Smaill reported the optimization of substituted quinazoline and pyrido[3,4-*d*]pyrimidine derivatives as orally active, irreversible inhibitors of the EGFR family [30]. The synthesis of 6-amide substituted quinazolines, pyrido[3,4-*d*]pyrimidines, and quinoline-3-carbonitriles is well precedented, and the present work proceeded along similar lines. The substituted acrylamides (**480–483**) of Scheme 3.57 were prepared by reaction of the 6-aminoquinazoline (**479**) with the respective acid chlorides (Scheme 3.57).

The 4-anilinoquinazolines (**488–519**) of Scheme 3.58 were prepared by aniline displacement of the 4-chloro substituent of quinazoline **487**. This array substrate was prepared from 7-fluoro-6-nitroquinazolone (**484**) by methoxide displacement of the activated fluorine to give **485**, followed by an iron dust reduction of the nitro group and acylation with propionic anhydride to give propanamide **486**. Chlorination in refluxing $POCl_3$ then gave **487** (Scheme 3.58). The quinazolines (**524–532**) of Scheme 3.59 were prepared by ethoxide displacement on **484** followed by $POCl_3$ mediated

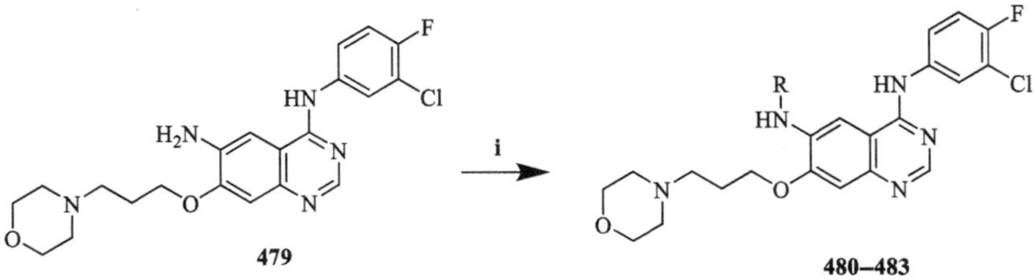

Scheme 3.57 Reagents and conditions: (i) acid chloride, Et_3N, THF, rt.

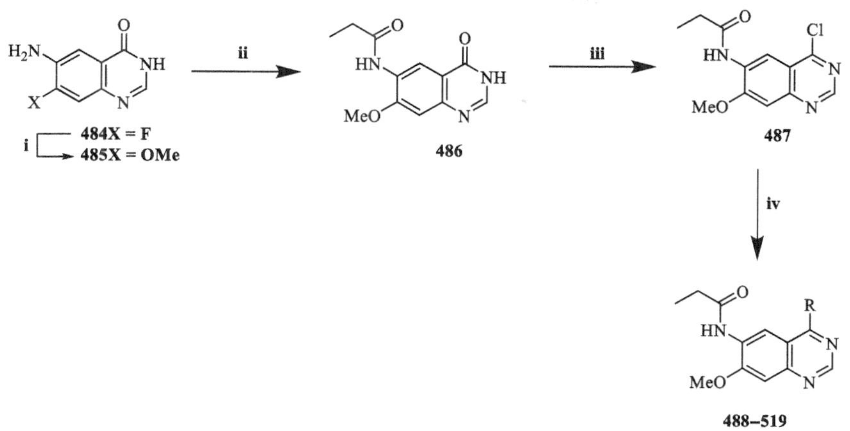

Scheme 3.58 Reagents and conditions: (i) NaOMe, MeOH, 100°C, 18 h; (ii) (a) Fe, AcOH, EtOH, water, 100°C, 1 h and (b) propionic anhydride, 120°C, 2 h; (iii) $POCl_3$, reflux, 90 min; and (iv) subst. aniline, *i*PrOH, c. HCl (cat.), 100°C, 0.5–1 h.

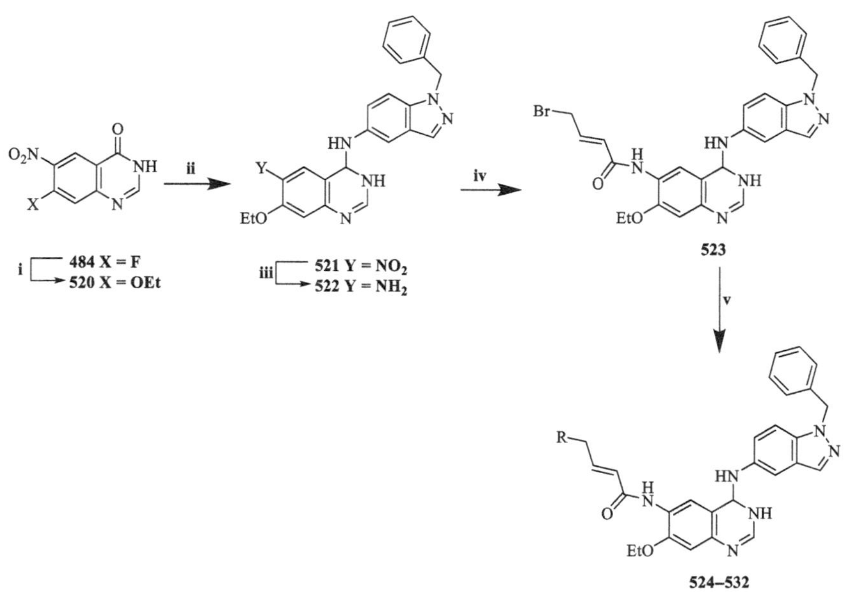

Scheme 3.59 Reagents and conditions: (i) NaOEt, EtOH, 50°C, 4 h; (ii) (a) $POCl_3$, reflux, 3 h and (b) 1-benzyl-1*H*-indazol-5-amine, *i*PrOH/CH_2Cl_2, 50°C, 15 h; (iii) Raney nickel, H_2, EtOAc/MeOH; (iv) (a) *E*-4-bromobut-2-enoic acid, oxalyl chloride, cat. DMF, CH_2Cl_2, rt and (b) acid chloride from (a), THF, −20°C, 0.75 h; and (v) amine, DMA, −20°C, 2 h.

chlorination and displacement with 1-benzyl-1*H*-indazol-5-amine to give **521**. Nitro group reduction via hydrogenation over Raney nickel then gave amine **522**, which was acylated with the freshly prepared acid chloride of *E*-4-bromobut-2-enoic acid to give bromide **523**. The low-temperature displacement of the bromide with a range of secondary amines then gave compounds **524–532** of Scheme 3.59.

The quinazolines (**534–550**) of Scheme 3.60 were prepared in an analogous manner to those of Scheme 3.59 described above. The pyrido[3,4-*d*] pyrimidines (**556–568**) were prepared similarly from 6-fluoropyrido[3,4-*d*] pyrimidin-4(3*H*)-one **551** by chlorination with $SOCl_2$ and displacement with the respective anilines to give compounds **552**. Fluorine displacement with 4-methoxybenzylamine then gave the PMB-protected amines (**553**), which were deprotected using TFA to give the 6-amino derivatives (**554**). Acylation of these with the freshly prepared acid chloride of *E*-4-bromobut-2-enoic acid gave the bromides (**555**). The low-temperature displacement of the bromides with a range of secondary amines then gave compounds **556–568** of (Scheme 3.60).

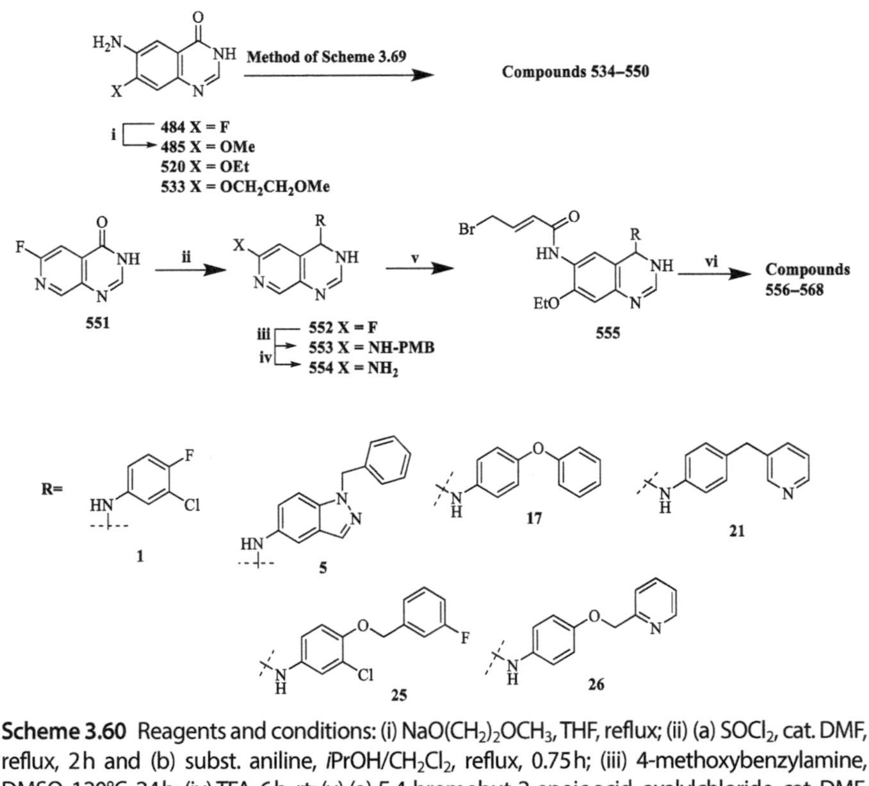

Scheme 3.60 Reagents and conditions: (i) $NaO(CH_2)_2OCH_3$, THF, reflux; (ii) (a) $SOCl_2$, cat. DMF, reflux, 2h and (b) subst. aniline, *i*PrOH/CH_2Cl_2, reflux, 0.75h; (iii) 4-methoxybenzylamine, DMSO, 120°C, 24h; (iv) TFA, 6h, rt; (v) (a) *E*-4-bromobut-2-enoic acid, oxalylchloride, cat. DMF, CH_2Cl_2, rt and (b) acid chloride from (a), THF, −20°C, 0.75h; and (vi) amine, DMA, −20°C, 2h.

The biological evaluation data of synthesized compounds is given in Chart 3.25. They have described an approach to the development of an irreversible inhibitor of erbB1, 2, and 4 with an improved pharmacokinetic and toxicity profile relative to canertinib. Compounds **535** (PF-00299804) and **560** were identified as clear lead candidates for advanced studies, having demonstrated selective, potent, irreversible inhibition of erbB1 in vitro and prolonged inhibition of erbB1 autophosphorylation in vivo (Chart 3.25). The sustained target modulation for **535** in vivo is consistent with dramatically improved pharmacokinetics relative to canertinib in rats, dogs, and monkeys, a consequence of the metabolic stability of the piperidinyl crotonamide Michael acceptor.

Jeff Bruce Smaill proposed that the Michael acceptor at the 6-position, particularly with a 3-chloro-4-fluoroanilino group at the 4-position, provides an optimum pharmacophore for irreversible inhibition of erbB1, 2, and 4 on both quinazoline and pyrido[3,4-*d*]pyrimidine scaffolds. Pharmacokinetic comparison of compounds **535** and **560** in rats, dogs, and monkeys identified compound **535** over (the more variable) compound **560** as the preferred clinical candidate.

The irreversible inhibition of erB1 was confirmed at the enzyme level, employing an ELISA-based enzyme assay, and at the cellular level by assaying the inhibition of erbB1 autophosphorylation following drug exposure time course and extensive wash-out in A431 cells. This was further supported by solution state protein mass spectrometry experiments and an erbB1 T790M X-ray cocrystal structure at 1.8A resolution, clearly showing the presence of compound **535** (PF-00299804) covalently linked to the target cysteine. Compound **535** PF-00299804 has been screened against a panel of 38 protein kinases to assess off-target interactions and was found to be particularly selective for the erbB family with only Lck, Src, and JAK3 returning isolated enzyme inhibitory $IC_{50s} < 10\,\mu M$ ($IC_{50} = 0.094$, 0.110, and 3.57 μM, respectively).

Coumar et al. reported the fast-forwarding hit to lead optimization approach for the development of *EGFR* inhibitors [31]. Compound **570**

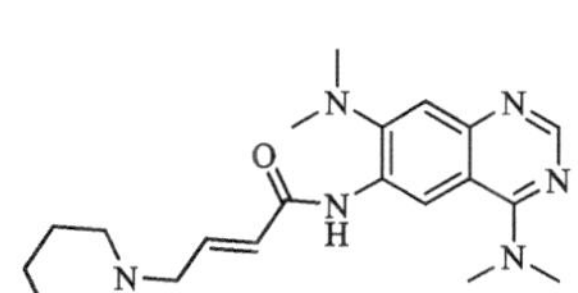

535
erbB1 enzyme $IC_{50} = 0.006\,\mu M$
erbB2 cellular $IC_{50} = 0.046\,\mu M$
erbB4 enzyme $IC_{50} = 0.074\,\mu M$
JAK3 enzyme $IC_{50} = 3.57\,\mu M$

Chart 3.25 In vitro EGFR inhibitory profile of compound 535.

was synthesized by nucleophilic substitution reaction of **569** with various amines in *n*-butanol (Scheme 3.61). The aniline functional group of **571** was converted into urea, amide, and sulfonamide functionality by acylation with phenyl isocyanate, benzoyl chloride, or sulfonyl chloride to give compounds **572–574** (Scheme 3.62). Similarly, the ester function of **575** was hydrolyzed to the carboxylic acid **576**, which was coupled with aniline to give the amide **577** (Scheme 3.63).

The Michael acceptor group of **581b** could covalently interact with the Cys773 residue by solving the X-ray cocrystal complex of **581b** with *EGFR* kinase. A clear covalent linkage between the Michael acceptor group of **581b** and Cys773 residue sulfur atom of EGFR protein was observed,

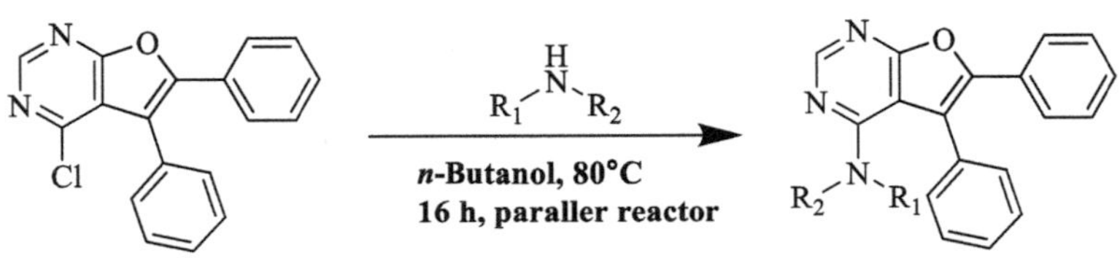

569 570

Scheme 3.61 Reagents and conditions: *n*-Butanol, reflux for 16 h at 80°C.

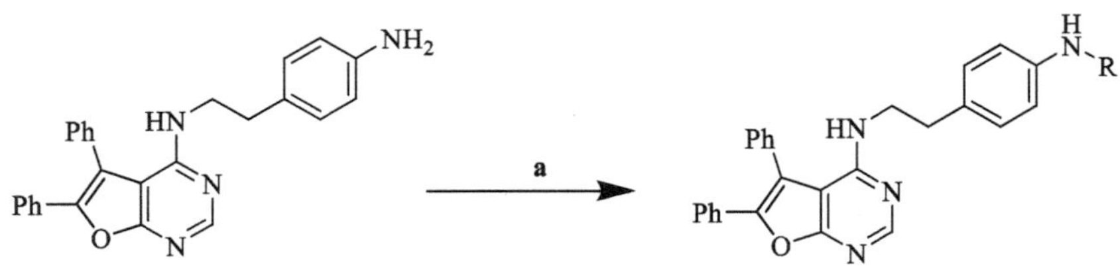

571

572 R = CONHPh
573 R = COPH
574 R = SO_2Ph

Scheme 3.62 Reagents and conditions: (a) LiOH, MeOH/H_2O (4:1), reflux, 2 h.

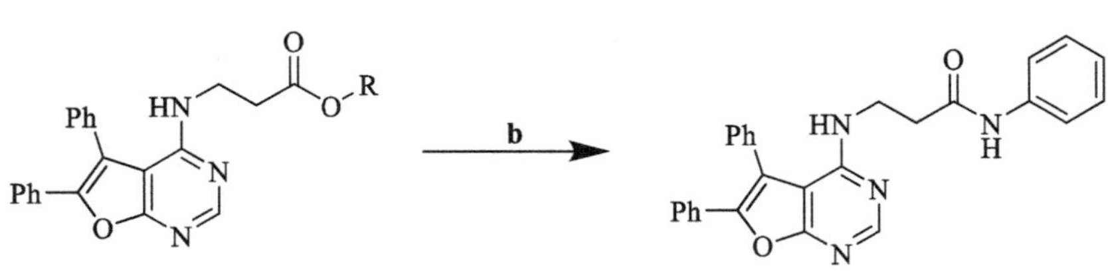

a 575 R = C_2H_5
576 R = H

577

Scheme 3.63 Reagents and conditions: (b) HOBt, EDC, DMF, aniline, rt, 16 h.

which would result in irreversible inhibition of EGFR kinase, explaining the potent activity of **581b**, compared to the hit **570s**.

The introduction of the acrylamide Michael acceptor group led to the synthesis of **581b**, which showed potent activity in both WT and mutant *EGFR* kinase, with potent antiproliferative activity in HCC827 lung cancer cell line (Scheme 3.64). Meanwhile the X-ray cocrystal complex of **581b** with EGFR kinase showed the presence of a covalent bond between the Michael acceptor region of the inhibitor and the Csy733 residue of EGFR, as predicted (Chart 3.26).

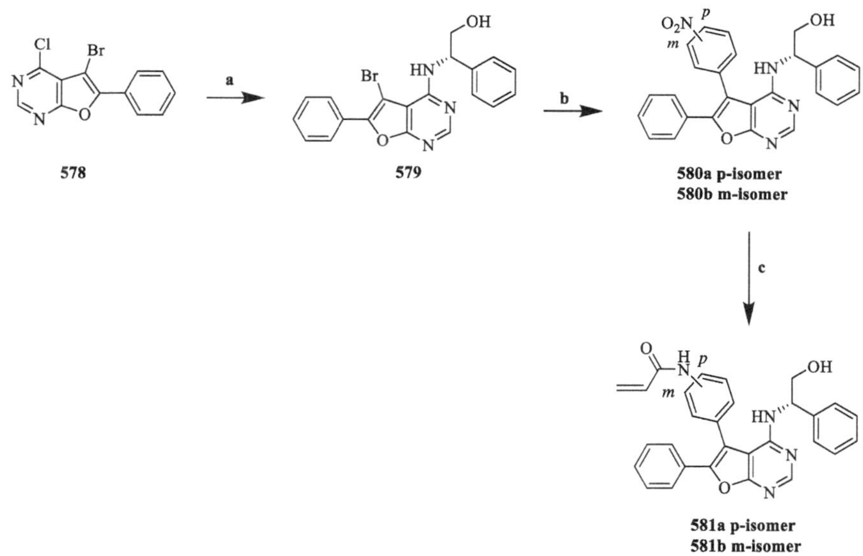

Scheme 3.64 Reagents and conditions: (a) phenylglycinol, *n*-butanol, 80°C, 16h; (b) $ArB(OH)_2$, $Pd(dppf)_2Cl_2$, Na_2CO_3, dioxane/H_2O, 95°C, 16h; and (c) H_2, Pd/C, EtOH, atm, 16h (i) acryloyl chloride, pyridine, Et_2O, rt, 16h.

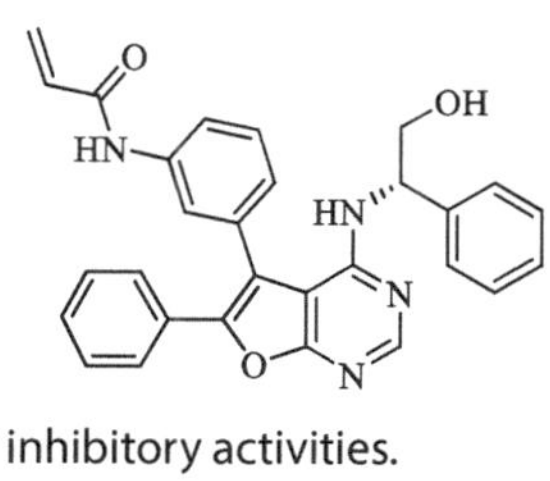

581b EGFR kinase
WT IC_{50} = 7.00 nM
DM IC_{50} = 28.00 nM
HCC827 IC_{50} = 8.00 nM

Chart 3.26 *EGFR* inhibitory activities.

REFERENCES

[1] T.S. Mok, Y.L. Wu, S. Thongprasert, Gefitinib or carboplatin-paclitaxel in pulmonary adenocarcinoma, N. Engl. J. Med. 361 (2009) 947–957.

[2] T. Mitsudomi, S. Morita, Y. Yatabe, S. Negoro, I. Okamoto, J. Tsurutani, Gefitinib versus cisplatin plus docetaxel in patients with non-small-cell lung cancer harbouring mutations of the epidermal growth factor receptor (WJTOG3405): an open label, randomized phase 3 trial, Lancet Oncol. 11 (2010) 121–128.

[3] M. Maemondo, A. Inoue, K. Kobayashi, S. Sugawara, S. Oizumi, H. Isobe, Gefitinib or chemotherapy for non-small-cell lung cancer with mutated EGFR, N. Engl. J. Med. 362 (2010) 2380–2388.

[4] R. Rosell, E. Carcereny, R. Gervais, A. Vergnenegre, B. Massuti, E. Felip, Erlotinib versus standard chemotherapy as first-line treatment for European patients with advanced EGFR mutation-positive non-small-cell lung cancer (EURTAC): a multicentre, open-label, randomised phase 3 trial, Lancet Oncol. 13 (2012) 239–246.

[5] L.V. Sequist, J.C. Yang, N. Yamamoto, K. O'Byrne, V. Hirsh, T. Mok, Phase III study of afatinib or cisplatin plus pemetrexed in patients with metastatic lung adenocarcinoma with EGFR mutations, J. Clin. Oncol. 31 (2013) 3327–3334.

[6] S. Xu, T. Xu, L. Zhang, Z. Zhang, J. Luo, Y. Liu, X. Lu, Z. Tu, X. Ren, K. Ding, Design, synthesis, and biological evaluation of 2-oxo-3,4 dihydropyrimido [4,5-d] pyrimidinyl derivatives as new irreversible epidermal growth factor receptor inhibitors with improved pharmacokinetic properties, J. Med. Chem. 56 (21) (2013) 8803–8813.

[7] Y. Zhang, K. Zhang, M. Zhao, L. Zhang, M. Qin, S. Guo, Y. Zhao, P. Gong, Discovery of a novel class anti-proliferative agents and potential inhibitors of EGFR tyrosine kinases based on 4-anilinotetrahydropyrido[4,3-d]pyrimidine scaffold: design, synthesis and biological evaluations, Bioorg. Med. Chem. 23 (15) (2015) 4591–4607.

[8] C. Han, Z. Huang, C. Zheng, L. Wan, Y. Lai, S. Peng, K. Ding, H. Ji, Y. Zhang, Nitric oxide donating anilinopyrimidines: synthesis and biological evaluation as EGFR inhibitors, Eur. J. Med. Chem. 66 (2013) 82–90.

[9] J. Engel, A. Richters, M. Getlik, S. Tomassi, M. Keul, M. Termathe, J. Lategahn, C. Becker, S. Mayer-Wrangowski, C. Grutter, N. Uhlenbrock, J. Krüll, N. Schaumann, S. Eppmann, P. Kibies, F. Hoffgaard, J. Heil, S. Menninger, S. Ortiz-Cuaran, J.M. Heuckmann, V. Tinnefeld, R.P. Zahedi, M.L. Sos, C. Schultz-Fademrecht, R.K. Thomas, S.M. Kast, D. Rauh, Targeting drug resistance in EGFR with covalent inhibitors: a structure-based design approach, J. Med. Chem. 58 (17) (2015) 6844–6863.

[10] C. Han, Z. Huang, C. Zheng, L. Wan, L. Zhang, S. Peng, K. Ding, H. Ji, J. Tian, Y. Zhang, Novel hybrids of phenylsulfonylfuroxan and anilinopyrimidine as potent and selective epidermal growth factor receptor inhibitors for intervention of non-small cell lung cancer, J. Med. Chem. 56 (11) (2013) 4738–4748.

[11] M.R.V. Finlay, M. Anderton, S. Ashton, P. Ballard, P.A. Bethel, M.R. Box, R.H. Bradbury, S.J. Brown, S. Butterworth, A. Campbell, C. Chorley, N. Colclough, D.A.E. Cross, G.S. Currie, M. Grist, L. Hassall, G.B. Hill, D. James, M. James, P. Kemmitt, T. Klinowska, G. Lamont, S.G. Lamont, N. Martin, H.L. McFarland, M.J. Mellor, J.P. Orme, D. Perkins, P. Perkins, G. Richmond, P. Smith, R.A. Ward, M.J. Waring, D. Whittaker, S. Wells, G.L. Wrigley, Discovery of a potent and selective EGFR inhibitor (AZD9291) of both sensitizing and T790M resistance mutations that spares the wild type form of the receptor, J. Med. Chem. 57 (20) (2014) 8249–8267.

[12] M. Qin, T. Wang, B. Xu, Z. Ma, N. Jiang, H. Xie, P.G.Y. Zhao, Novel hydrazone moiety-bearing amino pyrimidines as selective inhibitors of epidermal growth factor receptor T790M mutant, Eur. J. Med. Chem. 104 (2015) 115–126.

[13] R.A. Ward, M.J. Anderton, S. Ashton, P.A. Bethel, M. Box, S. Butterworth, N. Colclough, C.G. Chorley, C. Chuaqui, D.A.E. Cross, L.A. Dakin, J.E. Debreczeni, C. Eberlein, M.R.V. Finlay, G.B. Hill, M. Grist, T.C.M. Klinowska, C. Lane, S. Martin, J.P. Orme, P. Smith, F. Wang, M.J. Waring, Structure- and reactivity-based development of covalent inhibitors of the activating and gatekeeper mutant forms of the epidermal growth factor receptor (EGFR), J. Med. Chem. 56 (17) (2013) 7025–7048.
[14] Q. Xiao, R. Qu, D. Gao, Q. Yan, L. Tong, W. Zhang, J. Ding, H. Xie, Y. Li, Discovery of 5-(methylthio)pyrimidine derivatives as L858R/T790M mutant selective epidermal growth factor receptor (EGFR) inhibitors, Bioorg. Med. Chem. 24 (12) (2016) 2673–2680.
[15] W. Zhou, X. Liu, Z. Tu, L. Zhang, X. Ku, F. Bai, Z. Zhao, Y. Xu, K. Ding, H. Li, Discovery of pteridin-7(8H)-one-based irreversible inhibitors targeting epidermal growth factor receptor (EGFR) kinase T790M/L858R mutant, J. Med. Chem. 56 (20) (2013) 7821–7837.
[16] Y.H. Peng, H.Y. Shiao, C.H. Tu, P.M. Liu, J.T.A. Hsu, P.K. Amancha, J.S. Wu, M.S. Coumar, C.H. Chen, S.Y. Wang, W.H. Lin, H.Y. Sun, Y.S. Chao, P.C. Lyu, H.P. Hsieh, S.Y. Wu, Protein kinase inhibitor design by targeting the Asp-Phe-Gly (DFG) motif: the role of the DFG motif in the design of epidermal growth factor receptor inhibitors, J. Med. Chem. 56 (10) (2013) 3889–3903.
[17] W. Zhou, D. Ercan, P.A. Janne, N.S. Gray, Discovery of selective irreversible inhibitors for EGFR-T790M, Bioorg. Med. Chem. Lett. 21 (2) (2011) 638–643.
[18] S. Chan, K. Han, R. Qu, L. Tong, Y. Li, Z. Zhang, H. Cheng, X. Lu, A. Patterson, J. Smaill, X. Ren, J. Ding, H. Xie, K. Ding, 2,4-Diarylamino-pyrimidines as kinase inhibitors co-targeting IGF1R and EGFRL858R/T790M, Bioorg. Med. Chem. Lett. 25 (19) (2015) 4277–4281.
[19] R.P. Wurz, L.H. Pettus, K. Ashton, J. Brown, J.J. Chen, B. Herberich, F.-T. Hong, E. Hu-Harrington, T. Nguyen, D.J. St. Jean, S. Tadesse Jr., D. Bauer, M. Kubryk, J. Zhan, K. Cooke, P. Mitchell, K.L. Andrews, F. Hsieh, D. Hickman, N. Kalyanaraman, T. Wu, D.L. Reid, E.K. Lobenhofer, D.A. Andrews, N. Everds, R. Guzman, A.T. Parsons, S.J. Hedley, J. Tedrow, O.R. Thiel, M. Potter, R. Radinsky, P.J. Beltran, A.S. Tasker, Oxopyrido[2,3-d]pyrimidines as covalent L858R/T790M mutant selective epidermal growth factor receptor (EGFR) inhibitors, ACS Med. Chem. Lett. 6 (9) (2015) 987–992.
[20] S. Chang, L. Zhang, S. Xu, J. Luo, X. Lu, Z. Zhang, T. Xu, Y. Liu, Z. Tu, Y. Xu, X. Ren, M. Geng, J. Ding, D. Pei, K. Ding, Design, synthesis, and biological evaluation of novel conformationally constrained inhibitors targeting epidermal growth factor receptor Threonine790 → Methionine790 mutant, J. Med. Chem. 55 (6) (2012) 2711–2723.
[21] X. Ji, T. Peng, X. Zhang, J. Li, W. Yang, L. Tong, R. Qu, H. Jiang, J. Ding, H. Xie, H. Liu, Design, synthesis and biological evaluation of novel 6-alkenylamides substituted of 4-anilinothieno[2,3-d]pyrimidines as irreversible epidermal growth factor receptor inhibitors, Bioorg. Med. Chem. 22 (7) (2014) 2366–2378.
[22] C.H. Wu, M.S. Coumar, C.Y. Chu, W.H. Lin, Y.R. Chen, C.T. Chen, H.Y. Shiao, R. Shaik, S.Y. Wang, H. Hsu, C.H. Chen, C.Y. Chang, T.Y. Chang, T.W. Lien, M.Y. Fang, K.C. Yeh, C.P. Chen, T.K. Yeh, S.H. Hsieh, J.T.A. Hsu, C.C. Liao, Y.S. Chao, H.P. Hsieh, Design and synthesis of tetrahydropyridothieno[2,3-d]pyrimidine scaffold based epidermal growth factor receptor (EGFR) kinase inhibitors: the role of side chain chirality and Michael Acceptor Group for Maximal Potency, J. Med. Chem. 53 (20) (2010) 7316–7326.
[23] E.J. Hennessy, C. Chuaqui, S. Ashton, N. Colclough, D.A. Cross, J.E. Debreczeni, C. Eberlein, L. Gingipalli, T.C. Klinowska, J.P. Orme, L. Sha, X. Wu, Utilization of structure-based design to identify novel, irreversible inhibitors of EGFR harboring the T790M mutation, ACS Med. Chem. Lett. 7 (5) (2016) 514–519.

[24] H. Cheng, S.K. Nair, B.W. Murray, C. Almaden, S. Bailey, S. Baxi, D.C. Behenna, S.C. Schultz, D. Dalvie, D.M. Dinh, M.P. Edwards, J.L. Feng, R. Ferre, K.S. Gajiwala, M.D. Hemkens, A.J. Fisher, M. Jalaie, T.O. Johnson, R.S. Kania, S. Kephart, J.A. Lafontaine, E. Lunney, K.K.C. Liu, Z. Liu, J. Matthews, A. Nagata, S. Niessen, M.A. Ornelas, S.T.M. Orr, M. Pairish, S. Planken, S. Ren, D.T. Richter, K. Ryan, N.W. Sach, H. Shen, T. Smeal, J. Solowiej, S.C. Sutton, K. Tran, E. Tseng, W.F. Vernier, M. Walls, S. Wang, S.L. Weinrich, S. Xin, H. Xu, M.J. Yin, R. Zhou, M. Zientek, J. Kath, Discovery of 1-{(3R,4R)-3-[5-chloro-2-(1-methyl-1H-pyrazol-4-ylamino)-7H-pyrrolo [2,3-d]pyrimidin-4-yloxymethyl]-4-methoxypyrrolidin-1-yl}propenone (PF 06459988), a potent, WT sparing, irreversible inhibitor of T790M-containing EGFR mutants, J. Med. Chem. 59 (5) (2016) 2005–2024.

[25] C. Han, L. Wan, H. Ji, K. Ding, Z. Huang, Y. Lai, S. Peng, Y. Zhang, Synthesis and evaluation of 2-anilinopyrimidines bearing 3-aminopropamides as potential epidermal growth factor receptor inhibitors, Eur. J. Med. Chem. 77 (2014) 75–83.

[26] M.Y. Cha, K.O. Lee, S.J. Kang, Y.H. Jung, J.Y. Song, K.J. Choi, J.Y. Byun, H.J. Lee, G.S. Lee, S.B. Park, M.S. Kim, Synthesis and biological evaluation of pyrimidine-based dual inhibitors of human epidermal growth factor receptor 1 (HER-1) and HER-2 tyrosine kinases, J. Med. Chem. 55 (6) (2012) 2846–2857.

[27] Y. Hao, X. Wang, T. Zhang, D. Sun, Y. Tong, Y. Xu, H. Chen, L. Tong, L. Zhu, Z. Zhao, Z. Chen, J. Ding, H. Xie, Y. Xu, H. Li, Discovery and structural optimization of N5-substituted 6,7-dioxo-6,7-dihydropteridines as potent and selective epidermal growth factor receptor (EGFR) inhibitors against L858R/T790M resistance mutation, J. Med. Chem. 59 (15) (2016) 7111–7124.

[28] Y. Kawakita, H. Banno, T. Ohashi, T. Tamura, T. Yusa, A. Nakayama, H. Miki, H. Iwata, H. Kamiguchi, T. Tanaka, N. Habuka, S. Sogabe, Y. Ohta, T. Ishikawa, Design and synthesis of pyrrolo[3,2-d]pyrimidine human epidermal growth factor receptor 2 (HER2)/epidermal growth factor receptor (EGFR) dual inhibitors: exploration of novel back-pocket binders, J. Med. Chem. 55 (8) (2012) 3975–3991.

[29] G. Lelais, R. Epple, T. Marsilje, Y. Long, M. McNeill, B. Chen, W. Lu, J. Anumolu, S. Badiger, B. Bursulaya, M. DiDonato, R. Fong, J. Juarez, J. Li, M. Manuia, D.E. Mason, P. Gordon, T. Groessl, K. Johnson, Y. Jia, S. Kasibhatla, C. Li, J. Isbell, G. Spraggon, S. Bender, P.Y. Michellys, Discovery of (R,E)-N-(7-chloro-1-(1-[4-(dimethylamino) but-2-enoyl]azepan-3-Yl)-1H-benzo[D]imidazol-2-yl)-2-methylisonicotinamide (EGF816), a novel, potent, and WT sparing covalent inhibitor of oncogenic (L858R, Ex19del) and resistant (T790M) EGFR mutants for the treatment of EGFR mutant non-small-cell lung cancers, J. Med. Chem. 59 (14) (2016) 6671–6689.

[30] J.B. Smaill, A.J. Gonzales, J.A. Spicer, H. Lee, J.E. Reed, K. Sexton, I.W. Althaus, T. Zhu, S.L. Black, A. Blaser, W.A. Denny, P.A. Ellis, S. Fakhoury, P.J. Harvey, K. Hook, F.O.J. McCarthy, B.D. Palmer, F. Rivault, K. Schlosser, T. Ellis, A.M. Thompson, E. Trachet, R.T. Winters, H. Tecle, A. Bridges, Tyrosine kinase inhibitors. 20. Optimization of substituted quinazoline and pyrido[3,4-d]pyrimidine derivatives as orally active, irreversible inhibitors of the epidermal growth factor receptor family, J. Med. Chem. 59 (17) (2016) 8103–8124.

[31] M. Coumar, C. Chu, C. Lin, H. Shiao, Y. Ho, R. Reddy, W. Lin, C. Chen, Y. Peng, J. Leou, T.W. Lien, C.T. Huang, M.Y. Fang, S.H. Wu, J.S. Wu, S.K. Chittimalla, J.S. Song, J.T. Hsu, S.Y. Wu, C.C. Liao, Y.S. Chao, H.P. Hsieh, Fast-forwarding hit to lead: aurora and epidermal growth factor receptor kinase inhibitor lead identification, J. Med. Chem. 53 (13) (2010) 4980–4988.

CHAPTER 4

Current Approaches in the Development of Covalent Reversible EGFR Inhibitors

The advances in third-generation EGFR inhibitors have demonstrated the ability to overcome T790M-mediated resistance, which is one of the major constraints for first- and second-generation EGFR inhibitors [1]. Third-generation inhibitors AZD9291 (osimertinib) and CO1686 (rociletinib) both form a covalent bond with Cys797 in the active site and show activity against the resistance mutations (T790M/*L858R*, TMLR; T790M/del (746–750), TMdel) as well as the two initial activating mutations [2, 3]. However, recent reports of the *C797S* mutation emerging in the clinic that confer resistance to AZD9291 support the continued need for a T790M inhibitor that does not depend on a covalent interaction with Cys797 [4, 5].

4.1 RECENTLY REPORTED COVALENT REVERSIBLE EGFR INHIBITORS

Hanan et al. reported the discovery of selective and noncovalent diaminopyrimidine-based inhibitors of EGFR containing the T790M resistance mutation [6]. Compounds **592–594** were prepared as described in Scheme 4.1. 2-Bromo-4,5-diaminopyridine and pyrazole 4-carboxylate were converted to imidazopyridine **584** via a polyphosphoric acid-mediated cyclization. Subsequent bis-SEM protection afforded intermediate **585**. The corresponding pyrrolopyridine intermediates were prepared by starting with 2-bromo-4-aminopyridine **586**, which was iodinated and *N*-sulfonylated to provide **587**. Annulation with optionally SEM-protected 4-alkynylpyrazoles provided the pyrrolopyridines **588a** and **588b**. N1 alkylation or SEM protection of **588a–b** provided the key intermediates **589a–b**. The SNAr reaction of 4-methoxypiperidine with 2-chloro-4-aminopyrimidine **590** afforded intermediate **591.** Finally the palladium-catalyzed coupling of intermediates **585** or **589a–b** with **591** yielded compounds **592–594**. Schemes 4.2–4.4 describe the preparation

Third-Generation EGFR Inhibitors
https://doi.org/10.1016/B978-0-08-102661-8.00004-4

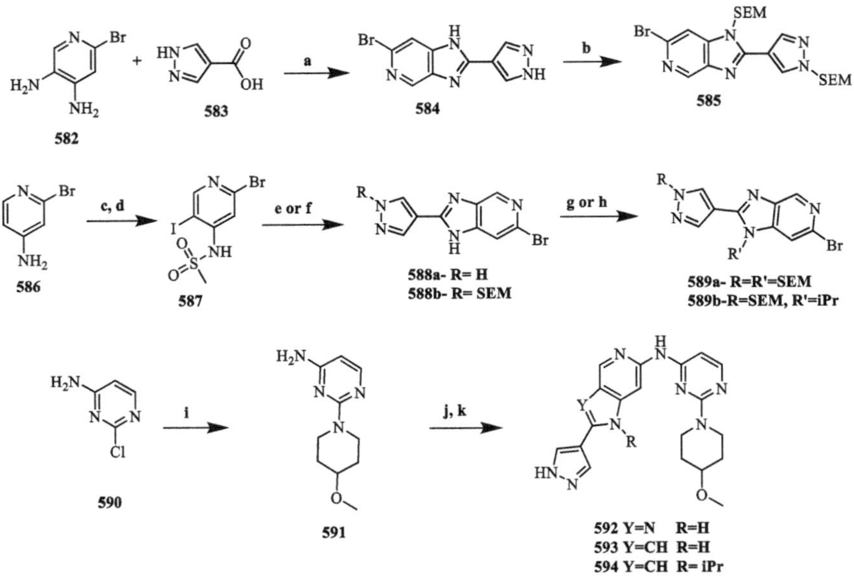

Scheme 4.1 Reagents and conditions: (a) polyphosphoric acid, 200°C, 18h, quant.; (b) sodium hydride, SEM-Cl, THF, DMF, rt., 2h, 81%; (c) iodine monochloride, NaOAc, HOAc, 70°C, 18h, 34%; (d) methanesulfonyl chloride, triethylamine, DCM, 0°C to rt., 2h, 46%; (e) 4-ethynyl 1Hpyrazole, $Pd(PPh_3)_2Cl_2$, CuI, triethylamine, DMF, 50–100°C, 33%; (f) 4-ethynyl-1-((2-(trimethylsilyl)ethoxy)methyl)-1H-pyrazole, $Pd(PPh_3)_2Cl_2$, CuI, triethylamine, DMF, 50–100°C, 34%; (g) NaH, SEM-Cl, THF, 0°C to rt., 5h, 36%; (h) 2-iodopropane, Cs_2CO_3, DMF, 90°C, 48h, 40%; (i) 4-methoxypiperidine hydrochloride, Cs_2CO_3, DMF, 120°C, 18h, 87%; (j) **585**, **589a**, or **589b**, Pd_2dba_3, X-Phos or Xantphos, Cs_2CO_3, 1,4-dioxane, 120°C, 28%–81%; (k) HCl, MeOH, rt. to 50°C, 22%–76%.

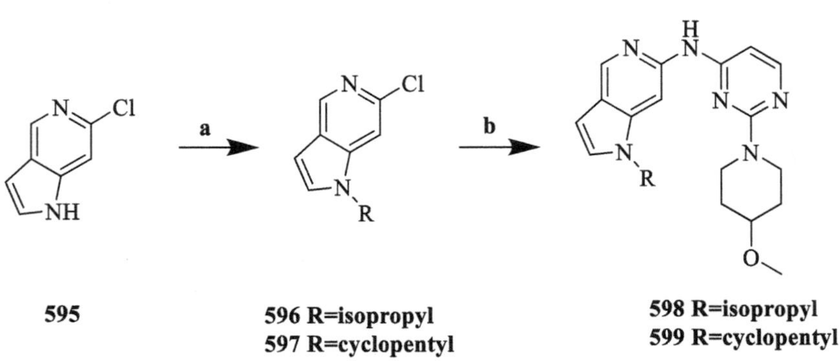

Scheme 4.2 Reagents and conditions: (a) 2-iodopropane or iodocyclopentane, Cs_2CO_3, DMF, 90°C, 18–20h, 56%–90%; (b) 647, chloro[[BrettPhos][2-(2-aminoethylphenyl) palladium(II)]]/[BrettPhos], NaOtBu, tBuOH, 90°C, 5h, 33%–49%.

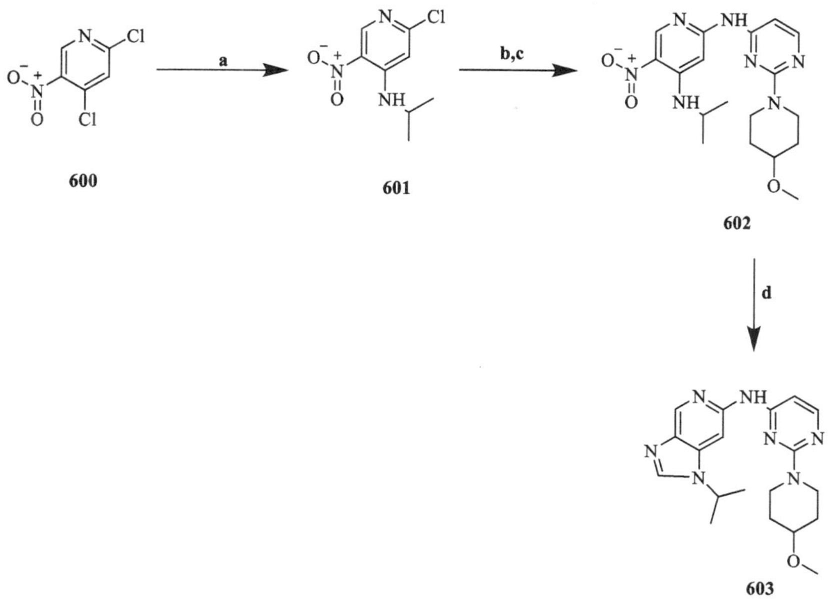

Scheme 4.3 Reagents and conditions: (a) isopropylamine, triethylamine, THF, rt., 3 h, quant.; (b) 647, Pd_2dba_3, X-Phos, Cs_2CO_3, 1,4-dioxane, reflux, 3 h, 72%; (c) Pd/C, EtOAc, MeOH, H_2, rt., 18 h, 92%; (d) trimethyl orthoformate, formic acid, 100°C, 1 h, 27%.

of additional bicyclic pyridine analogs. Pyrrolopyridines **598** and **599** were readily prepared, as shown in Scheme 4.2. Alkylation of commercially available 6-chloro-1*H*-pyrrolo[3,2-*c*]pyridine to give **596** and **597** followed by a palladium-catalyzed coupling to aminopyrimidine **591** provided compounds **598** and **599**. As described in Scheme 4.3, a nucleophilic displacement reaction with isopropylamine and 2,4-dichloro-5-nitropyridine provided chloropyridine **601**, which was coupled with the aminopyrimidine fragment **591** followed by the reduction of the nitro group to provide **602**. Closure of the imidazo ring with trimethyl orthoformate yielded **603**. Scheme 4.4 describes the synthesis of 3-aminopyrazolopyridine **611**, beginning with compound 2-bromo-5-iodopyridin-4-amine (**604**). A Sandmeyer reaction converted the 4-amino functionality to a 4-chloro group (**605**). Conversion of the 3-iodo group to a methyl ester proceeded via palladium-catalyzed carbonylation. Subsequent reduction to the aldehyde with diisobutylaluminum hydride produced compound **606**.

A significant benefit in removing this pyrazole was realized with respect to general selectivity over other kinases. Pyrazole containing compounds **592–594** exhibit undesirable activity against CDK2, with **594** also showing

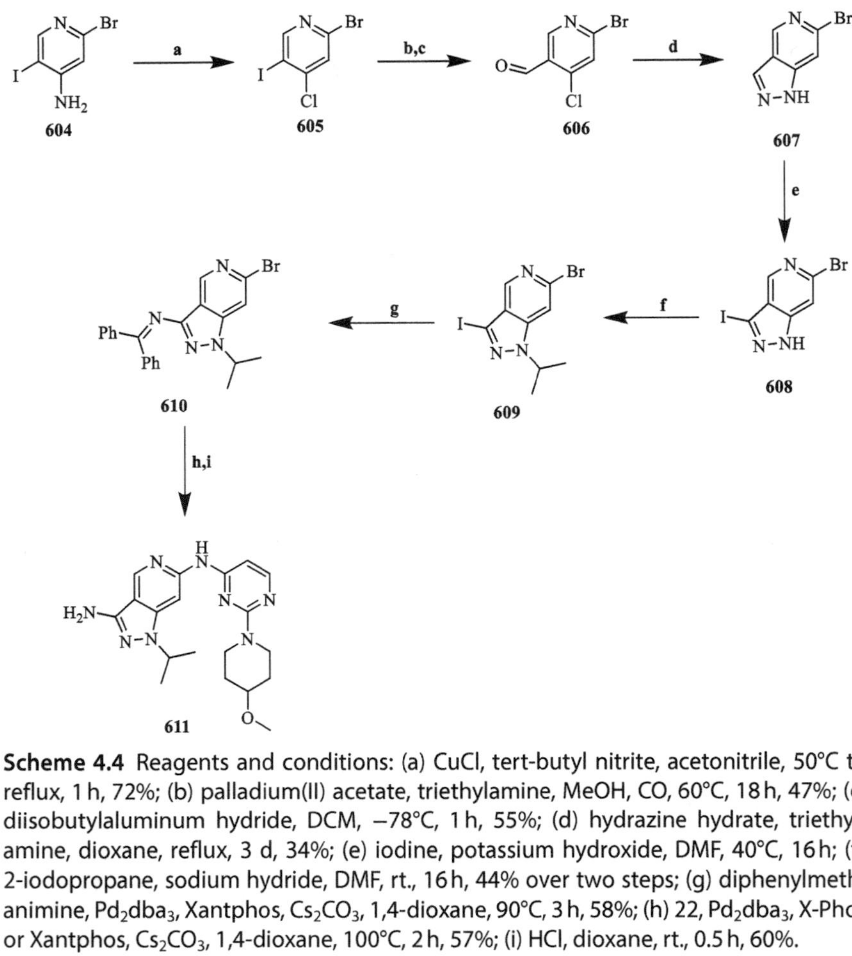

Scheme 4.4 Reagents and conditions: (a) CuCl, tert-butyl nitrite, acetonitrile, 50°C to reflux, 1 h, 72%; (b) palladium(II) acetate, triethylamine, MeOH, CO, 60°C, 18 h, 47%; (c) diisobutylaluminum hydride, DCM, −78°C, 1 h, 55%; (d) hydrazine hydrate, triethylamine, dioxane, reflux, 3 d, 34%; (e) iodine, potassium hydroxide, DMF, 40°C, 16 h; (f) 2-iodopropane, sodium hydride, DMF, rt., 16 h, 44% over two steps; (g) diphenylmethanimine, Pd_2dba_3, Xantphos, Cs_2CO_3, 1,4-dioxane, 90°C, 3 h, 58%; (h) 22, Pd_2dba_3, X-Phos or Xantphos, Cs_2CO_3, 1,4-dioxane, 100°C, 2 h, 57%; (i) HCl, dioxane, rt., 0.5 h, 60%.

decreased selectivity against Aurora B and KDR as compared to C, **592**, and **593**. Interestingly, while the imidazopyridine compound **592** ablates the Jak2 and Tyk2 activity exhibited by original hits A–C (<50% inhibition of Jak2 or Tyk2 activity observed at 1 μM), both pyrrolopyridine compounds **593** and **594** retain significant Jak family activity. NH pyrrolopyridine **593** is equipotent against TMLR and Jak2/Tyk2, while *N*-isopropyl **594** exhibits approximately 10-fold selectivity for TMLR over Jak2. The differential activity against Jak2 for **592** relative to **593** and **594** may be understood using a model of all three compounds bound in the Jak2 active site in the orientation observed for **592** and **593** in TMLR. In this orientation the five-membered heterocycle is adjacent to the gatekeeper residue and the hydrophobic side chain of Jak2's Met 929 gatekeeper can be expected to interact preferentially with the pyrrolo C3 methines of **593** and **594**, while the imidazo N2 of **592** introduces unfavorable polarity in this hydrophobic region. The selectivity of compound **594** can be improved with the removal of the terminal pyrazole moiety; as compared to **594**, compound **598** exhibits minimal activity against the off-targets Aurora B, KDR, and Jak2. This improvement in broader kinase selectivity is perhaps unsurprising given the alternative hinge-binding motif provided by the unsubstituted pyrazole. While **598** has >20-fold selectivity over Aurora B, KDR, and Jak2, we sought to improve its selectivity over CDK2 in order to drive an EGFR-dependent antiproliferative effect. Modification of the ribose-pocket group from an isopropyl to the slightly larger cyclopentyl (compound **599**) to fill a greater portion of the lipophilic ribose pocket does improve potency against TMLR and TMdel slightly (~2-fold). However, this change from isopropyl to cyclopentyl increases lipophilicity and does not substantially affect kinase selectivity. A comparison of X-ray crystal structures of **599** with TMLR and CDK2 suggested an opportunity to enhance selectivity in subsequent designs. The binding mode of **599** in TMLR is consistent with that observed for **594**, while in CDK2 the binding mode corresponds to the original binding mode observed for compounds **592** and **593**.

The gatekeeper residue in CDK2 is a phenylalanine (Phe80), and this bulky side chain precludes the binding of the ligand in the orientation observed in TMLR. Noting that the phenyl group of the Phe80 side chain provides hydrophobic contacts with the aromatic methines of the pyrrole ring, they hypothesized that introducing polarity in this region would disfavor interaction with CDK2 Phe80 and would improve the selectivity, similar to the selectivity benefit seen with **592**, as compared with **593**. Indeed the addition of an extra ring of nitrogen in imidazopyridine compound **603** provides an improvement in selectivity over all key off-targets as compared to **598** and **599**, with a 24-fold reduction in activity against CDK2. Aurora B, KDR, and Jak2 also have lipophilic gatekeeper residues (Aurora B = Leu170, KDR = Val914, Jak2 = Met929), and the concurrent reduction in all of these off-target activities suggests that the binding mode observed for **599** in CDK2 may be shared among these other off-targets, consistent with the Jak2 model. When tested at 1 μM in a broader panel of 219 kinases, **603** inhibits only 5 kinases at >80%. Compared with **598**, compound **603** also has reduced lipophilicity, while potency is maintained and thus provides an increased LLE.

Additionally the 3-aminopyrazolopyridine compound **611** is 3–8 times more potent than **603** with a TMLR Ki of 12 nM and TMdel Ki of 37 nM.

This compound reached our initial objective of simultaneously achieving good potency with selectivity against WT EGFR (8–26 times) and with acceptable physicochemical properties that are substantially improved over those of compounds A–C (Log D < 2.5, kinetic solubility > 50 μM). Critically, **611** has good selectivity over a broad panel of kinases, inhibiting only 5 out of 219 at >80% when tested at 1 μM. Other bicyclic systems were tested, but they did not result in further improvement over 10. 2-Oxoimidazole **613** was similarly anticipated to insult the CDK2 and Jak2 active sites and provide good kinase selectivity; however, compared with **611**, this compound has decreased activity against TMLR and was therefore not pursued. The dihydropyrrole **614** was designed to increase the percentage of sp3 centers in an attempt to improve the solubility of pyrrolopyridine **598**. This compound does exhibit an improvement in kinetic aqueous solubility (increased from 8 to 69 μM), but it has significantly reduced TMLR potency. While the structure of **599** with TMLR suggests that the sp3 carbons of **614** should be accommodated sterically, a greater energetic benefit may be provided by the aliphatic-aromatic contacts of pyrroles **598** and **599** sandwiched between Leu718 and Gly796, as compared to the aliphatic-aliphatic contacts afforded by the dihydropyrrole of **614**. Finally, pyridone **615** was synthesized to evaluate 6–6 bicyclic systems, but the lack of potency and poor solubility led to no additional pursuits.

Treatment with hydrazine affected the ring closure to pyrazolopyridine **607**. Iodination at the 3-position to **608** followed by *N*-alkylation provided key intermediate **609**. Sequential palladium-catalyzed couplings first at the iodo position and then at the bromo position, followed by deprotection, thus producing compound **611**.

Beginning with potent inhibitors of Jak2/Tyk2, they identified a novel series of noncovalent diaminopyrimidine based T790M EGFR inhibitors. The replacement of the biarylamide motif with bicyclic substructures provided a significant reduction in lipophilicity. The bicyclic scaffolds identified in Chart 4.1 (**598**, **603**, and **611**) enable a substantial improvement in LLE and are promising leads in the search for selective and noncovalent inhibitors of T790M-containing EGFR mutants [6].

Bryan et al. reported the pyridones as highly selective, noncovalent inhibitors of T790M double mutants of EGFR [7]. A synthetic route was developed, allowing for rapid exploration at C3 and C4 (Scheme 4.5). For diversification at C4, pyridine **616** was first oxidized to carboxylic acid **617**. Amide bond formation using various coupling conditions installed R1 at C3 and demethylation produced the pyridone. The final step diversification

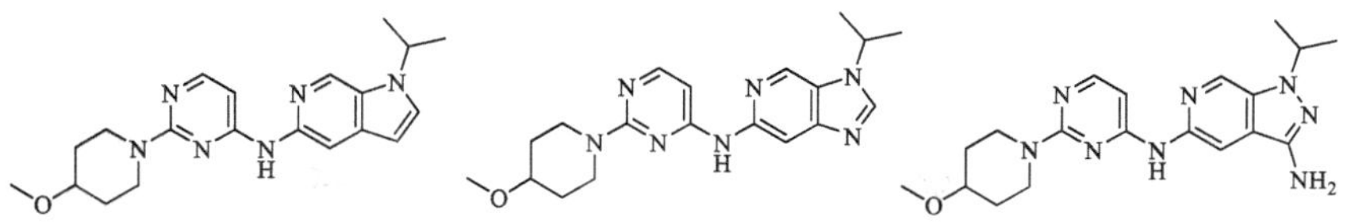

598	603	611
TMLR LLE = 4.8	TMLR LLE = 4.9	TMLR LLE = 5.6
TMLR Kiapp = 33 nM	TMLR Kiapp = 40 nM	TMLR Kiapp = 12 nM
TMdel Kiapp = 62 nM	TMdel Kiapp = 211 nM	TMdel Kiapp = 67 nM
WT EGFR Kiapp = 1100 nM	WT EGFR Kiapp = 559 nM	WT EGFR Kiapp = 317 nM

Chart 4.1 Antiproliferative and in vitro EGFR inhibitory activities.

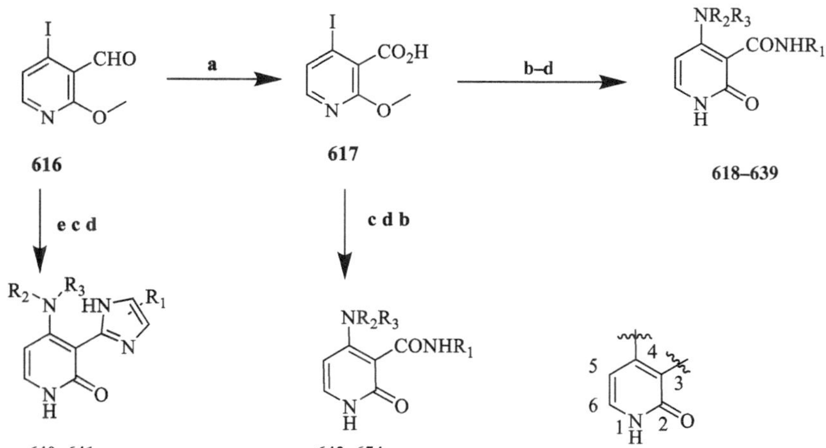

Scheme 4.5 Reagents and conditions: (a) 2-methyl-2-butene, $NaClO_2$, $NaPO_4H_2$-H_2O; (b) HBTU or oxalyl or thionyl chloride, NH_2R_1; (c) HCl; (d) NHR_2R_3, Δ; (e) $R_1C_1NH_2C_2NH_2$, MeOH, Δ.

at C4 with R2 and R3 occurred via nucleophilic aromatic substitution (SNAr) to give **618–639**. Diversification at C3 followed similar steps to those described above for compounds **642–654**. From intermediate **617**, demethylation followed by SNAr at C4 gave the pyridone late stage intermediate. The amide bond was then formed via the acid chloride. Benzimidazoles **640–641** were formed upon the condensation of **616** with the required diamine prior to pyridone formation and coupling to HNR_2R_3.

The biological activity of all the synthesized compounds is given in Chart 4.2. The replacement of the phenyl ring with six-membered heteroaryls then occurred. 2-Pyridine **631** and 2,4-pyrimidine **632** were equipotent to phenyl **622**. The addition of an amine to the C3 position of **632**, designed to target polar interactions with the side chains of catalytic Lys745 and Glu762, generated the first single-digit nanomolar compound of the

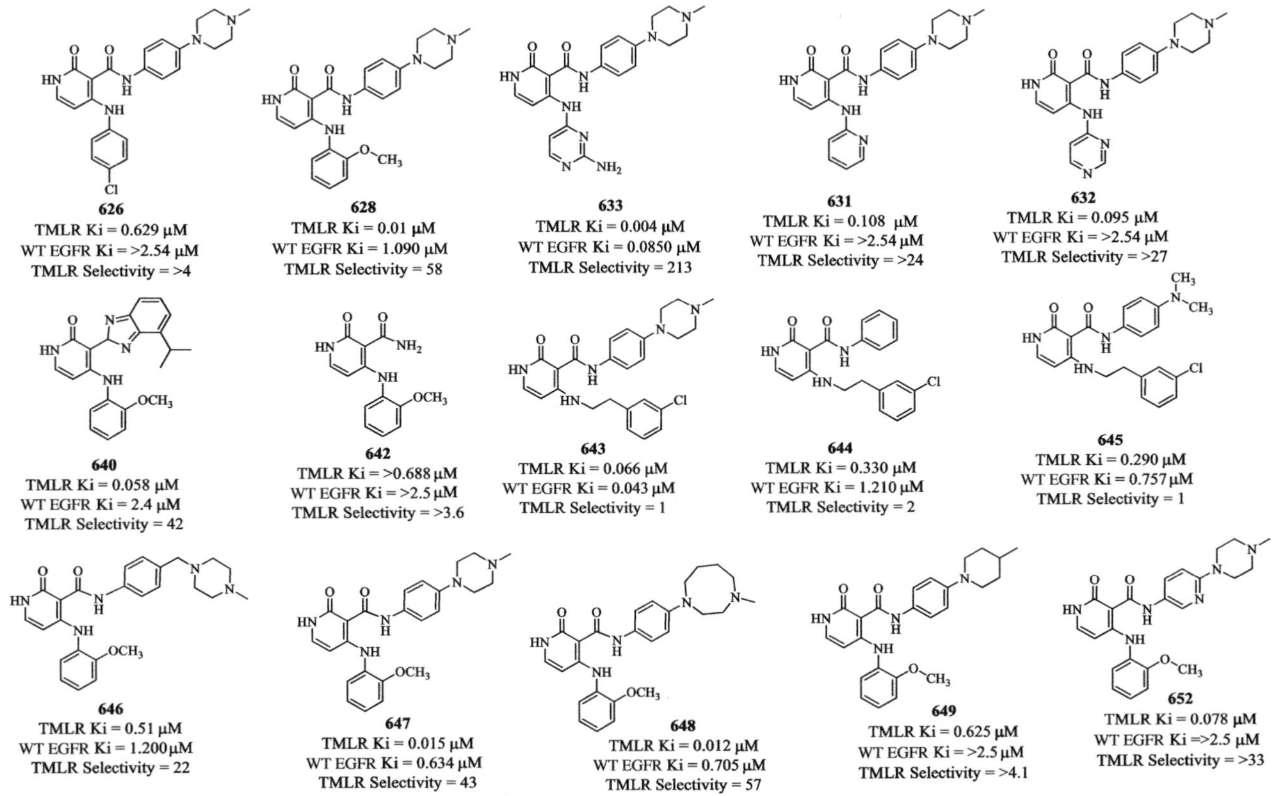

Chart 4.2 Antiproliferative and in vitro EGFR inhibitory activities. Yang et al. reported the structural optimization and structure–activity relationships of N^2-(4-(4-methylpiperazin-1-yl)phenyl)-N^8-phenyl-9*H*-purine-2,8-diamine derivatives, a new class of reversible kinase inhibitors targeting both EGFR-activating and resistance mutations [8]. Structural optimization was carried out by focusing on the *N*-9-(*H*), C-2 (anilino group), and C-8 positions (anilino group) of the purine scaffold (Chart 4.3).

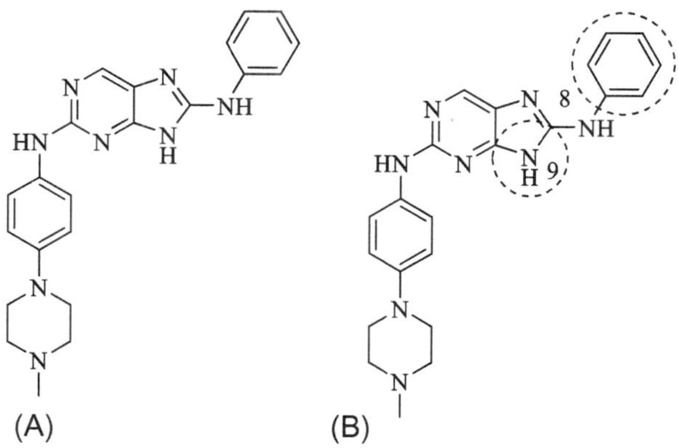

Chart 4.3 (A) Structure of hit compound **A**. (B) Moieties in compound **A** targeted for modification in this study.

series, with >200× selectivity over WT EGFR (compound **633**). However, the removal of one of the pyrimidine nitrogens of **633** erased this gain in potency (pyridine **626**); this is potentially due to the diaminopyridine being positively charged at physiological pH, unlike the pyrimidine, and its clashing with Lys745 (Chart 4.2).

In order to understand the minimal motifs necessary for inhibition a number of abridged amides were generated (Chart 4.2). Truncation to the primary amide gave a significant loss in activity (**642**) when compared to **628**. Similar reductions in potency were found with the unsubstituted phenyl ring analog (**644**) and the dimethylaniline (**645**), which lost 4–5 fold, when compared to the fully elaborated **643**. Efforts to enforce the conformational restriction of the amide led to benzimidazole **640**, which maintained selectivity over WT EGFR with only a modest decrease in potency relative to compound **628**. The isopropyl group on the phenyl ring was predicted to occupy the lipophilic space within the ribose pocket.

Further, Bryan et al. concluded that several changes could be tolerated on piperazine ring such as the extension of the linker between the aryl ring and the piperazine (**646**), replacement of the linking nitrogen with carbon (**647**) and increase in the ring size (**647**) were all tolerated by TMLR. Interestingly the replacement of the terminal nitrogen with carbon (**648**) was not tolerated, but the replacement with a sulfone (**649**) was with less than two-fold loss in potency relative to **628**; this suggests that polarity was beneficial in this region, though a positive charge was not required.

Heteroaromatic replacements at C3 were also evaluated to reduce the lipophilicity of compound **628** (Log D 7.4 = 3.5). While pyridine **652** suffered a modest reduction in activity relative to compound **628**, it did improve overall lipophilicity (Log D 7.4 = 2.6) and kinetic solubility (65 μM versus 4 μM for compound **628**).

Scheme 4.6 depicts the synthetic route to the 2,8-dianilinopurine derivatives **663a–h**, in which the *N*-9-(*H*) position of the purine scaffold is substituted by various groups. Commercially available 4-fluoronitrobenzene (**656**) reacted with 1-methylpiperazine in the presence of K_2CO_3 in DMSO to produce the intermediate (**657**). The catalytic hydrogenation of the nitro functionality in (**657**) with palladium on carbon (Pd/C) provided the desired aniline (**658**) in an excellent yield. The reaction of the commercially available 2,4-dichloro-5-nitropyrimidine (**659**) with various amines,

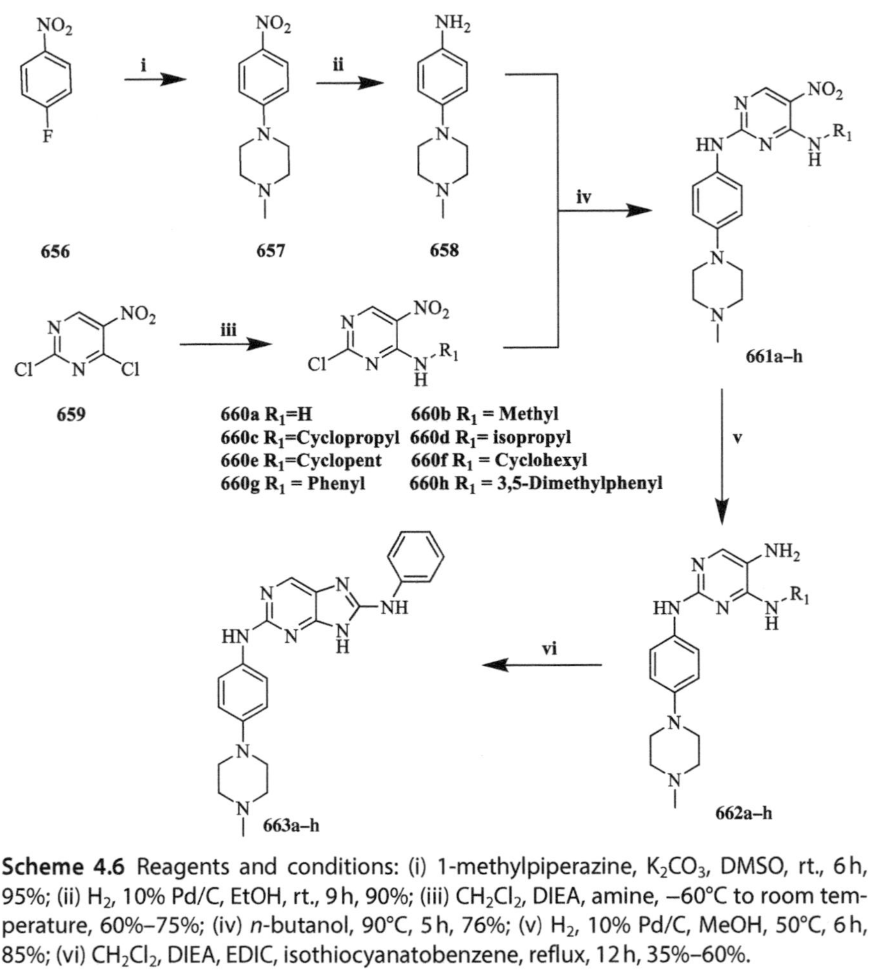

Scheme 4.6 Reagents and conditions: (i) 1-methylpiperazine, K_2CO_3, DMSO, rt., 6 h, 95%; (ii) H_2, 10% Pd/C, EtOH, rt., 9 h, 90%; (iii) CH_2Cl_2, DIEA, amine, −60°C to room temperature, 60%–75%; (iv) *n*-butanol, 90°C, 5 h, 76%; (v) H_2, 10% Pd/C, MeOH, 50°C, 6 h, 85%; (vi) CH_2Cl_2, DIEA, EDIC, isothiocyanatobenzene, reflux, 12 h, 35%–60%.

including ammonia, methanamine, cyclopropanamine, propan-2-amine, cyclopentylamine, cyclohexylamine, aniline, and 3,5-dimethylaniline, produced intermediates **660a–h**. The nucleophilic aromatic substitution of the 2-chloro position in **660a–h** with **658** in *n*-butanol at 90°C yielded compounds **661a–h**. The subsequent hydrogenation of the nitro group in **661a–h** using Pd/C as a catalyst provided a good yield of the desired intermediates **662a–h**, which were treated with isothiocyanatobenzene to produce the final 2,8-dianilinopurine derivatives **663a–h**.

Scheme 4.7 shows a general route for the synthesis of 2,8-dianilino-9-cyclopentylpurine derivatives **668a–e** bearing different substituents on the phenyl ring of the 2-anilino group. The fluorine in 4-fluoronitrobenzene was substituted via conventional nucleophilic aromatic substitution using morpholine, 4-methylpiperidine, and ethanolamine in the presence of K_2CO_3 in DMSO to yield the corresponding substituted nitrobenzenes **664a**, **664b**, and **664c**, respectively. As before the nitro functional groups in compounds **664a–c** were reduced via catalytic hydrogenation to generate the corresponding anilines **665a–c** in high yield. These intermediates,

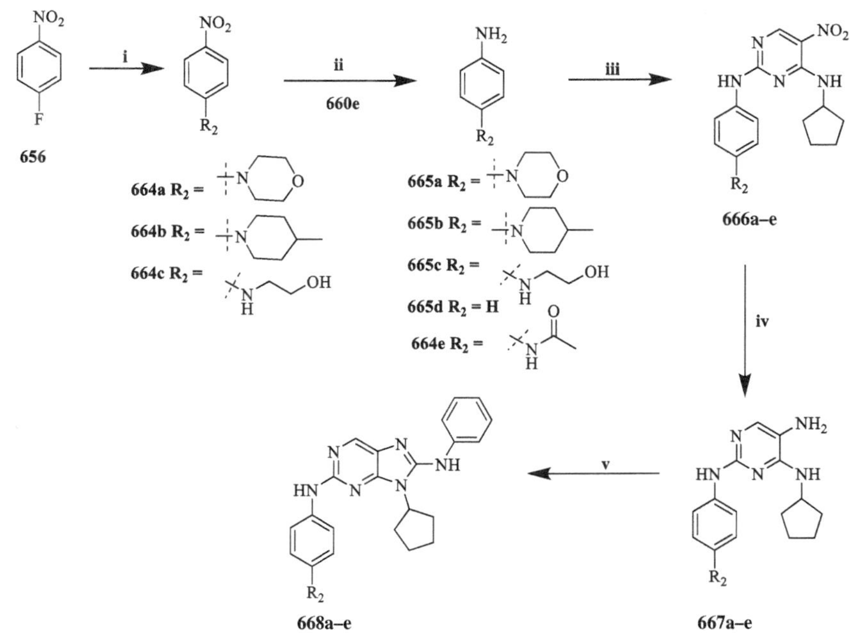

Scheme 4.7 Reagents and conditions: (i) morpholine, 4-methylpiperidine, ethanolamine, K_2CO_3, DMSO, rt., 6h, 70%–85%; (ii) H_2, 10% Pd/C, EtOH, rt., 9h, 90%; (iii) *n*-butanol, 90°C, 5h, 76%; (iv) H_2, 10% Pd/C, MeOH, 50°C, 6h, 85%; (v) CH_2Cl_2, DIEA, EDIC, isothiocyanatobenzene, reflux, 12h, 35%–60%.

665a–c, along with the commercially available compounds **665d** and **665e** were treated with compound **660e** in *n*-butanol at 90°C to produce **666a–e**. These compounds were then hydrogenated with catalytic Pd/C, and the resultant intermediates (**667a–e**) were treated with isothiocyanatobenzene to yield the final 2,8-dianilino-9-cyclopentylpurine derivatives **668a–e**.

Scheme 4.8 depicts the synthetic route to 2,8-dianilino-9-cyclopentylpurine derivatives **673a–j** with different substituents on the phenyl ring of the 8-anilino group. Substituted anilines **669a–j** reacted with CS_2 and DABCO in ethanol at room temperature to yield the intermediate compounds **670a–j**, which were directly treated with triphosgene at 35–40°C in dichloromethane to produce substituted phenyl isothiocyanates **671a–j** with yields of 45%–60%. Compounds **671a–j** were treated with compound **662e** and EDCI in the presence of *N*,*N*-diisopropylamine in dichloromethane to yield the final 2,8-dianilino-9-cyclopentylpurine derivatives **673a–j** in acceptable yields.

The biological evaluation data of the synthesized compound is given Chart 4.4. They described the structural optimization of a hit compound, *N*2-(4-(4-methylpiperazin-1-yl)phenyl)-*N*8-phenyl-9*H*-purine-2,8-diamine (**A**) (Chart 4.4.), which exhibited poor binding affinity (Kd) for both EGFR-activating and drug-resistant mutants. SAR studies centering

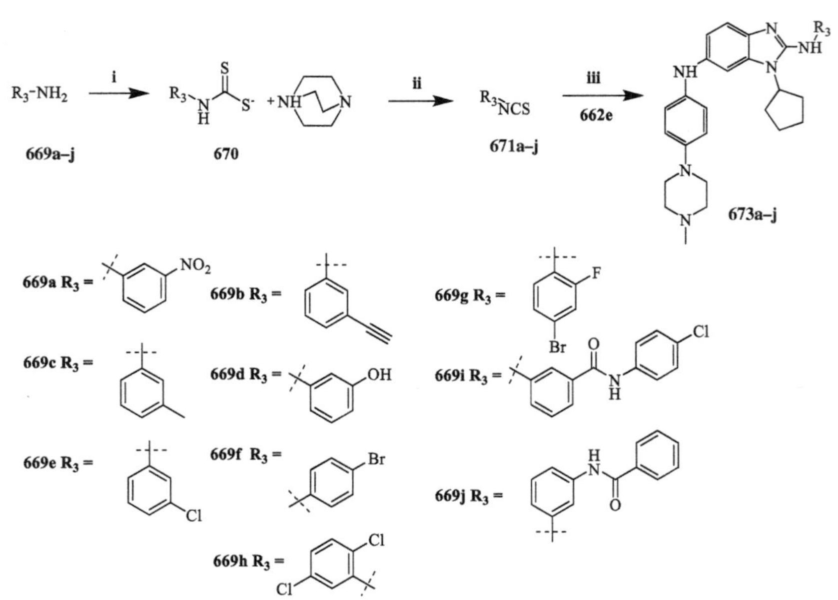

Scheme 4.8 Reagents and conditions: (i) EtOH, CS_2, DABCO, rt., 12h, 60%; (ii) CH_2Cl_2, triphosgene, 0°C, 1h, reflux, 4h, 65%; (iii) CH_2Cl_2, DIEA, EDIC, reflux, 12h, 35%–60%.

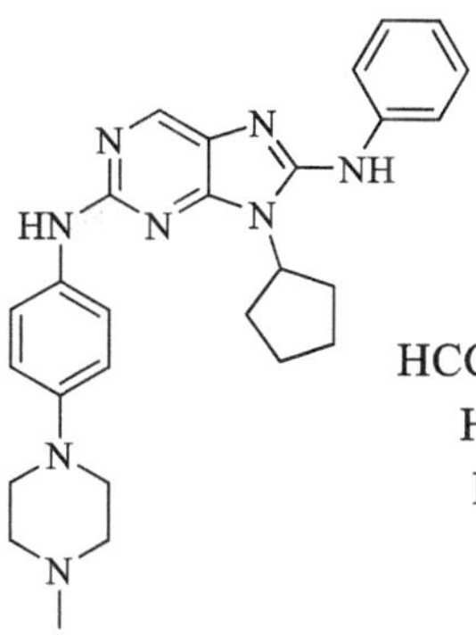

663e
HCC827 IC_{50} = <0.00001 μM
H1975 IC_{50} = 0.36 μM
HepG2 IC_{50} = >5 μM

Chart 4.4 Antiproliferative and in vitro EGFR inhibitory activities.

on the *N*-9 (H), C-2 (anilino), and C-8 (anilino) groups of the purine scaffold led to the discovery of compound **663e** (Chart 4.4).

This compound exhibited significantly increased the antiviability potency against the HCC827 cell line, which harbors an activating mutation in EGFR, and the H1975 cell line, which contains the drug-resistance EGFR mutation T790M. Compound **663e** displayed only weak inhibition against HepG2, a cell line that is not EGFR driven.

Further in vitro assays indicated that compound **663e** has a high potency against both EGFR-activating and drug-resistant EGFR mutants, as well as a good kinase spectrum selectivity; it also inhibits several other oncokinases.

Western blot analysis showed that compound **663e** significantly inhibited EGFR phosphorylation and the activation of downstream signaling proteins. In vivo antitumor activity assays showed that an oral once-daily dose of compound **663e** at 5 mg/kg for 15 days led to a complete tumor regression in the HCC827 xenograft model, and that a once-daily dose of compound **663e** at 50 mg/kg exhibited considerable tumor inhibition in the H1975 xenograft model.

Heald et al. reported the noncovalent mutant selective EGFR inhibitors. The preparation of *N*-substituted aza-benzimidazole analogs **684–691**, **692–694, 695, 703**, and **698** was carried out as outlined in Scheme 4.9 [9]. Regiospecific SNAr, followed by reduction and cyclization with trimethyl orthoformate, afforded azabenzimidazole template **677**.

A final Buchwald-Hartwig coupling furnished targets **684–687, 689–690, 691, 692–694**, and **695**. Chiral resolution of racemate **683** afforded **ent-699** and **ent-700**. Compound **688** was prepared by installing the RHS unit on intermediate **675** via Buchwald-Hartwig coupling, followed by the reduction and subsequent cyclization to generate the final target (sequence **682–683–688**).

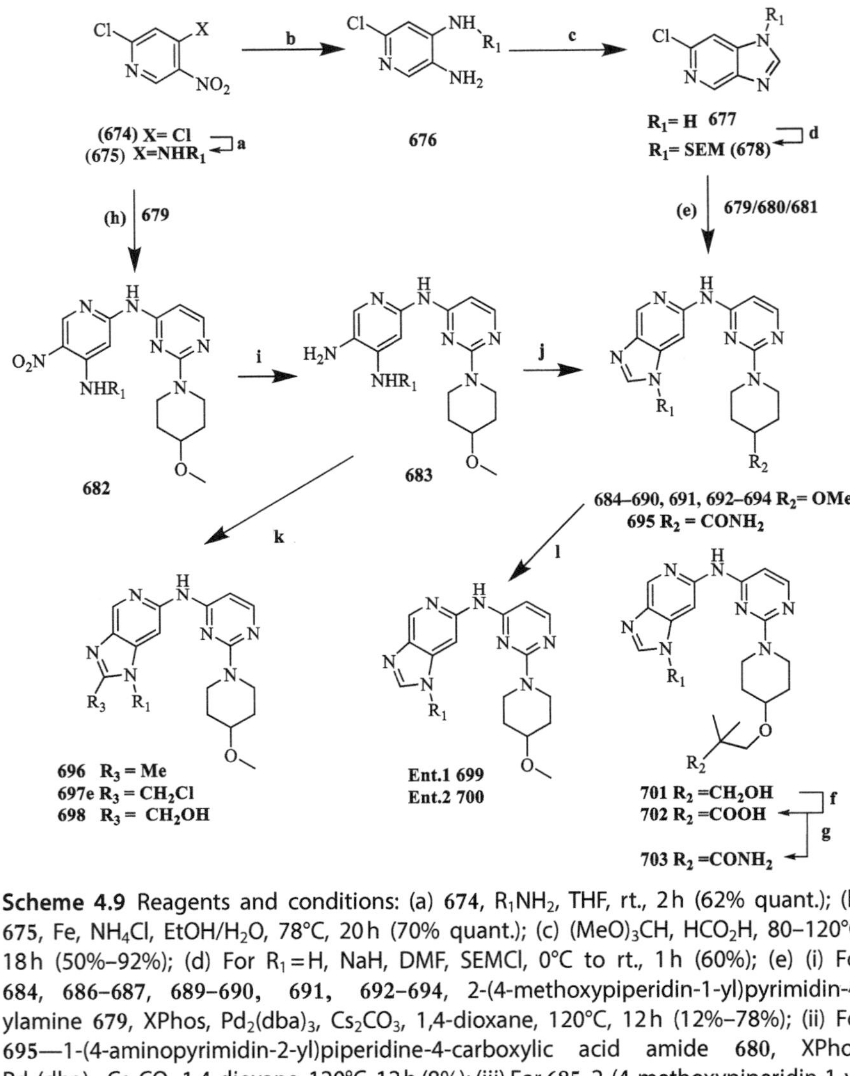

Scheme 4.9 Reagents and conditions: (a) **674**, R_1NH_2, THF, rt., 2h (62% quant.); (b) **675**, Fe, NH_4Cl, EtOH/H_2O, 78°C, 20h (70% quant.); (c) $(MeO)_3CH$, HCO_2H, 80–120°C, 18h (50%–92%); (d) For R_1=H, NaH, DMF, SEMCl, 0°C to rt., 1h (60%); (e) (i) For **684, 686–687, 689–690, 691, 692–694**, 2-(4-methoxypiperidin-1-yl)pyrimidin-4-ylamine **679**, XPhos, $Pd_2(dba)_3$, Cs_2CO_3, 1,4-dioxane, 120°C, 12h (12%–78%); (ii) For **695**—1-(4-aminopyrimidin-2-yl)piperidine-4-carboxylic acid amide **680**, XPhos, $Pd_2(dba)_3$, Cs_2CO_3,1,4-dioxane, 120°C, 12h (8%); (iii) For **685**, 2-(4-methoxypiperidin-1-yl) pyrimidin-4-ylamine **679**,chloro{[BrettPhos][2-(2aminoethylphenyl] palladium(II)]}/ [BrettPhos], NaO*t*Bu, *t*BuOH, 120°C, 2h (6%); (iv) For **703**—2,2-dimethylpropane-1,3-diol **681**, XantPhos $Pd_2(dba)_3.CHCl_3$, Cs_2CO_3, 1,4-dioxane, 100°C, 1h (55%); (f) **701**, PDC, DMF, rt., 24h (19%); (g) **702**, CDI, NH_4OH, DMF, rt., 18h (20%); (h) **675e**, 2-(4-methoxypiperidin-1-yl)pyrimidin-4-ylamine **679**, XPhos, $Pd_2(dba)_3$, Cs_2CO_3, 1,4-dioxane, 100°C, 3h (72%); (i) **682e**, H_2, Pd/C (10% *w/w*), EtOAc/MeOH, rt., 18h (92%); (j) **683e**, $(MeO)_3CH$, HCO_2H, 100°C, 1h (27%); (k) (i) For R_3=Me **683–696**: $MeC(OEt)_3$, HCO_2H, μW, 220°C, 5.5h (90%); (ii) For R_3=CH_2OH **698**: (i) **683**, 2-chloro-1,1,1-trimethoxyethane, p-TSA, 100°C, 1h (86%); (ii) **697e**, KOAc, DMF, 100°C, 45min (52%); (l) chiral resolution of **690**, ChiralPak IA column (20mm×250mm), 10% EtOH, 10% TBME and 0.5% DEA in heptane (isocratic), flow rate 18mL/min (**699**, ent. 1 84%, > 99% ee; **700**, ent. 2 78%, 93% ee).

Variation of R_3 was facilitated by condensation of **683** with the appropriate orthoester derivative (**696–698**).

In order to fully probe the SAR and facilitate diversification at the pyrimidine 2-substituent, late stage intermediates **712/713** were targeted as outlined in Scheme 4.10. Bromination of **704**, followed by regiospecific SNAr displacement, reduction, and cyclization with the requisite orthoester/acid chloride derivative gave **708–710**. A final Buchwald-Hartwig coupling

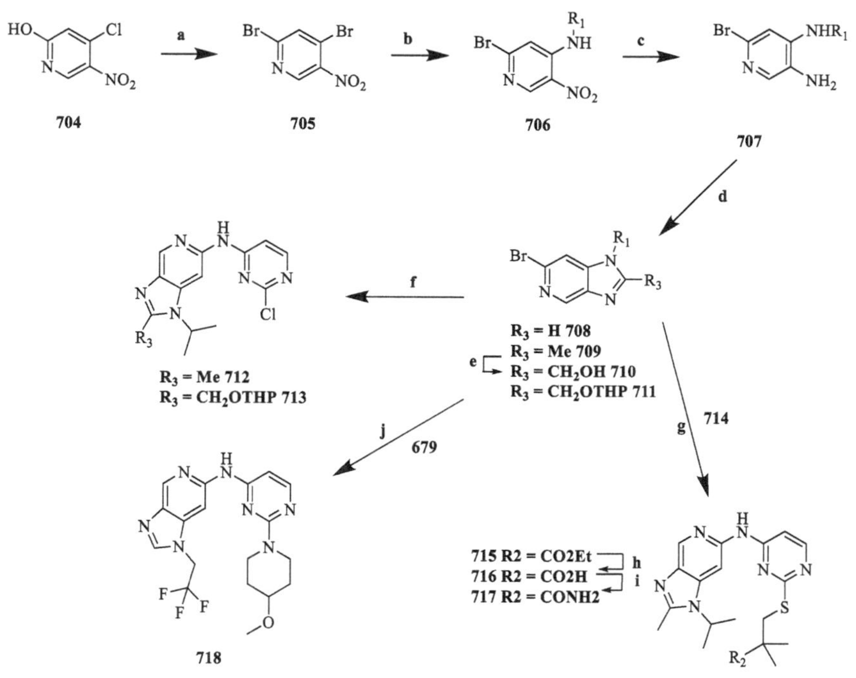

Scheme 4.10 Reagents and conditions: (a) $POBr_3$, CH_3CN, 80°C, 2h (41%); (b) **705**, *i*$PrNH_2$, THF, rt., 1h (quant.); **705**, $CF_3CH_2NH_2$, THF, rt., 18h, then 35°C, 18h (79%); (c) **707e**, **706**, Fe, EtOH/AcOH, 80°C, 4h (R_1 = *i*Pr (94%)); **707h**, **706**, Fe, EtOH/AcOH, rt., 18h (R1 = CF_3CH_2 (94%)); (d) For **708h**, **707h**, $(MeO)_3CH$, HCO_2H, 110°C, 18h (54%); For **709e**—(i) **707e**, $(EtO)_3CMe$, AcOH, 80°C, 24h (88%), (ii) K_2CO_3, DMF, 100°C, 24h (47%); for **710e**—(i) **707e**, $MeCO_2CH_2COCl$, Et_3N, DCM, 0°C, 1h (67%), (ii) K_2CO_3, DMF, 120°C, 24h (57%); (e) **710e**, 3,4-dihydro-2H-pyran, p-TSA,THF, 60°C, 12h (51%); (f) For **709e–712**, 2-chloropyrimidin-4-ylamine, XantPhos, $Pd_2(dba)_3.CHCl_3$, Cs_2CO_3, 1,4-dioxane, 100°C, 4h (26%); for **711e–713**, 2-chloropyrimidin-4-ylamine, XantPhos, $Pd_2(dba)_3$, Cs_2CO_3, 1,4-dioxane, 100°C, 1h (91%); (g) **709e**, 3-(4- aminopyrimidin-2-ylsulfanyl)-2,2-dimethylpropionic acid ethyl ester **714**, CuI, trans-*N,N'*-dimethylcyclohexane-1,2-diamine, Cs_2CO_3, 1,4-dioxane, 90°C 1.5h (30%); (h) **715**, LiOH, THF/H_2O, rt., 4h then, 50°C, 20h then, μW, 100°C, 4h (quant.); (i) **716**, NH_4Cl, HATU, DMAP (cat.), DIPEA, DMF, rt., 3h (46% over two steps); (j) **708h**, 2-(4-methoxypiperidin-1-yl)pyrimidin-4-ylamine **679**, XPhos, $Pd_2(dba)_3$, Cs_2CO_3, 120°C, 18h (22%).

with 2-chloropyrimidin-4-ylamine afforded key intermediates **712/713**. For **710**, wherein R = CH_2OH, this intermediate was further derivatized to the corresponding THP protected analog **711** prior to coupling.

Direct SNAr displacement of the chloride in **712/713** with appropriately functionalized primary/secondary amines, primary alcohols, or Pd-mediated couplings with the requisite vinyl boronate were then effected to afford target compounds **720–721**, **723–726**, **728**, **754**, **751**, **745**, **733**, **734**, and **736–743**, as outlined in Scheme 4.11. Alternatively, as revealed in Scheme 4.11, the intact pyrimidinyl unit could be installed directly via a Pd-catalyzed arylation; this was the strategy adopted for compounds **718** and **717**. Fluoropiperidine analogs **763** and **764** were identified and initially profiled as a mixture of diastereoisomers (cis-rel-OMe/F and OH/F respectively). The azabenzimidazole intermediate **758** was prepared from (*S*)-**7561** via cyclisation with the corresponding imidate generated in situ from (*R*)-lactamide/Meerwein's salt followed by protection of the secondary alcohol. A subsequent Buchwald-Hartwig coupling with (±)-**759**/(±)-**760**, followed by THP deprotection, afforded cis-rel **763** and **764**. The individual diastereoisomers **763a/b** and **764a/b** were in turn prepared by performing the latter coupling with the appropriate homochiral piperidinyl-pyrimidine unit, as shown in Scheme 4.12.

The hydroxy piperidine analog **720** was of comparable potency to **696**, and was much more stable in HLM (3 cf. 14 mL/min/kg) (Chart 4.5). However, lower stability was observed in hepatocytes presumably due to Phase II conjugation of the hydroxyl group of **720**. The LLE values based on measured Log D indicate that both **696** and **720** represent a marginal improvement in potency/lipophilicity balance. Two of these analogs, **726** and **728**, showed a retention of potency with increased LLE, indicating that the added polar groups, although suboptimal, were achieving a positive interaction with the protein. Significantly, compounds **726** and **728** also have improved HLM stability relative to compound **696** (Clhep = 2 and 6 respectively compared to 14 mL/min/kg).

Basu et al. described the structure-based design and synthesis of covalent-reversible inhibitors to overcome drug resistance in EGFR [10]. They synthesized a focused library of EGFR inhibitors utilizing WZ4002 as lead structure by utilizing a generic route to obtain the common intermediate **772** (Scheme 4.13). Subsequently, they altered the Michael acceptor by the introduction of various covalent reversible warheads and further derivatized the β-carbon of the resulting 2-substituted acrylamide with suitable aliphatic and aromatic groups to explore the associated structure

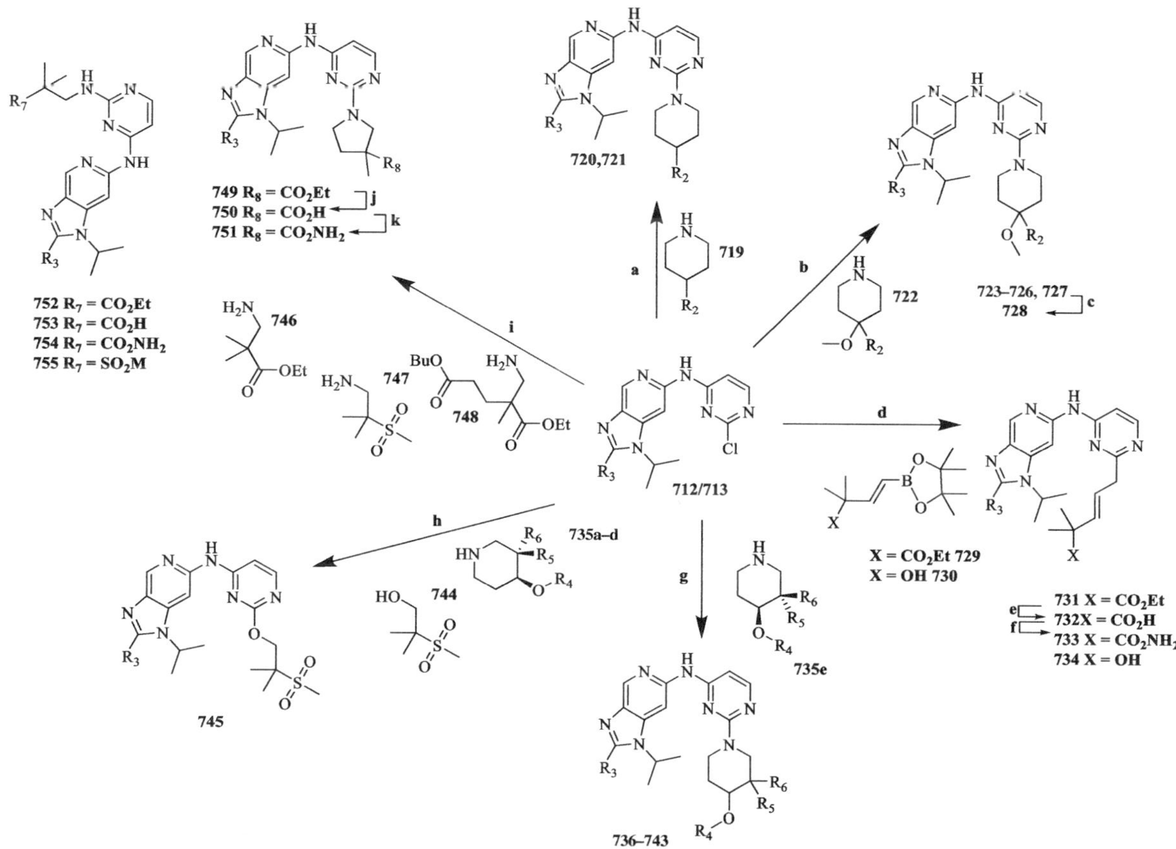

Scheme 4.11 Reagents and conditions: (a) **712**, **719a** ($R_1 = OH$)/**719b** ($R_1 = SO_2Me$), Et_3N, IPA, μW, 150°C, 0.5–0.75 h (29%–80%); (b) **712**, **722a** ($R_1 = CH_2NMe_2$)/**722b** ($R_1 = CH_2OMe$)/**722c** ($R_1 = CH_2OH$)/**722d** ($R_1 = CH_2CH_2OH$)/**722e** ($R_1 = CH_2CH_2N_3$), Et_3N, IPA, μW, 150°C, 1 h (14%–72%); (c) For **727–728**, H_2, Pd/C, IMS, rt., 16 h (68%); (d) (i) For **712–731**, **729**, Pd(dppf)Cl_2.DCM, 2 M Na_2CO_3, DME, 100°C, 16 h (42%); (ii) For **712–734**, **730**, Pd(dppf)Cl_2.DCM, 2 M Na_2CO_3, DME, 100°C, 64 h (79%); (e) (i) **731**, LiOH·H_2O, dioxane, μW, 100°C, 1 h (quant.); (f) **732**, NH_4Cl, HATU, DMAP, DIPEA, DMF, rt., 1 h (33%); (g) (i) **735** (**a** $R_4 = Me$, $R_5 = F$, $R_6 = H$; **b** $R_4 = H$, $R_5 = F$, $R_6 = H$; **c** $R_4 = Me$, $R_5 = F$, $R_6 = F$, **d** $= R_4 = H$, $R_5 = F$, $R_6 = F$; **e** $= R_4 = Me$, $R_5 = H$, $R^6 = F$), **712/713**, iPr2NEt, IPA, μW, 150°C, 30–60 min. (**736**, **738**, **741**, **742**, 42%–70%); (ii) 1.25 M HCl in MeOH, rt., 1 h (**737**, **739**, **740**, **743**, 36%–60% over two steps); (h) **712**, **744**, K_2CO_3, IPA, μW, 250°C, 4.5 h (52%); (i) **746** or **747** or **748**, DIPEA, IPA, μW, 150°C, 1–10 h (50% Quant.); For **752–754**; (j) **752**, 6 M HCl, μW, 120°C, 1 h (97%); (k) **753**, NH_4Cl, HATU, iPr2NEt, DMF, rt., 2.5 h (41%); for **749–751**—(j) **749**, 3 M HCl, μW, 110°C, 1 h (92% over two steps from **712**); (k) **750**, NH_4Cl, HATU, iPr2NEt, DMF, rt., 1.5 h (70%).

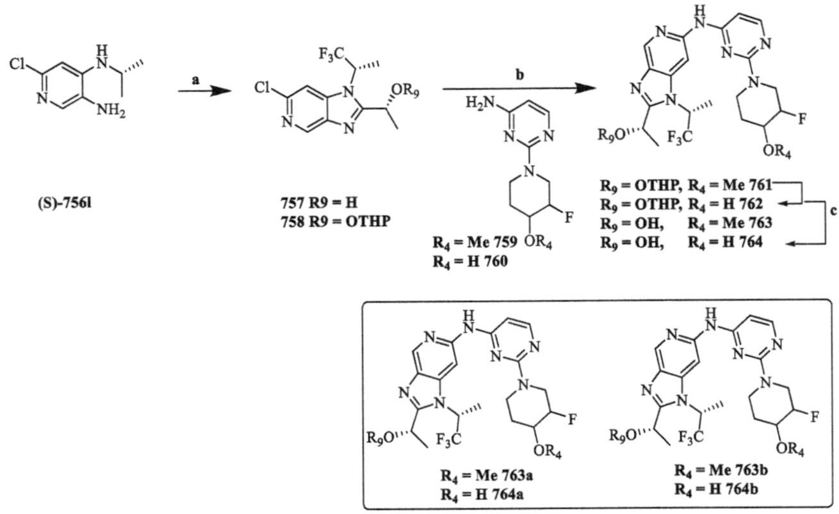

Scheme 4.12 Reagents and conditions: (a) (i) (*R*)-Lactamide, Et3O + BF_4, THF, rt., 2.0 h; (ii) (*S*)-**756l**, EtOH, 75°C, 3 h (65% over two steps); (iii) 3,4-dihydro-2H-pyran, p-TSA, THF, 65°C, 16 h (87%); (b) (i) For **758–761**, (±)-cis-2-(3-fluoro-4-methoxypiperidin-1-yl)pyrimidin-4-ylamine **759**, XPhos, $Pd_2(dba)_3$, Cs_2CO_3, 1,4-dioxane, 100 °C, 4 h (70%); (ii) For **758–762**, (±)-cis-1-(4-aminopyrimidin-2-yl)-3- fluoropiperidin-4-ol **760**, XPhos, $Pd_2(dba)_3$, Cs_2CO_3, 110°C, 5 h (49%); (iii) For **761a–758**, (−)-2 ((3*R*,4*S*)-3-fluoro-4-methoxypiperidin-1-yl) pyrimidin-4-ylamine **759a**, XPhos, $Pd_2(dba)_3.CHCl_3$, Cs_2CO_3, 1,4-dioxane, 100°C, 1.5 h (53%); (iv) For **761b–758**, (+)-2-((3*S*,4*R*)-3- fluoro-4-methoxypiperidin-1-yl)pyrimidin-4-ylamine **769b**, XPhos, $Pd_2(dba)_3.CHCl_3$, Cs_2CO_3, 1,4-dioxane, 100°C, 1.5 h (48%); (v) For **762a–758**, (−)-(3*R*,4*S*)-1-(4-aminopyrimidin-2-yl)-3- fluoropiperidin-4-ol **760a**, XPhos, $Pd_2(dba)_3.CHCl_3$, Cs_2CO_3, 1,4-dioxane, 100°C, 1 h (55%); (vi) For **762b–758**, (+)-(3*S*,4*R*)-1-(4-aminopyrimidin-2-yl)-3-fluoropiperidin-4-ol **760b**, XPhos, $Pd_2(dba)_3.CHCl_3$, Cs_2CO_3, 1,4-dioxane, 100°C, 1 h (44%); c) For **763a/b, 764, 764a/b–761a/b, 762, 762a/b**, 1.25 M HCl in MeOH, rt., 1–2 h (32%–64%); for **761–763**, 4 M HCl in dioxane, rt., 12 h (73%).

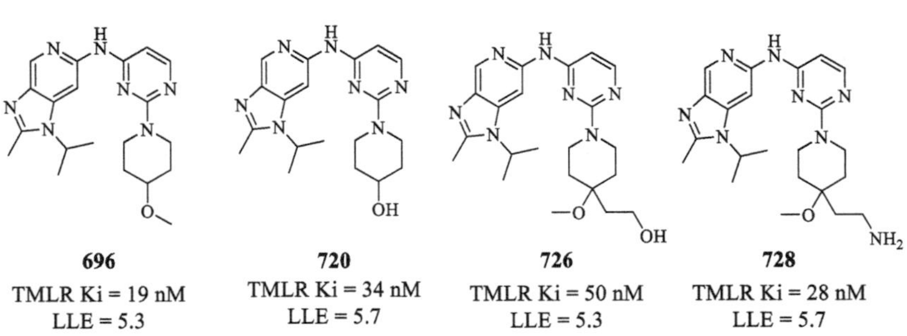

Chart 4.5 Antiproliferative and in vitro EGFR inhibitory activities.

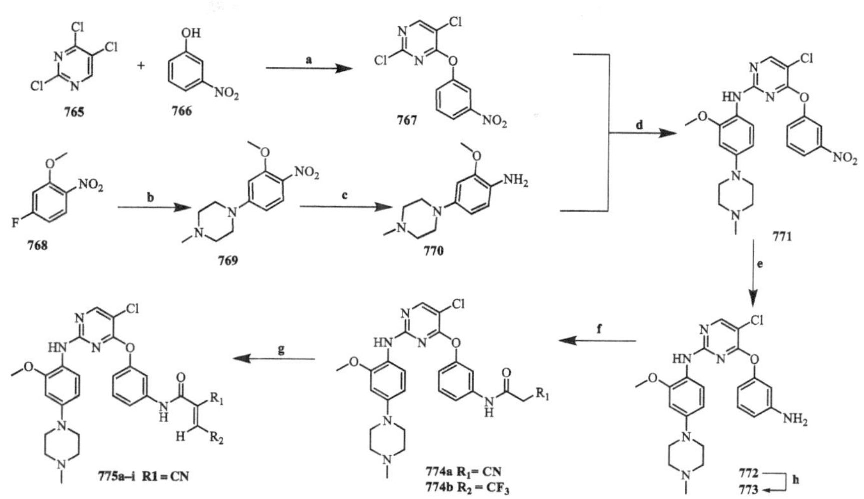

Scheme 4.13 Reagents and conditions: (a) K_2CO_3, DMF, 80°C, 8 h, 86%; (b) *N*-methylpiperizine, K_2CO_3, DMF, rt., 6 h, 98%; (c) Zn, NH_4Cl, THF, MeOH, rt., 18 h, 74%; (d) TFA, 2-BuOH, 100°C, 18 h, 65%; (e) Zn, NH_4Cl, MeOH, THF, rt., 18 h, 91%; (f) For R_1 = CN, cyanoacetic acid, EDC, HOBt, *N,N*-diisopropylethylamine, DCM, rt., 18 h, 89%; R_2 = CF_3, 3,3,3-trifluoropropanoic acid, EDC, HOBt, *N,N*-diisopropylethylamine, DMF, rt., 18 h, 72%; (g) Compound **813a**, ArCHO, piperidine, EtOH, 80°C, 8–36 h, 70%–82%; (h) acetic anhydride, Et_3N, DCM, 2 h, 86%.

activity relationships (SARs). Initially, 3-nitrophenol was regio-selectively attached at the 4 position of 2,4,5-trichloropyrimidine to obtain aryl ether **767** (Scheme 4.13). At the same time, 5-fluoro-2-nitroanisole was decorated with *N*-methylpiperizine at the 5 position. The following reduction of the nitro group led to compound **770**, which was coupled under acidic condition with aryl ether **767** to obtain the core intermediate **771**. The nitro group of **771** was reduced to obtain the corresponding amine **772**, which was coupled with cyanoacetic acid and 3,3,3-trifluoropropanoic acid to the corresponding amide using standard coupling conditions (EDC, HOBt, DIPEA) to obtain key intermediates **774a** and **774b**, respectively. An aldol condensation of compound **774a** with various commercially available aldehydes to generate our designed covalent reversible inhibitors (CRIs). The aromatic aldehydes underwent efficient aldol condensation with compound **774a**, using piperidine as base in refluxing ethanol to afford the desired products **775a–i** predominantly as trans-olefins. The tool compound **773** was prepared by treating compound **772** with acetic anhydride and DIPEA. Although aldol condensations using aromatic aldehydes were successful, analogous reactions using aliphatic aldehydes did not result in the desired products. Constitutive studies revealed that strong dehydrating reaction conditions are required for the complete formation of the olefin from the respective intermediate.

Having successfully synthesized the 3-substituted-2-cyanoacrylamides, Basu et al. further synthesized a novel class of CRIs disclosing a trifluoromethyl- instead of cyano-functionality as an electron withdrawing group.

The aldol condensation of aromatic aldehydes with compound **774b** to produce the desired products by applying the previously used method was likewise unsuccessful. Hence **775n–p** was synthesized in a two-step process from amine precursor **772** (Scheme 4.14). Initially, the 3-aryl-2-trifluoromethylacrylic acids were prepared by the aldol condensation of appropriate aromatic aldehydes with 3,3,3-trifluoropropanoic acid using $TiCl_4$ in DCM as solvent. Subsequently the 3-aryl-2-trifluoromethylacrylic acids were coupled to the amine **772** using EDCI and HOBt in DMF to obtain the desired products **775n–p** in good yields (80%–90%). Tool compound **781** was likewise prepared in two steps. Therefore the condensation product **780** of malonic acid and cyclopropane carboxaldehyde was coupled to amine **772** by utilizing EDC/HOBt coupling conditions (Scheme 4.15). Similarly, cyanoacetic acid was condensed with cyclopropane carboxaldehyde to obtain compound **783**, which was further subjected to hydrogenation to obtain compound **784**. The latter was coupled with amine **772** to obtain compound the reversible analog **785** (Scheme 4.15).

A focused library of covalent reversible inhibitors (CRIs) was synthesized and biologically evaluated by activity-based studies (Chart 4.6). The entity of CRIs was found to be a highly mutant selective for EGFR *L858R* and EGFR *L858R*/T790M while sparing the inhibition of WT EGFR. Furthermore, CRIs comprising cyano functionality (**775a** and **775e**) as an electron-withdrawing element were, in general, more potent

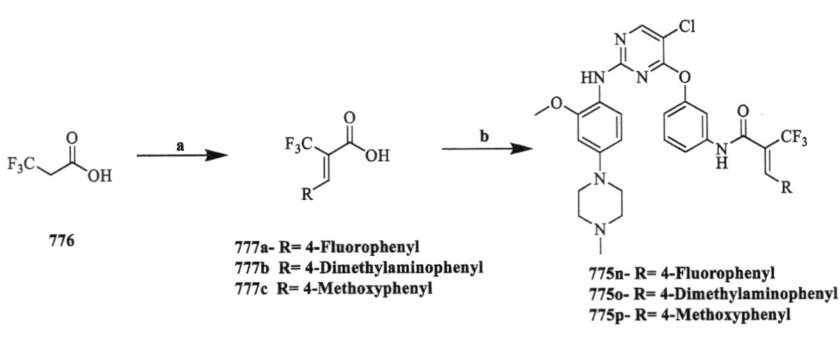

Scheme 4.14 Reagents and conditions: (a) Aromatic aldehyde, $TiCl_4$, DCM, THF, rt., 80%–90%; (b) 772, EDCI, HOBt, DIPEA, DMF, rt., 18h, 80%–90%.

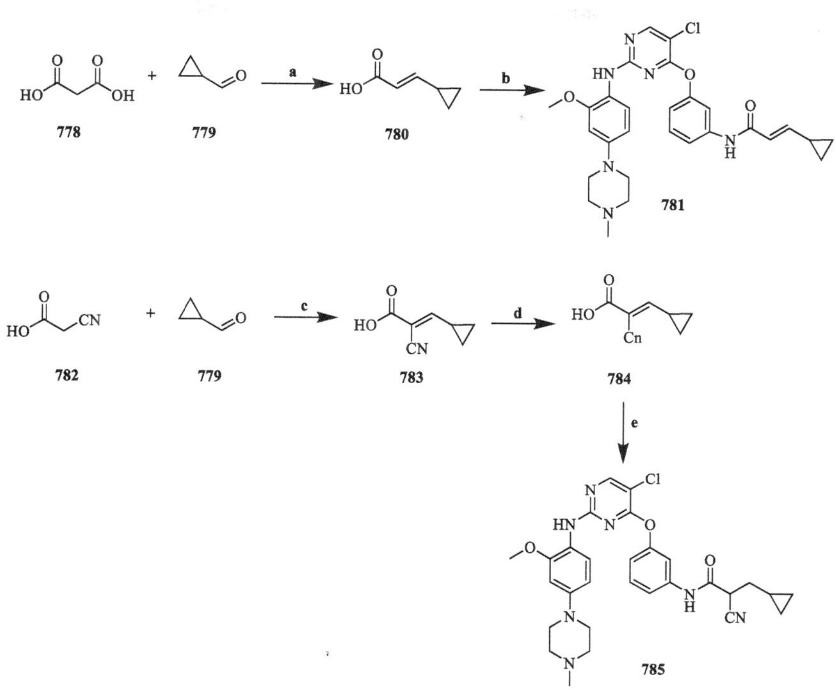

Scheme 4.15 Reagents and conditions: (a) piperidine, pyridine, 80°C, 2h, 60%; (b) 772, EDC, HOBT, DIPEA, DCM, rt., 18h, 65%; (c) piperidine, EtOH, 80°C, 2h, 85%; (d) H_2, cat. Pd/C, MeOH, rt., 2h, quant.; (e) EDC, HOBT, DIPEA, DCM, rt., 18h, 82%.

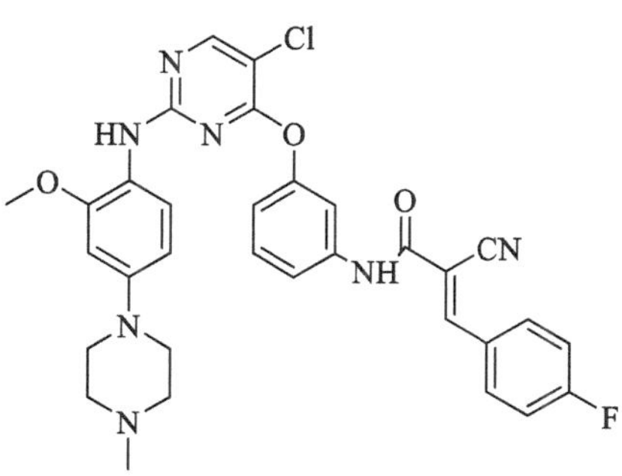

775a

WT IC_{50} = 7.0 μM
L858R IC_{50} = 0.54 μM
L858R/T790M IC_{50} = 0.12 μM

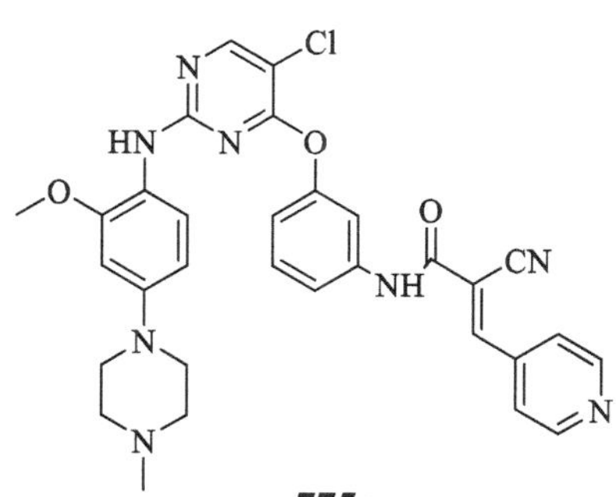

775g

WT IC_{50} = 8.7 μM
L858R IC_{50} = 0.15 μM
L858R/T790M IC_{50} = 0.037 μM

Chart 4.6 Antiproliferative and in vitro EGFR inhibitory activities.

than their trifluoromethyl decorated counterparts (**775n** and **775o**). Among 3-substituted heterocycles of 2-cyanoacrylamide CRIs, **775g** decorated with a 4-pyridyl group was found to be the most potent CRI, most likely due to a charged interaction of the protonated pyridyl group with Asn800 of EGFR.

Kaspersen et al. identified the new 4-*N*-substituted 6-aryl-7*H*-pyrrolo[2,3-*d*] pyrimidine-4-amines as highly potent EGFR-TK inhibitors with Src-family activity [11]. To enable efficient synthesis of a compound library, two different but complementary routes were used. The starting material 6-aryl-4-chloropyrrolopyrimidine **786a–d** derivatives were synthesized from 1-aryl-2-bromoethanones in three steps. Amination chemistry in *n*-butanol gave the 4,6-disubstituted pyrrolopyrimidines **811–834**, with varying R, R1 and R2 substituents as shown in Scheme 4.16. Derivatives **811–832** were in most cases prepared as their pure (*R*)-enantiomer, except **814a, 818a, 824a, 825a**, and **832b**, which were racemates. First, as shown in Scheme 4.17, the commercially available compound **835** was converted to **837** by standard chlorination and protection. The following iodination has previously been performed by others, but preparative HPLC was needed to purify the iodo compound **838**.

The pyrrolopyrimidine **839** obtained by deprotection could be transformed to give the precursors **841–844** in 56%–86% a nonoptimized yield. Suzuki coupling at C-6 using XPhos and 2nd generation XPhos precatalyst was then performed using selected arylboronic acids giving most of the targeted products. However, applying ortho- and meta-hydroxyphenylboronic

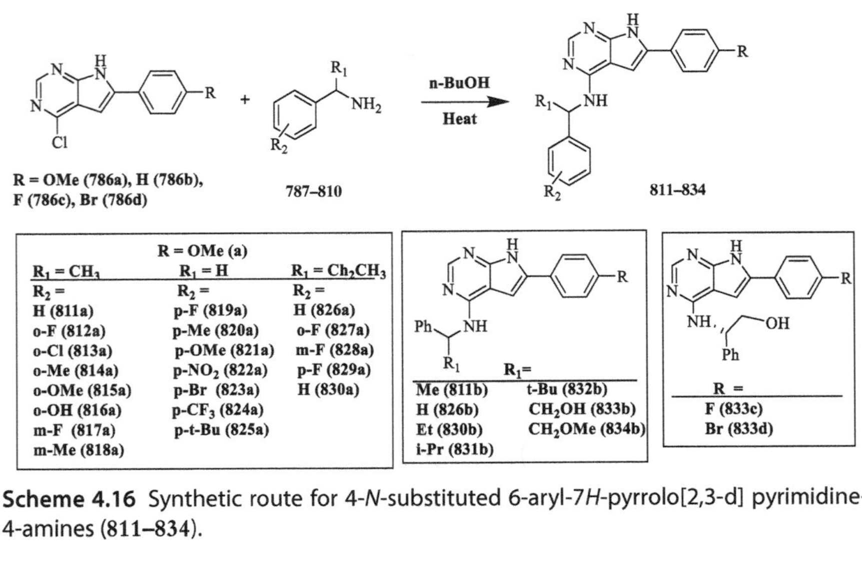

Scheme 4.16 Synthetic route for 4-*N*-substituted 6-aryl-7*H*-pyrrolo[2,3-d] pyrimidine-4-amines (811–834).

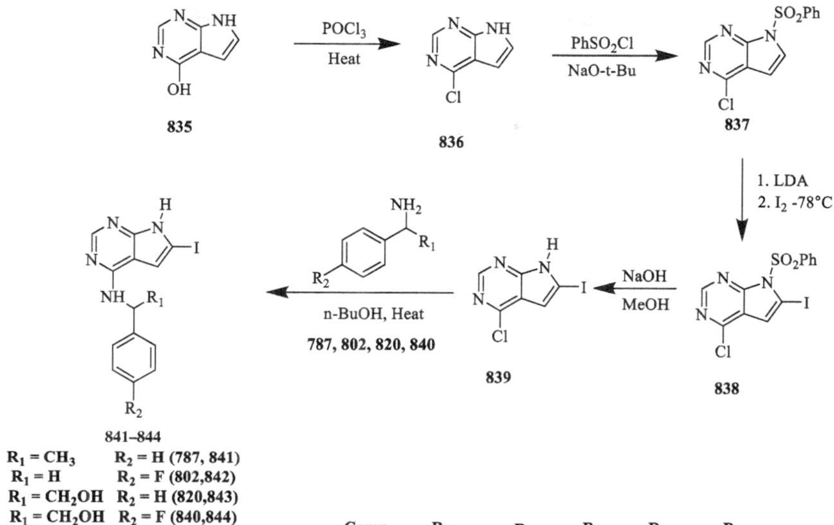

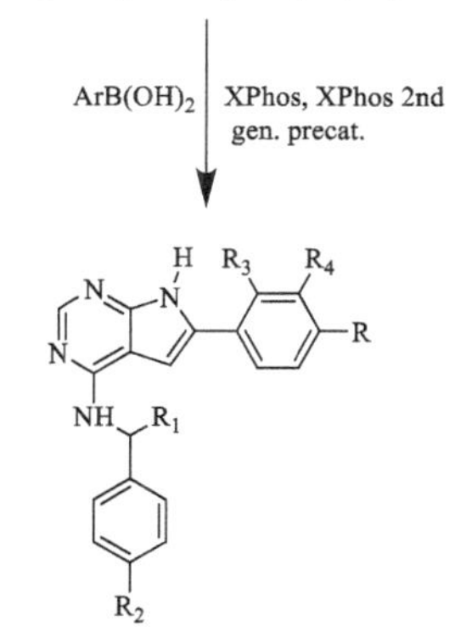

	Comp	R_1	R	R_2	R_3	R_4
(a)	811e	CH_3	H	H	OMe	H
	811f	CH_3	H	H	OH	H
(a)	811g	CH_3	H	H	H	OMe
	811h	CH_3	H	H	H	OH
	811j	CH_3	H	H	CH_2OH	H
	811k	CH_3	H	H	H	CH_2OH
	811l	CH_3	CH_2OH	H	H	H
(b)	811m	H	H	CHO	OMe	H
	811n	H	CH_2OH	H	OMe	H
(b)	826m	H	CHO	H	OMe	H
	826n	H	CH_2OH	H	OMe	H
	833e	CH_2OH	H	H	OMe	H
	833l	CH_2OH	CH_2OH	H	H	H
(b)	833m	CH_2OH	CHO	H	OMe	H
	833n	CH_2OH	CH_2OH	H	OMe	H
	833o	CH_2OH	F	H	OMe	H
(b)	845m	CH_2OH	CHO	F	OMe	H
	845n	CH_2OH	CH_2OH	F	OMe	H

Scheme 4.17 Reagents and conditions: Pyrrolopyrimidines synthesized by Suzuki coupling on **841–844** giving (*R*)-**811e–h**, (*R*)-**811j–n**, **826m–n**, (*S*)-**833e**, (*S*)-**833l–n** and (*S*)-**833m–n**. (a) BBr_3 and (b) $NaBH_4$ in MeOH.

acids, the coupling only reached 50% conversion in 40 h. Instead, these derivatives were made by deprotection of the methoxy analogues (*R*)-**811e** and (*R*)-**811g**. The para-hydroxymethyl derivatives (*R*)-**811n**, **826n**, (*S*)-**833n**, and (*S*)-**845n** were obtained by sodium borohydride reduction of the corresponding aldehyde derivatives (*R*)-**811m**, **826m**, (*S*)-**833m**, and (*S*)-**845m**.

Eight of the new pyrrolopyrimidine based EGFR-TK inhibitors were evaluated in two genetically modified model cells Ba/F3-673 EGFRL858R and Ba/F3-EGFR T790M, and four human cancer cell lines including A-431 (epidermoid carcinoma), AU-565 (breast adeno-675 carcinoma), K-562 (leukaemia) and C-33A (cervix carcinoma).

The Ba/F3-EGFR cells display growth and survival depending on EGFR signaling. Cell proliferation study with Ba/F3-EGFR *L858R* cells using the XTT assay proved to be a highly sensitive test system for evaluating the EGFR-TK inhibitors. Of the eight compounds assayed, seven were highly active, with IC_{50} values in the range of 80–172 nM, while the benzylamine based compound **826n** was less potent than the other derivatives. Compounds **(*R*)-811n**, **(*S*)-833n** and **(*S*)-845n** sharing the same aryl moiety as well as the para hydroxymethyl derivative (*R*)-261 were found more active than erlotinib. None of the compounds inhibited the proliferation of the Ba/F3-EGFR T790M cell line, indicating that the cyto-688 toxicity is mainly due to EGFR *L858R* inhibition.

All the compounds assayed were found to be very potent toward both the EGFR *L858R* and EGFR L861Q mutants. On average, as compared to the wild-type EGFR-TK assay, somewhat higher IC_{50} values were seen toward EGFR *L858R*, while mostly similar IC_{50} values were noted in the case of EGFR L861Q. As expected for reversible inhibitors, testing toward the EGFR T790M mutant only revealed activity in the μM range. Compounds **(*R*)-811l** and **826n** showed lower inhibitory potency toward EGFR T790M as compared to the other drug candidates, as seen in Chart 4.7.

Hanan et al. reported the 4-aminoindazolyl-dihydrofuro[3,4-*d*]pyrimidines as noncovalent inhibitors of mutant EGFR TK [12].

Noncommercial 3-aminoindazoles **848** were obtained through the reaction of ortho-fluorobenzonitriles **846** with hydrazine **847**. The SNAr reaction of these aminoindazoles **848** with 2,4-dichloro-7,7-dimethyl-5,7-dihydrofuro[3,4-*d*]pyrimidine **849** afforded the penultimate intermediates. Suzuki coupling with aryl boronates provided the final test compounds **852a–j**, **853–855**, and **856–860** as shown in Scheme 4.18.

Structure activity relationship indicate that further modification of pyrazole **852b** lost most of the activity owe to imidazole and could not be recovered through the use of dichloro-substitution as a phenyl isostere (compound **852c**), as shown in Chart 4.8. Saturation of the phenyl ring, which would be desirable to reduce planarity and improve overall physiochemical properties, was also not tolerated (compound **852d**), as shown in Chart 4.8. Indole **852e** was slightly less active than the indazole (17 nM versus 2 nM); however, benzoisothiazole **852f** lost a nearly 400-fold activity. This data suggests that both the hydrogen bond donor and the acceptor of the indazole are important for binding, with a more significant contribution from the hydrogen-bond donor.

Benzothiazole **852g** improved HLM and RLM stability, but it was completely inactive. They attempted to modulate the metabolism of the indazole

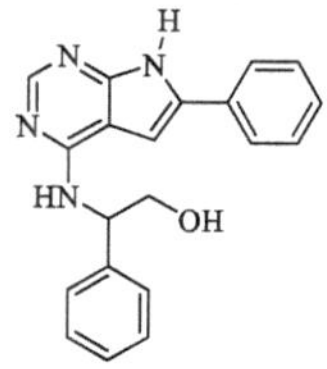

833
EGFR-TK IC_{50} = 0.7 nM

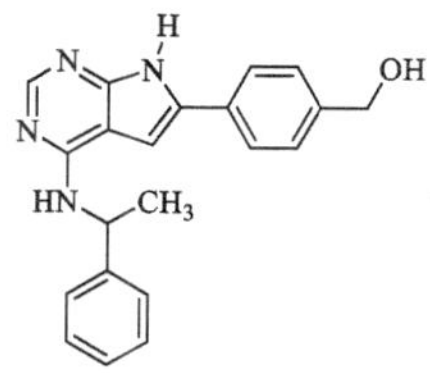

811l
EGFRL858R IC_{50} = 0.8 nM
EGFRL861Q IC_{50} = 0.3 nM
EGFRT790M (% inhibition) = 44

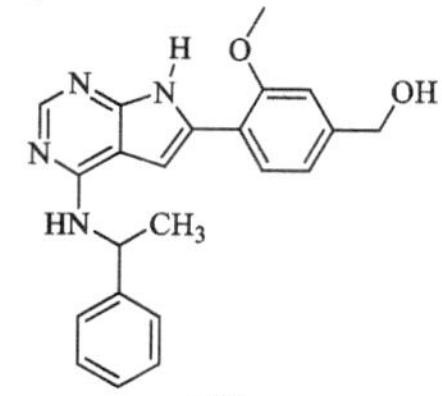

811n
EGFRL858R IC_{50} = 0.4 nM
EGFRL861Q IC_{50} = 0.2 nM
EGFRT790M (% inhibition) = 95

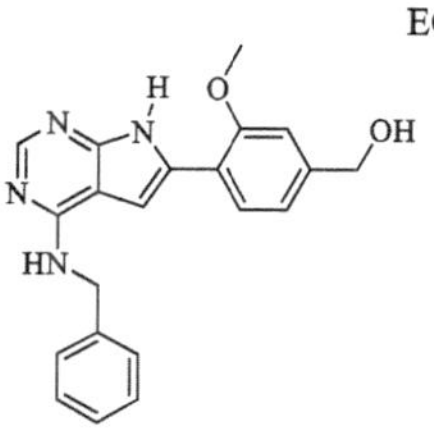

826n
EGFRL858R IC_{50} = 0.8 nM
EGFRL861Q IC_{50} = 0.4 nM
EGFRT790M (% inhibition) = 66

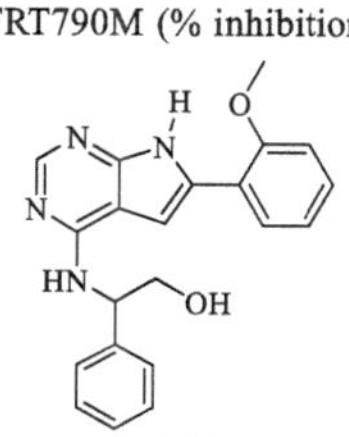

833e
EGFRL858R IC_{50} = 0.1 nM
EGFRL861Q IC_{50} = 0.2 nM
EGFRT790M (% inhibition) = 94

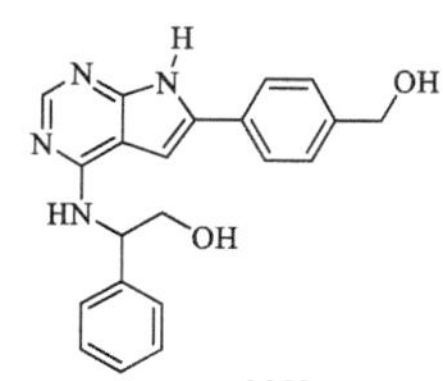

833l
EGFRL858R IC_{50} = 0.4 nM
EGFRL861Q IC_{50} = 0.2 nM
EGFRT790M (% inhibition) = 84

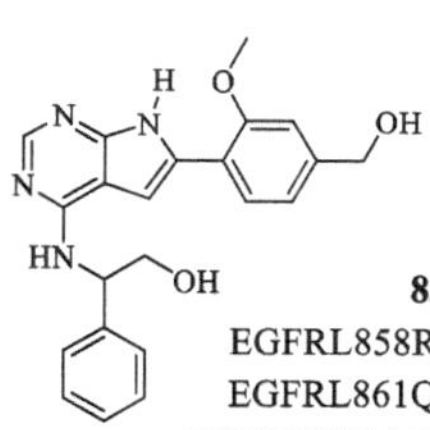

833n
EGFRL858R IC_{50} = 0.4 nM
EGFRL861Q IC_{50} = 0.3 nM
EGFRT790M (% inhibition) = 97

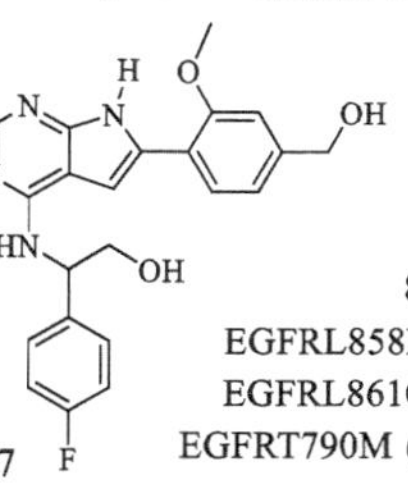

845n
EGFRL858R IC_{50} = 0.1 nM
EGFRL861Q IC_{50} = 0.2 nM
EGFRT790M (% inhibition) = 94

Chart 4.7 Antiproliferative and in vitro EGFR inhibitory activities.

ring through the introduction of an additional heteroatom to reduce electron density; however, pyrazolopyrimidine **852h** lost all activity, as shown in Chart 4.8. The substitution of the indazole ring with electron-withdrawing substituents was more fruitful; compounds **852i**, **852j**, **852k**, and **852l** showed that fluoro substitution was tolerated with a minimal loss of activity, with substitution at the 6 or 7 position affording compounds equipotent with starting point **852a**. Unfortunately, mono-fluoro substitution at any position failed to provide an improvement in HLM or RLM stability. Larger groups such as chloro or trifluoromethyl were moderately tolerated only at the 7 position (**852m** and **852o**), with larger drops in potency observed for substitution at the 6 position (**852n** and **852p**). No improvement in HLM or RLM stability was observed with either chloro or trifluoromethyl substitution. Disubstitution with electron withdrawing groups was evaluated

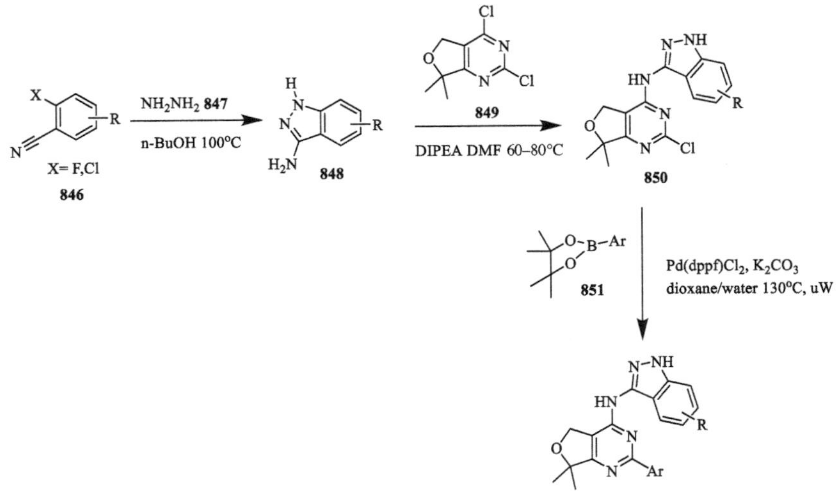

852 (a–j), 853–855, 856–860

Scheme 4.18 Synthetic route for 4-aminoindazolyl-dihydrofuro[3,4-d]pyrimidines **852a–j, 853–855, 856–860.**

(compounds **852q**), and while activity was retained the clearance predicted by HLM and RLM remained in the moderate to high range.

Compounds with single fluoro or chloro substitution (**852k** and **852o**) were observed to have a high clearance similar to the starting point **852a** and were above the expected clearance based on liver microsome and hepatocyte incubations. Disubstituted compound **852r** resulted in only a slight reduction in clearance (Cl = 60 mL/min/kg). While 7-trifluoromethyl substitution (**852m**) afforded a reduction in plasma clearance to 27 mL/min/kg; unfortunately, oral bioavailability remained low (F = 5%). Conversely, 6-trifluoromethyl substitution (**852n**) resulted in high clearance.

Shao et al. reported the 6-oxooxazolidine–quinazolines as noncovalent inhibitors with the potential to target mutant forms of EGFR. The synthesis of the anilinoquinazolines **865** is depicted in Scheme 4.19 [13]. Briefly, 2-amino-4-fluorobenzoic acid **860** was converted to the quinazolinone **865** in a 50% yield in six steps, according to the reported procedure. The reaction sequence was initiated using commercially available 2-amino-4-fluorobenzoic acid **860** and formamidine acetate in 2-methoxyethanol at reflux to yield 7-fluoroquinazolin-4(3*H*)-one **861**, which was nitrified with concentrated H_2SO_4 and fuming HNO_3. The resulting 7-fluoro-6-nitro-3*H*-quinazolin-4-one **862** was chlorinated with excessive thionyl

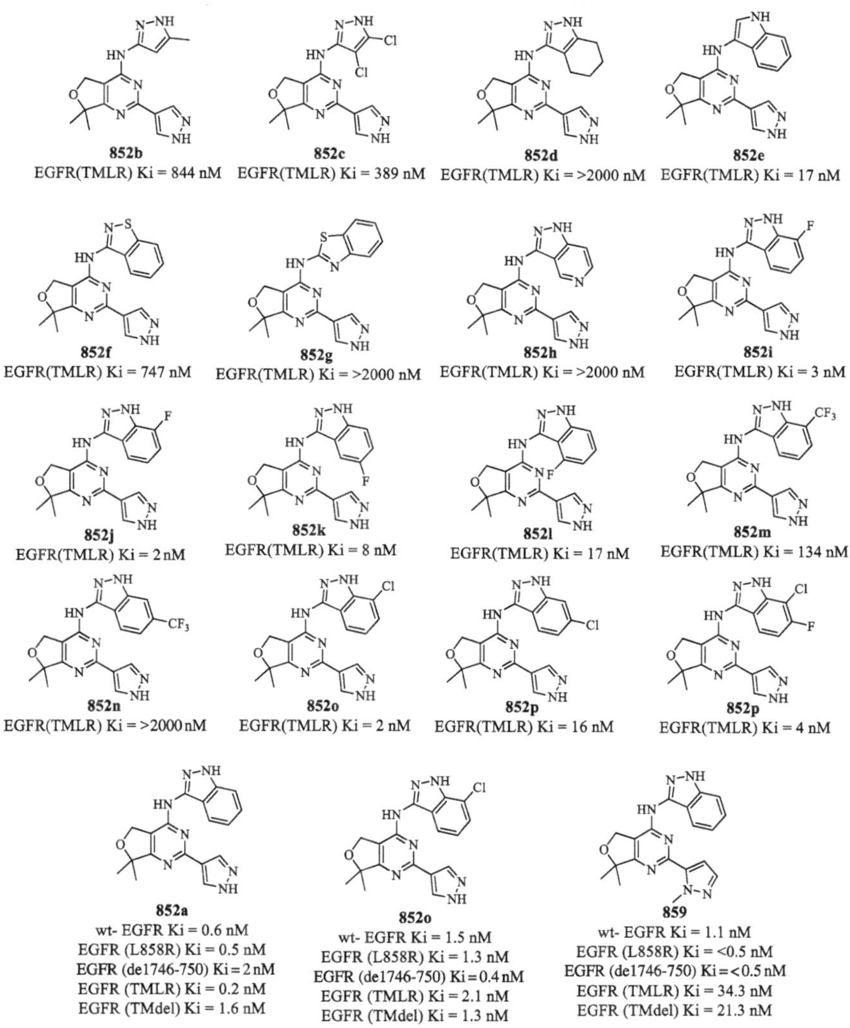

Chart 4.8 Antiproliferative and in vitro EGFR inhibitory activities.

chloride in the presence of a catalytic amount of DMF for the preparation of 4-chloroquinazoline with subsequent aniline substitution to provide 4-anilino-7-fluoro-6-nitroquinazoline **863** (Scheme 4.20).

The fluorine atom was substituted by a variety of alcohols using t-BuOK in DMSO yielding the intermediate **864**. Reduction of the 6-nitro group with $NaBH_4$ and $NiCl_6 \cdot H_2O$ in methanol gave the desired product **865**. Various oxo-oxazolidine carboxylic acid groups, such as compounds **867a–d** were synthesized and incorporated into the 6 position of the anilinoquinazoline **865** via amide bond formation to give the

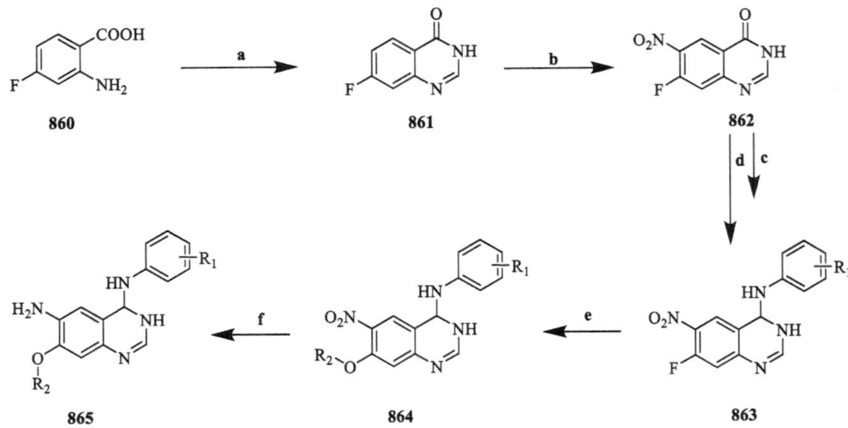

Scheme 4.19 Reagents and conditions: (a) formamidine acetate, 2-methoxyethanol, reflux, 18h; (b) HNO_3, H_2SO_4, 110°C, 2h; (c) $SOCl_2$, DMF, reflux, 4h; (d) aniline, DCM, rt., 30min; (e) aliphatic alchol, t-BuOK, DMSO, rt., 30min; (f) $NaBH_4$, $SnCl_2$ $6H_2O$, DCM–MeOH, ice-bath, 30min.

corresponding compounds **868–871** (Schemes 4.19 and 4.21). The purities of all compounds were evaluated by HPLC–MS. The stereocenter of the products was not affected during the synthesis procedure, as reported.

The inhibitory activities of the selected compounds against different types of kinases (WT EGFR, EGFR *L858R*, and EGFR T790M) were evaluated using the homogeneous time-resolved fluorescence (HTRF)

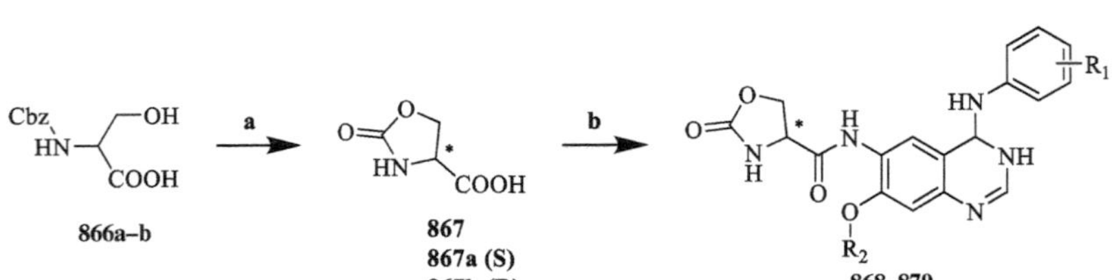

868a: (S), R_1=3-Cl-4-F, R_2=Me	**868o:** (S), R_1=3-Cl-4-F, R_2=((R)-4-methylmorpholin-3-yl)methyl
868b: (S), R_1=3-Cl-4(3-fluorobenzyloxy), R_2=Et	**868p**: (S), R_1=3-Cl-4-F, R_2=2-methoxyethyl
868c: (S), R_1=3-Br, R_2=Me	**868q**: (S), R_1=3-Cl-4-FR_2=2-(piperidin-1-yl)ethyl
868d: (S), R_1=3-CF_3, R_2=Me	**868r**: (S), R_1=3-Cl-4-F, R_2=3-morpholinopropyl
868e: (S), R_1=3-Cl-4-(pyridin-2-ylmethoxy), R_2=Et	**869a**: (R), R_1=3-Cl-4-F, R_2=Me
868g: (S), R_1=(3-fluorobenzyl)-1 H-indazol-5-yl, R_2=Et	**869b**: (R), R_1=3-Cl-4(3-fluorobenzyloxy), R_2=Et
868h: (S), R_1=3-Cl-4-(6-methylpyridin-3-yl)oxy, R_2=Et	**869c**: (R), R_1=3-Br, R_2=Me
868i: (S), R_1=3-Cl-4-cyclopropylmethoxy, R_2=Et	**869e**: (R), $R_1$1=3-Cl-4-(pyridin-2-ylmethoxy), R_2=Et
868j: (S), R_1=4-methoxy, R_2=Me	**869g**: (R), R_1=(3-fluorobenzyl)-1 H-indazol-5-yl, R_2=Et
868k: (S), R_1=3-Cl-4-F, R_2=(S)-tetrahydrofuran-3-yl	**869h**: (R), R_1=3-Cl-4-F, R_2=(S)-tetrahydrofuran-3-yl
868l: (S), R_1=3-Cl-4-F, R_2=(S)-1-methylpyrrolidin-3-yl	**869i**: (R), R_1=3-Cl-4-F, R_2=(1-methylpiperidin-4-yl)methyl
868m: (S), R_1=3-Cl-4-F, R_2=(1-methylpiperidin-4-yl)methyl	**869j**: (R),R_1=3-Cl-4-F, R_2=2-methoxyethyl
868n: (S), R_1=3-Cl-4-F, R_2=2,2,2-trifluoroethyl	**869l**: (R), R_1=3-Cl-4-F, R_2=3-morpholinopropyl

Scheme 4.20 Reagents and conditions: (a) NaOH-H_2O, 0°C-RT; (b) Compound 865, ($CoCl_2$), Et_3N, DMF, DCM, ice-bath to rt, overnight.

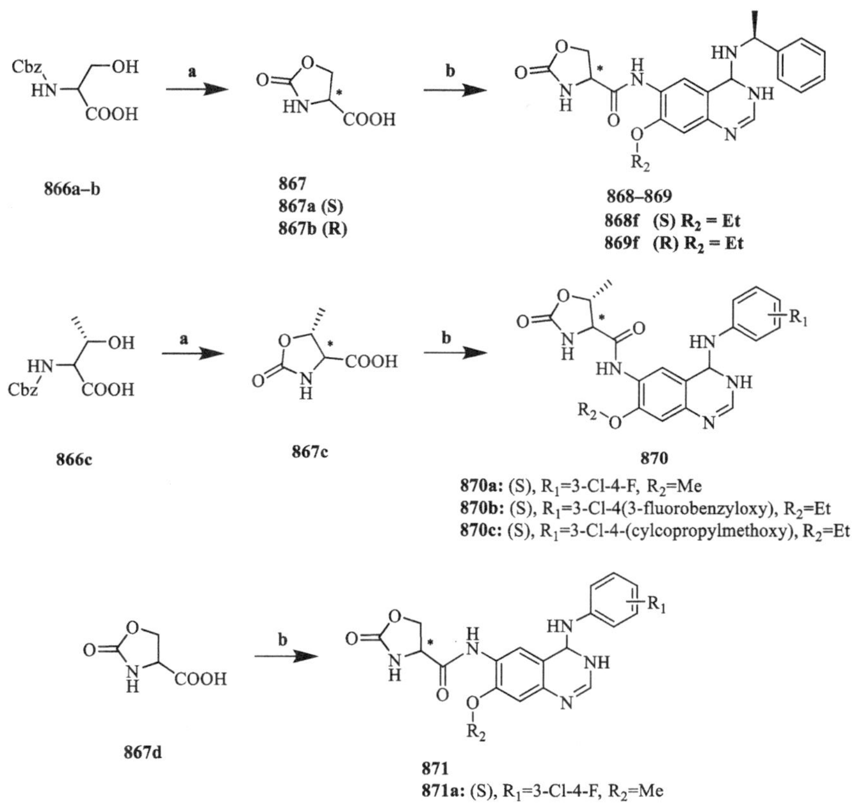

Scheme 4.21 Reagents and conditions: (a) NaOH–H_2O, 0°C–rt.; (b) compound **865**, $(COCl)_2$, Et_3N, DMF, DCM, ice-bath to rt., overnight.

method, and gefitinib was employed as a positive control. Among the tested compounds, some exhibited more potency than gefitinib, especially on EGFR T790M. Notably, compounds **868a**, **868r**, and **870a** ($IC_{50} < 100$ nM) with a 3-chloro-4-fluoro anilino moiety at the 4-position of quinazoline ring displayed a higher inhibitory activity on EGFR T790M than gefitinib ($IC_{50} = 173$ nM).

The replacement of aniline with chiral alkyl amine (**868f**) led to a loss in inhibitory activity (for EGFR T790M, $IC_{50} > 1000$ nM), the bulky substituents (compounds **868b**, **869b**, and **870b**) at the 4 position was not favorable, leading a 2–5 fold decrease on the EGFR T790M inhibitory activity. Among them, the most potent compound **868a** exhibited excellent inhibition against EGFR *L858R* and EGFR T790M, with an IC_{50} at 0.31 nM and 26.33 nM, respectively (Chart 4.9).

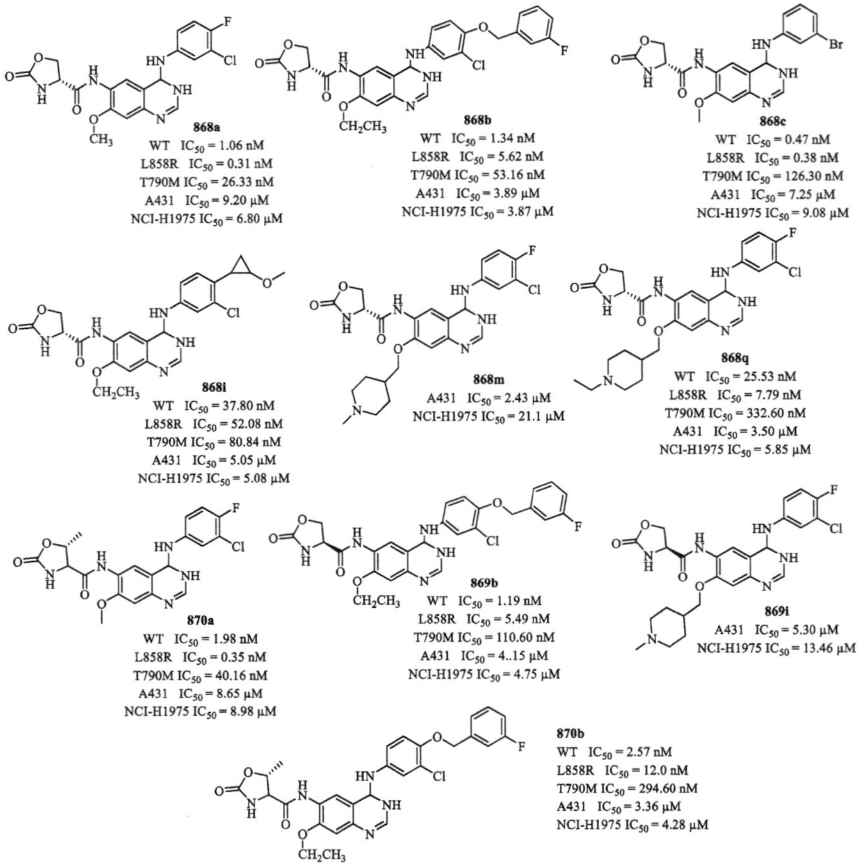

Chart 4.9 Antiproliferative and in vitro EGFR inhibitory activities.

4.2 IRREVERSIBLE VERSUS REVERSIBLE INHIBITORS

To overcome the enhanced ATP binding in T790M mutants, the next-generation drug must bind to the T790M mutant with a higher affinity than that of the first-generation drugs. Irreversible EGFR inhibitors containing a Michael acceptor functional group have been developed to overcome this enhanced ATP binding by irreversibly alkylating a cysteine residue (C797) in the ATP binding site of the EGFR [13, 14].

However, except for a few examples, such as BIBW2992 [15], PF00299804 [16], and WZ-4002 [17], these irreversible inhibitors have thus far shown a limited clinical efficacy. Inherent weaknesses, including relatively high toxicity and a decreased binding velocity to the mutant kinase, may be responsible for the lack of clinical efficacy [15–17]. Hence the development of next-generation reversible EGFR inhibitors that can inhibit the drug-resistant T790M-bearing mutants and remain potent in other activating mutants is a highly attractive goal that has received much attention in recent years.

For covalent third-generation inhibitors a very reactive warhead can both help achieve high potency against the target, but it can also cause idiosyncratic toxicity by reacting with off-target proteins. Therefore, it is desirable to increase the binding affinity to achieve high potency, while reducing the intrinsic chemical reactivity of the warhead in order to minimize idiosyncratic toxicity. Reducing the warhead reactivity relative to the second-generation EGFR covalent inhibitors was one of the medicinal chemistry goals for the discovery of osimertinib. Guided by a novel in vitro irreversible kinetics assay that defines binding affinity and irreversible inactivation, the warhead in PF-06747775 was optimized to demonstrate minimal intrinsic chemical reactivity, hence exhibiting considerably reduced proteome-wide reactivity relative to earlier inhibitors. Human clinical data will delineate the degree of efficacy and safety benefit the excellent proteome-wide selectivity provides for PF-06747775 [18–21].

REFERENCES

[1] Y. Kim, J. Ko, Z. Cui, A. Abolhoda, J.S. Ahn, S.H. Ou, The EGFR T790M mutation in acquired resistance to an irreversible second-generation EGFR inhibitor, Cancer Ther. 11 (2012) 784.

[2] H.A. Yu, M.E. Arcila, N. Rekhtman, C.S. Sima, Analysis of tumor specimens at the time of acquired resistance to EGFR-TKI therapy in 155 patients with EGFR-mutant lung cancers, Clin. Cancer Res. 19 (2013) 2240.

[3] A.O. Walter, R.T. Sjin, H.J. Haringsma, K. Ohashi, J. Sun, K. Lee, A. Dubrovskiy, M. Labenski, Z. Zhu, Discovery of a mutant-selective covalent inhibitor of EGFR that overcomes T790M-mediated resistance in NSCLC, Cancer Discov. 3 (2013) 1404.

[4] D.A. Cross, S.E. Ashton, S. Ghiorghiu, C. Eberlein, C.A. Nebhan, P.J. Spitzler, J.P. Orme, AZD9291, an irreversible EGFR TKI, overcomes T790M-mediated resistance to EGFR inhibitors in lung cancer, Cancer Discov. 4 (2014) 1046.

[5] M.R.V. Finlay, M. Anderton, S. Ashton, P. Ballard, P.A. Bethel, M.R. Box, R.H. Bradbury, S.J. Brown, Discovery of a potent and selective EGFR inhibitor (AZD9291) of both sensitizing and T790M resistance mutations that spares the wild type form of the receptor, J. Med. Chem. 57 (2014) 8249.

[6] E.J. Hanan, C. Eigenbrot, M.C. Bryan, D.J. Burdick, B.K. Chan, Y. Chen, J. Dotson, R.A. Heald, P.S. Jackson, H. La, M.D. Lainchbury, S. Malek, H.E. Purkey, G. Schaefer, S. Schmidt, E.M. Seward, S. Sideris, C. Tam, S. Wang, S.K. Yeap, I. Yen, J. Yin, C. Yu, I. Zilberleyb, T.P. Heffron, Discovery of selective and noncovalent diaminopyrimidine-based inhibitors of epidermal growth factor receptor containing the T790M resistance mutation, J. Med. Chem. 57 (23) (2014) 10176–10191.

[7] M.C. Bryan, D.J. Burdick, B.K. Chan, Y. Chen, S. Clausen, J. Dotson, C. Eigenbrot, R. Elliott, E.J. Hanan, R.A. Heald, P. Jackson, H. La, M. Lainchbury, S. Malek, S.E. Mann, H.E. Purkey, G. Schaefer, S. Schmidt, E. Seward, S. Sideris,

S. Wang, I. Yen, C. Yu, T.P. Heffron, Pyridones as highly selective, noncovalent inhibitors of T790M double-mutants of EGFR, ACS Med. Chem. Lett. 7 (1) (2016) 100–104.
[8] J. Yang, L.J. Wang, J.J. Liu, L. Zhong, R.L. Zheng, Y. Xu, P. Ji, C.H. Zhang, W.J. Wang, X.D. Lin, L.L. Li, Y.Q. Wei, S.Y. Yang, Structural optimization and structure-activity relationships of N2-(4-(4-methylpiperazin-1-yl) phenyl)-N8-phenyl-9H-purine-2,8-diamine derivatives, a new class of reversible kinase inhibitors targeting both EGFR-activating and resistance mutations, J. Med. Chem. 55 (23) (2012) 10685–10699.
[9] R. Heald, K.K. Bowman, M.C. Bryan, D. Burdick, B. Chan, E. Chan, Y. Chen, S. Clausen, C. Eigenbrot, R. Elliott, E.J. Hanan, P. Jackson, J. Knight, H. La, M. Lainchbury, S. Malek, S. Mann, M. Merchant, K. Mortara, H. Purkey, G. Schaefer, S. Schmidt, E. Seward, S. Sideris, L. Shao, S. Wang, K. Yeap, I. Yen, C. Yu, T.P. Heffron, Noncovalent mutant selective epidermal growth factor receptor inhibitors: a lead optimization case study, J. Med. Chem. 58 (22) (2015) 8877–8895.
[10] D. Basu, A. Richters, D. Rauh, Structure-based design and synthesis of covalent-reversible inhibitors to overcome drug resistance in EGFR, Bioorg. Med. Chem. 23 (12) (2015) 2767–2780.
[11] S.J. Kaspersen, J. Han, K.G. Norsett, L. Rydsa, E. Kjobli, B.H. Hoff, Identification of new 4-N-substituted 6-aryl-7H-pyrrolo[2,3-d] pyrimidine-4-amines as highly potent EGFR-TK inhibitors with Src-family activity, Eur. J. Pharm. Sci. 59 (2014) 69–82.
[12] E.J. Hanan, M. Baumgardner, M.C. Bryan, Y. Chen, C. Eigenbrot, P. Fan, X.H. Gu, H. La, S. Malek, H.E. Purkey, G. Schaefer, S. Schmidt, S. Sideris, I. Yen, C. Yu, T.P. Heffron, 4-Aminoindazolyl-dihydrofuro[3,4-d]pyrimidines as non-covalent inhibitors of mutant epidermal growth factor receptor tyrosine kinase, Bioorg. Med. Chem. Lett. 26 (2) (2016) 534–539.
[13] J. Shao, E. Chen, K. Shu, W. Chen, G. Zhang, Y. Yu, 6-Oxooxazolidine–quinazolines as noncovalent inhibitors with the potential to target mutant forms of EGFR, Bioorg. Med. Chem. 24 (16) (2016) 3359–3370.
[14] L.V. Sequist, B. Besse, T.J. Lynch, V.A. Miller, K.K. Wong, B. Gitlitz, K. Eaton, C. Zacharchuk, A. Freyman, C. Powell, R. Ananthakrishnan, S. Quinn, J.C. Soria, Neratinib an irreversible pan-ErbB receptor tyrosine kinase inhibitor: results of a phase II trial in patients with advanced non-small-cell lung cancer, J. Clin. Oncol. 28 (2010) 3076–3083.
[15] Y. Kim, J. Ko, Z. Cui, A. Abolhoda, J.S. Ahn, S.H. Ou, M.J. Ahn, K. Park, The EGFR T790M mutation in acquired resistance to an irreversible second-generation EGFR inhibitor, Mol. Cancer Ther. 11 (2012) 784–791.
[16] M.L. Sos, H.B. Rode, S. Heynck, M. Peifer, F. Fischer, S. Kluter, V.G. Pawar, C. Reuter, J.M. Heuckmann, J. Weiss, L. Ruddigkeit, M. Rabiller, M. Koker, J.R. Simard, M. Getlik, Y. Yuza, T.H. Chen, H. Greulich, R.K. Thomas, D. Rauh, Chemogenomic profiling provides insights into the limited activity of irreversible EGFR inhibitors in tumor cells expressing the T790M EGFR resistance mutation, Cancer Res. 70 (2010) 868–874.
[17] J.C. Soria, L.V. Sequist, S. Gadgeel, J. Goldman, H. Wakelee, A. Varga, P. Fidias, A.J. Wozniak, J.W. Neal, R.C. Doebele, E.B. Garon, S. Jaw-Tsai, L. Caunt, P. Kaur, L. Rolfe, A. Allen, D.R. Camidge, First-in-human evaluation of CO-1686, an irreversible, highly selective tyrosine kinase inhibitor of mutations of EGFR (activating and T790M), in: 15th World Conference on Lung Cancer, Sydney, Australia, 2013.
[18] M. Ranson, Preliminary results from a phase I study with AZD9291: an irreversible inhibitor of epidermal growth factor receptor (EGFR) activating and resistance mutations in non-small-cell lung cancer (NSCLC), in: European Cancer Congress, 2013. Abstract #33LBA.
[19] D.A.E. Cross, S.E. Ashton, S. Ghiorghiu, C. Eberlein, C.A. Nebhan, P.J. Spitzler, J.P. Orme, M.R.V. Finlay, R.A. Ward, M.J. Mellor, G. Hughes, A. Rahi, V.N. Jacobs, M.R. Brewer, E. Ichihara, J. Sun, H. Jin, P. Ballard, K. Al-Kadhimi, R. Rowlinson,

T. Klinowska, G.H.P. Richmond, M. Cantarini, D.W. Kim, M.R. Ranson, W. Pao, AZD9291, an irreversible EGFR TKI, overcomes T790M-mediated resistance to EGFR inhibitors in lung cancer, Cancer Discov. 4 (2014) 1046–1061.

[20] R.T.T. Sjin, K. Lee, A.O. Walter, A. Dubrovskiy, M. Sheets, T.S. Martin, M.T. Labenski, Z. Zhu, R. Tester, R. Karp, A. Medikonda, P. Chaturvedi, Y. Ren, H. Haringsma, J. Etter, M. Raponi, A.D. Simmons, T.C. Harding, D. Niu, M. Nacht, W.F. Westlin, R.C. Petter, A. Allen, J. Singh, In vitro and in vivo characterization of irreversible mutant -selective EGFR inhibitors that are wild-type sparing, Mol. Cancer Ther. 13 (2014) 1–12.

[21] P.A. Janne, S.S. Ramaligam, J.C.H. Yang, M.J. Ahn, D.W. Kim, S.W. Kim, D. Planchard, Y. Ohe, E. Felip, C. Watkins, M. Cantarini, S. Ghiorghiu, M. Ranson, Clinical activity of the mutant-selective EGFR inhibitor osimertinib (AZD9291) in patients (pts) with EGFR inhibitor resistant non-small-cell lung cancer (NSCLC), J. Clin. Oncol. 32 (2014) 8009.

CHAPTER 5
Mechanism of Resistance to Third-Generation Inhibitors

Epidermal growth factor receptor (EGFR) mutations are more commonly seen in NSCLC tumors with adenocarcinoma histology, specifically in women, Asian patients, and in nonsmokers [1–3]. EGFR mutations are seldom found in squamous cell carcinomas of the lung, small cell lung cancer, or other epithelial malignancies; thus activating somatic EGFR mutations are a unique feature of a subclass of NSCLC.

Epidermal growth factor receptor (EGFR) activating mutations are found in exons 18–21 of the EGFR gene, which is part of the gene coding for the tyrosine kinase domain of the EGFR protein. In patients diagnosed with advanced NSCLC the most common activating mutations observed prior to treatment with a tyrosine kinase inhibitor (TKI) are exon 19 deletions and an L858R point mutation in exon 21. The most prevalent EGFR mutations consist of small inframe deletions around the conserved LREA motif of exon 19 (corresponding to amino acid residues 747–750) and a point mutation (L858R) in exon 21 [4, 5], which account for more than 90% of all EGFR kinase mutations. These EGFR mutations activate the EGFR signaling pathway and promote EGFR-mediated prosurvival and antiapoptotic signals through downstream targets, such as AKT-PI3K, ERK, and STAT [6].

5.1 SECONDARY MUTATIONS: EGFR T790M

The first identified mechanism of acquired resistance to EGFR TKIs (first-generation and second-generation inhibitors) was the EGFR T790M mutation in 2005, which was observed in patients with NSCLC harboring either exon 19 deletions or the L858R mutation that progressed after a period of response to gefitinib or erlotinib. In postprogression biopsies the original EGFR mutation and the novel T790M in exon 20 were identified (Figs. 5.1 and 5.2) [7].

Third-Generation EGFR Inhibitors
https://doi.org/10.1016/B978-0-08-102661-8.00005-6

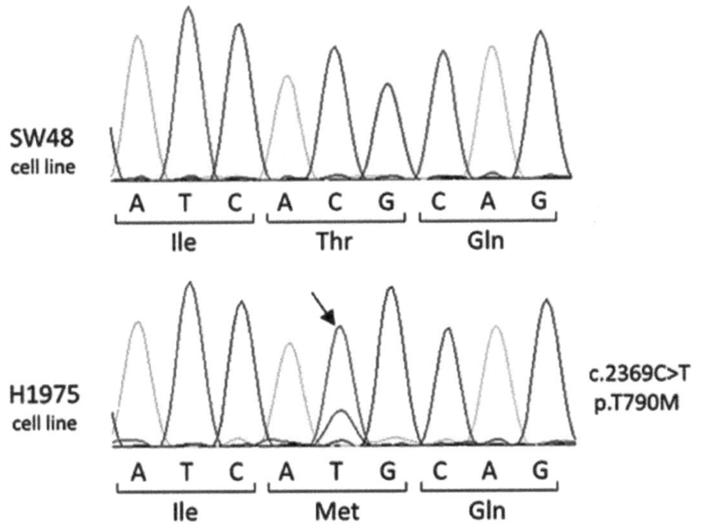

Fig. 5.1 Detection of the T790M mutation by DNA sequencing, where threonine amino acid is mutated by methionine. The mutated nucleotide is indicated with an *arrow*.

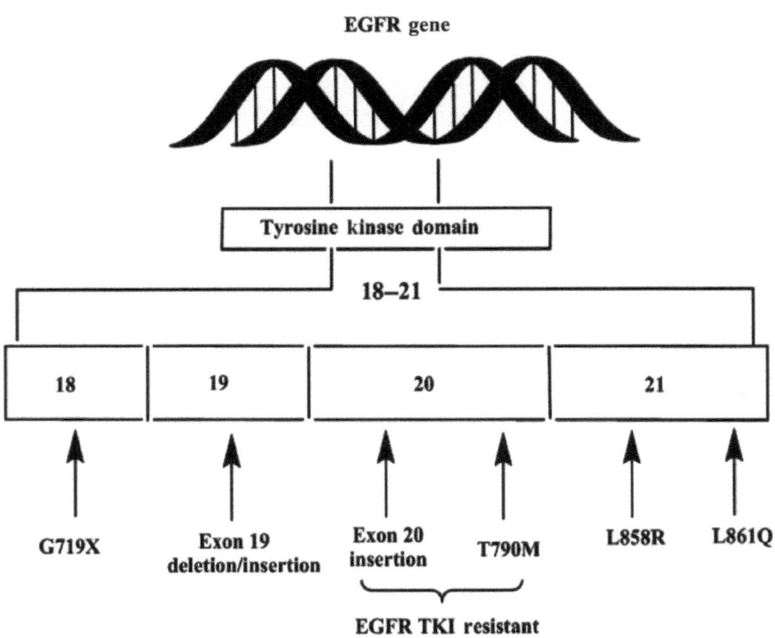

Fig. 5.2 EGFR tyrosine kinase mutation.

When T790M was introduced in vitro to sequences containing wild-type EGFR, exon 19 deletion–EGFR, or L858R-EGFR, the resulting proteins were significantly more resistant to gefitinib in the constructs containing T790M [8]. The inhibitory concentrations to erlotinib and gefitinib in T790M-containing constructs exceeded 5 μM, which is a concentration more than 100-fold higher than that required to inhibit exon 19 deletions or L858R-EGFR [9]. A NSCLC cell line with the L858R-T790M mutation (H1975) was significantly more resistant to gefitinib or erlotinib than lines with L858R (H3255) and an exon 19 deletion (HCC827) [10]. The T790M mutation is most often seen in cis; however, it can occur in trans to L858R or exon 19 deletions [10]. EGFR-mutated cell lines that have L858R-T790M or exon 19 deletions-T790M continue to be dependent on EGFR because alternative EGFR inhibitors (which will be described in greater detail below) halt cell proliferation and lead to apoptosis [11]. Some EGFR-mutated NSCLC cell lines (H3255 and PC-9) that are exposed to incremental concentrations of gefitinib in culture end up acquiring T790M [12]; mouse lung cancer models of L858R-T790M confirmed that these tumors are resistant to gefitinib and erlotinib [13]. This preclinical and clinical data supports T790M as a main mechanism of resistance to EGFR TKIs. How T790M affects the hypersensitivity of activating EGFR mutations is still not completely clear. It was initially speculated, based on the crystallographic structure of the kinase domain of EGFR, that the bulkier methionine residue of the "gatekeeper" T790M changed the ATP-binding pocket of the kinase, therefore blocking the engagement of erlotinib or gefitinib [13].

However, it was more recently demonstrated that T790M affected minimally the binding of Gefitinib to L858R-EGFR [13]. Instead, L858R-T790M-EGFR had increased affinity to ATP when compared with L858R alone, which is predicted to decrease the binding of gefitinib and erlotinib because these drugs are ATP-competitive kinase inhibitors [9]. These findings will certainly affect the development of the next generation of EGFR inhibitors with the ability to overcome T790M (Figs. 5.1 and 5.2).

While the third-generation EGFR TKIs are highly active, progression still occurs. Several mechanisms of acquired resistance to osimertinib and rociletinib have been described in vitro and in the clinical setting. The C797S mutation located within the tyrosine kinase domain of EGFR was reported to be a leading mechanism of resistance to the third-generation irreversible EGFR inhibitors targeting T790M mutation [14–17]. HER2 amplification [14], CMET amplification [14], and EGFR

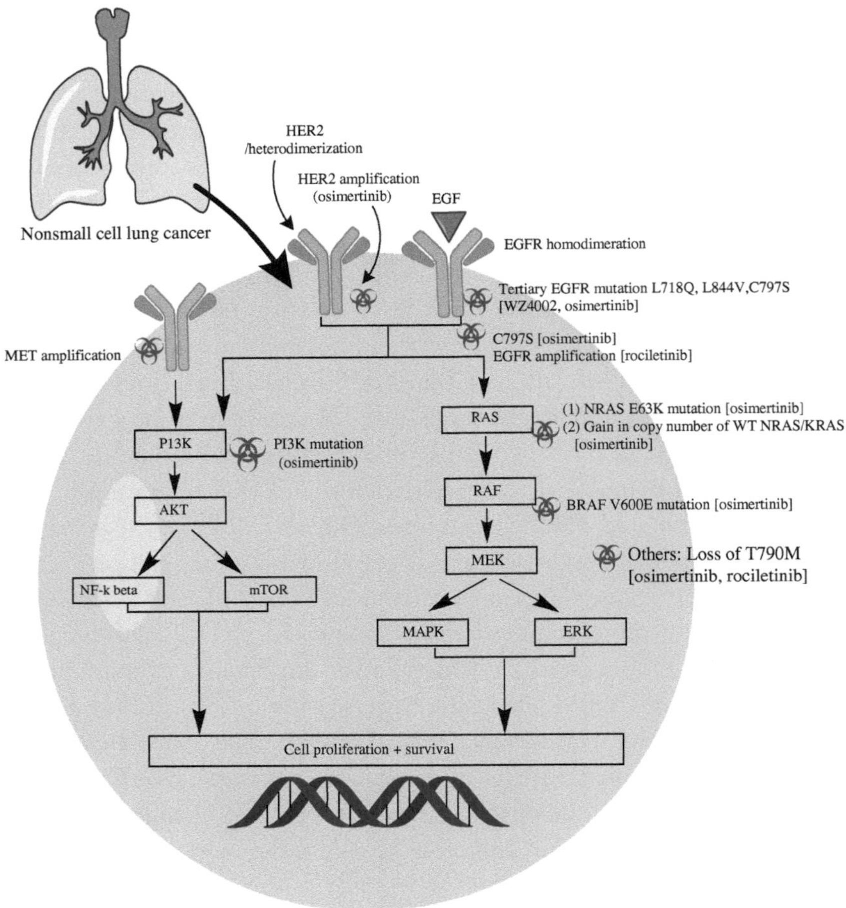

Fig. 5.3 Mechanism of resistance to third-generation EGFR tyrosine kinase inhibitors.

L718Q mutation [18] were also responsible for acquired resistance to third-generation EGFR inhibitors without a T790M or C797S mutation. Amplifications of MET, HER2, or KRAS G12S mutation was recently identified in NSCLC patients who progressed with treatments of osimertinib (AZD9291) or rociletinib (CO1686) [19], as shown in Fig. 5.3.

5.2 EGFR L718Q MUTATION

Bersanelli et al. presented a case of advanced NSCLC with a coexisting primary L585R and secondary T790M EGFR mutations that acquired resistance to AZD9291 (osimertinib) due to the occurrence of the tertiary

L718Q mutation. With the emergence of third-generation EGFR-tyrosine kinase inhibitors, such as osimertinib (AZD9291) and rociletinib (CO1686), new mechanisms of drug resistance are developing, like C797S and L884V EGFR mutations, in patients with EGFR T790M-positive nonsmall cell lung cancer. This is the first clinical report for this new mutation as an EGFR-dependent mechanism of resistance to osimertinib (AZD9291) (Fig. 5.3) [18].

5.3 L718Q, L844V, AND C797S MUTATIONS

Ercan et al. identified three major drug resistance mutations. EGFR L718Q, L844V, and C797S cause resistance to both WZ4002 and rociletinib (CO1686), while in contrast only EGFR C797S leads to osimertinib (AZD9291) resistance. Cells containing an EGFR sensitizing mutation, Del 19 or L858R, in conjunction with L718Q, L844V, or C797S retain sensitivity to the quinazoline-based EGFR inhibitors gefitinib and afatinib. The C797S mutation, in the presence of Del 19 or L858R and T790M, causes resistance to all current EGFR inhibitors, but L858R/T790M/C797S remains partially sensitive to cetuximab, which leads to disruption of EGFR dimerization (Fig. 5.3) [20].

5.4 C797S MUTATION

5.4.1 C797S Mutation Mediates Resistance to Osimertinib (AZD9291)

In the first in-human phase I/II AURA trial of osimertinib (AZD9291), systemic progression in NSCLC patients was seen after treatment for a median of 9.6 months [21]. The characterization of resistance mechanisms in 22 patients who became resistant to osimertinib (AZD9291) was reported [22]. These patients with progression on (osimertinib) AZD9291 in the AURA trial had paired pretreatment and posttreatment plasma samples.

Through the study of T790M-positive patients with an acquired resistance to osimertinib (AZD9291), three molecular subtypes of osimertinib (AZD9291) resistance were revealed: (1) Acquired C797S mutation together with a T790M mutation [23]. This novel mutation is a "tertiary" substitution mutation at the osimertinib (AZD9291) binding site, changing cysteine 797 to serine (EGFR C797S), which results in the blocking of drug binding; (2) T790M mutation without a C797S mutation; and (3) loss of the T790M mutation without a C797S mutation. The report also discovered that in some cases, two different nucleotide mutations (T to A and G to C) leading to C797S amino acid mutation occurred in the same patients. Since only 6 out of 15 cases acquired C797S mutation, additional mechanisms of resistance to osimertinib (AZD9291) must be present (Fig. 5.3) [24].

5.4.2 C797S Mutation Mediates Resistance to Olmutinib (HM61713)

Song et al. reported the first case of C797S mutation as a resistance mechanism of olmutinib (HM61713) that blocks EGFR inhibition and promotes resistance to third-generation EGFR TKIs. This case also highlights the importance of repeat biopsies to evaluate the underlying resistance mechanism in patients whose resistance to third-generation EGFR TKIs develops, as well as to make treatment decisions. (Fig. 5.3) [25].

5.4.3 Cis and Trans C797S

Niederst et al. demonstrated that the allelic context in which C797S was acquired may predict responsiveness to alternative treatments. If the C797S and T790M mutations are in trans, cells will be resistant to third-generation EGFR TKIs, but will be sensitive to a combination of first- and third-generation TKIs. If the mutations are in cis, no EGFR TKIs alone or in combination can suppress activity. If C797S develops in wild-type cells for T790 (when third-generation TKIs are administered in the first line setting), then the cells are resistant to third-generation TKIs, though they retain sensitivity to first-generation TKIs. The context in which the C797S develops with respect to the other EGFR alleles impacts the efficacy of subsequent treatments (Fig. 5.3) [26].

5.5 HER2 AND MET AMPLIFICATION MEDIATES RESISTANCE TO OSIMERTINIB (AZD9291)

As some patients who progressed on osimertinib (AZD9291) proved negative for the C797S mutation, additional resistance mechanisms must be present. In a case report a 54-year-old male with stage IV adenocarcinoma was found to have acquired a T790M mutation after progression from second-line treatment with gefitinib [27]. The disease progressed after 12 months of osimertinib (AZD9291) treatment on the AURA trial; HER2 amplification was identified without a C797S mutation from the tumor biopsy.

The second patient from the same report was a 60-year-old female, a nonsmoker who was diagnosed with stage IV adenocarcinoma with pleural metastasis [28]. Mutation analysis revealed the known EGFR-activating mutation in exon 21, L858R. She received erlotinib and gefitinib sequentially for 12 months. After disease progression, T790M was identified by NGS, and the patient was treated with osimertinib (AZD9291) on the AURA trial [27]. The patient had partial tumor regression and remained progression-free for 10 months. A re-biopsy of the osimertinib (AZD9291)-resistant tumor identified an EGFR activating mutation and CMET amplification without T790M or C797S mutation. These two cases indicated that in refractory NSCLC without T790M or C797S mutations, additional gene mutations or amplifications of tyrosine kinases other than EGFR can be the mechanisms of resistance. Additional treatment targeting the mutations will be needed (Fig. 5.3) [27].

5.6 RAS SIGNALING DEPENDENCE OF OSIMERTINIB (AZD9291)

Eberlein et al. reported an increasing dependence on RAS signaling that includes novel NRAS mutation (E63K), the gain of copy number of WT NRAS and KRAS in preclinical studies of acquired resistance to osimertinib. MEK inhibition (with selumetinib) in addition to osimertinib resulted in tumor regression in the transgenic model (Fig. 5.3) [29].

REFERENCES

[1] H. Shigematsu, L. Lin, T. Takahashi, et al., Clinical and biological features associated with epidermal growth factor receptor gene mutations in lung cancers, J. Natl. Cancer Inst. 97 (2005) 339–346.
[2] L.V. Sequist, D.W. Bell, T.J. Lynch, et al., Molecular predictors of response to epidermal growth factor receptor antagonists in non-small-cell lung cancer, J. Clin. Oncol. 25 (2007) 587–595.
[3] G.J. Riely, K.A. Politi, V.A. Miller, et al., Update on epidermal growth factor receptor mutations in non-small cell lung cancer, Clin. Cancer Res. 12 (2006) 7232–7241.
[4] S.V. Sharma, D.W. Bell, J. Settleman, et al., Epidermal growth factor receptor mutations in lung cancer, Nat. Rev. Cancer 7 (2007) 169–181.
[5] L.V. Sequist, V.A. Joshi, P.A. Janne, et al., Epidermal growth factor receptor mutation testing in the care of lung cancer patients, Clin. Cancer Res. 12 (2006) 4403s–4408s.
[6] D.B. Costa, B. Halmos, A. Kumar, et al., BIM mediates EGFR tyrosine kinase inhibitor-induced apoptosis in lung cancers with oncogenic EGFR mutations, PLoS Med. 4 (2007) 1669–1679.
[7] S. Kobayashi, T.J. Boggon, T. Dayaram, et al., EGFR mutation and resistance of non-small-cell lung cancer to gefitinib, N. Engl. J. Med. 352 (2005) 786–792.
[8] W. Pao, V.A. Miller, K.A. Politi, et al., Acquired resistance of lung adenocarcinomas to gefitinib or erlotinib is associated with a second mutation in the EGFR kinase domain, PLoS Med. 2 (2005) e73.

[9] T.A. Carter, L.M. Wodicka, N.P. Shah, et al., Inhibition of drug-resistant mutants of ABL, KIT, and EGF receptor kinases, Proc. Natl. Acad. Sci. 102 (2005) 11011–11016.

[10] S. Kobayashi, H. Ji, Y. Yuza, et al., An alternative inhibitor overcomes resistance caused by a mutation of the epidermal growth factor receptor, Cancer Res. 65 (2005) 7096–7101.

[11] E. Avizienyte, R.A. Ward, A.P. Garner, Comparison of the EGFR resistance mutation profiles generated by EGFR targeted tyrosine kinase inhibitors and the impact of drug combinations, Biochem. J. 415 (2008) 197–206.

[12] J.A. Engelman, T. Mukohara, K. Zejnullahu, et al., Allelic dilution obscures detection of a biologically significant resistance mutation in EGFR-amplified lung cancer, J. Clin. Investig. 116 (2006) 2695–2706.

[13] E.L. Kwak, R. Sordella, D.W. Bell, et al., Irreversible inhibitors of the EGF receptor may circumvent acquired resistance to gefitinib, Proc. Natl. Acad. Sci. 102 (2005) 7665–7670.

[14] D. Planchard, Y. Loriot, F. Andre, A. Gobert, N. Auger, L. Lacroix, EGFR independent mechanisms of acquired resistance to AZD9291 in EGFR T790M-positive NSCLC patients, Ann. Oncol. 26 (10) (2015) 2073–2078.

[15] N. Godin-Heymann, L. Ulkus, B. Brannigan, U. McDermott, J. Lamb, S. Maheswaran, J. Settleman, D. Haber, The T790M "gatekeeper" mutation in EGFR mediates resistance to low concentrations of an irreversible EGFR inhibitor, Mol. Cancer Ther. 7 (4) (2008) 874–879.

[16] P.A. Schwartz, P. Kuzmic, J. Solowiej, S. Bergqvist, B. Bolanos, C. Almaden, A. Nagata, K. Ryan, J. Feng, D. Dalvie, J.C. Kath, M. Xu, R. Wani, B.W. Murray, Covalent EGFR inhibitor analysis reveals importance of reversible interactions to potency and mechanisms of drug resistance, Proc. Natl. Acad. Sci. 111 (1) (2014) 173–178.

[17] H.A. Yu, S.K. Tian, A.E. Drilon, L. Borsu, G.J. Riely, M.E. Arcita, Acquired resistance of EGFR-mutant lung cancer to a T790M-specific EGFR inhibitor: emergence of a third mutation (C797S) in the EGFR tyrosine kinase domain, JAMA Oncol. 1 (7) (2015) 982.

[18] M. Bersanelli, R. Minari, P. Bordi, L. Gnetti, C. Bozzetti, A. Squadrilli, C.A. Lagrasta, L. Bottarelli, G. Osipova, E. Capelletto, M. Mor, M. Tiseo, L718Q mutation as new mechanism of acquired resistance to AZD9291 in EGER-mutated NSCLC, J. Thorac. Oncol. 11 (10) (2016) 121–123.

[19] S. Ortiz-Cuaran, M. Scheffler, D. Plenker, L. Dahmen, A.H. Scheel, L. Fernandez-Cuesta, L. Meder, C.M. Lovly, T. Persigehl, S. Merkelbach-Bruse, M. Bos, S. Michels, R. Fischer, K. Albus, K. König, H.U. Schildhaus, J. Fassunke, M.A. Ihle, H. Pasternack, C. Heydt, C. Becker, J. Altmüller, H. Ji, C. Müller, A. Florin, J.M. Heuckmann, P. Nuernberg, S. Ansen, L.C. Heukamp, J. Berg, W. Pao, M. Peifer, R. Buettner, J. Wolf, R.K. Thomas, M.L. Sos, Heterogeneous mechanisms of primary and acquired resistance to third-generation EGFR inhibitors, Clin. Cancer Res. 22 (19) (2016) 4837–4847.

[20] D. Ercan, H.G. Choi, C.H. Yun, M. Capelletti, T. Xie, M.J. Eck, N.S. Gray, P.A. Janne, EGFR mutations and resistance to irreversible pyrimidine based EGFR inhibitors, Clin. Cancer Res. 21 (17) (2015) 3913–3923.

[21] H.J. Lee, G. Schaefer, T.P. Heffron, L. Shao, X. Ye, S. Sideris, S. Malek, E. Chan, M. Merchant, H. La, S. Ubhayakar, R.L. Yauch, V. Pirazzoli, K. Politi, J. Settleman, Noncovalent wild-type-sparing inhibitors of EGFR T790M, Cancer Discov. 3 (2) (2013) 168–181.

[22] J. Singh, R.C. Petter, T.A. Baillie, A. Whitty, The resurgence of covalent drugs, Nat. Rev. Drug Discov. 10 (4) (2011) 307–317.

[23] S.B. Goldberg, G.R. Oxnard, S. Digumarthy, A. Muzikansky, D.M. Jackman, I.T. Lennes, L.V. Sequist, Chemotherapy with erlotinib or chemotherapy alone in advanced non small cell lung cancer with acquired resistance to EGFR tyrosine kinase inhibitors, Oncologist 18 (11) (2013) 1214–1220.

[24] S.S. Ramalingam, J. Yang, C. Lee, T. Kurata, D.W. Kim, T. John, N. Nogami, Y. Ohe, Y. Rukazenkov, P. Frewer, M. Cantarini, S. Ghiorghiu, P.A. Janne, MINI 16.07: AZD9291 in treatment-naïve EGFRm advanced NSCLC: AURA first-line cohort, J. Thorac. Oncol. 10 (2015) S319.
[25] K. Park, J.S. Lee, K.H. Lee, J.H. Kim, Y.J. Min, J.Y. Cho, J.Y. Han, B.S. Kim, J.S. Kim, D.H. Lee, J.H. Kang, E.K. Cho, I.J. Jang, J. Jung, H.Y. Kim, H.J. Sin, J. Son, J.S. Woo, D.W. Kim, Updated safety and efficacy results from phase I/II study of HM61713 in patients (pts) with EGFR mutation positive non-small cell lung cancer (NSCLC) who failed previous EGFR-tyrosine kinase inhibitor (TKI) [ASCO abstract 8084], J. Clin. Oncol. 33 (Suppl) (2015).
[26] A.B. Cortot, C.E. Repellin, T. Shimamura, M. Capelletti, K. Zejnullahu, D. Ercan, J.G. Christensen, K.K. Wong, N.S. Gray, P.A. Jänne, Resistance to irreversible EGF receptor tyrosine kinase inhibitors through a multistep mechanism involving the IGF1R pathway, Cancer Res. 73 (2) (2013) 834–843.
[27] Z. Piotrowska, M.J. Niederst, C.A. Karlovich, H.A. Wakelee, J.W. Neal, M. Mino-Kenudson, L. Fulton, A.N. Hata, E.L. Lockerman, A. Kalsy, S. Digumarthy, A. Muzikansky, M. Raponi, A.R. Garcia, H.E. Mulvey, M.K. Parks, R.H. DiCecca, D. Dias-Santagata, A.J. Iafrate, A.T. Shaw, A.R. Allen, J.A. Engelman, L.V. Sequist, Heterogeneity underlies the emergence of EGFR T790 wild-type clones following treatment of T790M positive cancers with a third generation EGFR inhibitor, Cancer Discov. 5 (7) (2015) 713–722.
[28] T. Wilson, J. Fridlyand, Y. Yan, E. Penuel, L. Burton, E. Chan, J. Peng, E. Lin, Y. Wang, J. Sosman, A. Ribas, J. Li, J. Moffat, D.P. Sutherlin, H. Koeppen, M. Merchant, R. Neve, J. Settleman, Widespread potential for growth-factor-driven resistance to anticancer kinase inhibitors, Nature 487 (7408) (2012) 505–509.
[29] M. Saki, M. Toulany, H.P. Rodemann, Acquired resistance to cetuximab is associated with the overexpression of ras family members and the loss of radiosensitization in head and neck cancer cells, Radiother. Oncol. 108 (3) (2013) 473–478.

CHAPTER 6

Discovery of Fourth-Generation Allosteric *C797S* Inhibitors

The T790M mutation is the most common mechanism for resistance to first- and second-generation tyrosine kinase inhibitors (TKI) for epidermal growth factor receptor (EGFR) [1]. Several third-generation EGFR mutant selective TKIs are being explored to conquer this resistance. AZD9291 (osimertinib) has been approved for treatment of the metastatic EGFR T790M mutation-positive nonsmall cell lung cancer [2]. *C797S* mutation was reported to be a major mechanism for resistance to T790M-targeting EGFR inhibitors AZD9291 (osimertinib) [3]. To combat this resistance, Jia et al. recently reported the allosteric inhibitors [4].

To search for an allosteric inhibitor that binds to EGFR away from the ATP-binding site, Jia et al. obtained human EGFR mutant peptides spanning residues 696–1022 (including wild-type, L858R, L858R/T790M, T790M, and T790M/V948R mutant) [4]. The purified EGFR L858R/T790M mutant enzyme was used to screen a library of 2.5 million compounds at 1 μM ATP concentration. There were 1322 top hits found in this round of screening. The top hits were then assayed for their IC_{50} values at both 1 μM and 1 mM ATP to separate the ATP competitive from noncompetitive compounds. The top hits were also screened against the wild-type EGFR to select out those with more specificity for the mutant EGFR.

The first compound, EAI001 (EGFR allosteric inhibitor), was discovered with potency and selectivity for mutant EGFR (at 1 mM ATP, IC50 0.024 μM for L858R T790M, IC50 > 50 μM for wild-type EGFR). However, it only had modest potency against individual L858R and T790M mutants. After a medicinal chemistry-based optimization of EAI001, the EAI045 inhibitor was found to have high potency and selectivity for L858R/T790M mutation. The profound selectivity for the EGFR mutant was confirmed through profiling a

Third-Generation EGFR Inhibitors
https://doi.org/10.1016/B978-0-08-102661-8.00006-8

panel of 250 protein kinases. EAI045 was therefore confirmed to be an allosteric and non-ATP competitive inhibitor of mutant EGFR [4].

EAI045 decreased but did not completely abolish the EGFR autophosphorylation in L858R/T790M-mutant NSCLC cell line H1975. In stably transfected NIH-3T3 cells harboring the L858R/T790M EGFR mutant, EAI045 showed the same activity. In L858R-mutant H3255 cells, EAI045 exhibited moderate activity [4]. In the HaCaT cells, which have a keratinocyte cell line with wild-type EGFR, EAI045 did not show any activity of inhibiting EGFR phosphorylation. These again confirmed the selectivity of EAI045 for mutant EGFR. Since dimerization of EGFR is required for its activation, these investigators hypothesized that the allosteric inhibitor was inactive for those asymmetric dimers/dimers between wild-type and mutant EGFR peptides [4]. The investigators confirmed that EAI045 was markedly more active in dimerization-defective EGFR mutants. When combined with cetuxirnab, a monoclonal antibody that can block EGFR dimerization by preventing EGF ligand binding, EAI045 markedly inhibited the proliferation of Ba/F3 cells bearing the L858R/T790M mutation. These in vitro studies proved that EA1045 is active and selective for T790M-harboring EGFR mutants that are in a monomer state [4].

In a genetically engineered mouse model of *L858R*/T790M-mutant-driven lung cancer the efficacy of EAI045 was tested both alone and in combination with cetuximab. Remarkable tumor regression was observed in *L858R*/T790M-mutant mice treated with the combination of EAI045 and cetuximab. There was no response observed in those mice treated with EAI045 alone; however, the same effect was observed in both *L858R*/T790M/*C797S*-engineered Ba/F3 cells and in mice carrying the *L858R*/T790M/*C797S* tumor xenografts. These assays clearly showed that EAI045 is able to overcome resistance from acquired T790M and *C797S* mutations [4].

6.1 U TO Y ALLOSTERIC STRATEGY TO COMBAT *C797S* EGFR RESISTANCE

The cocrystal structure (PDB code 3IKA and 4ZAU) of the EGFR T790M kinase domain in complex with WZ4002 and osimertinib (AZD9291)

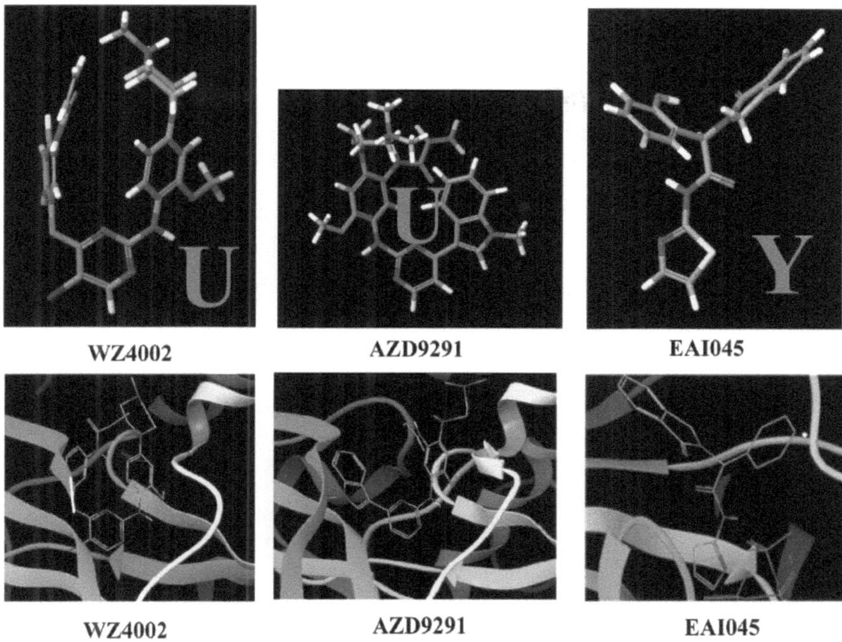

Fig. 6.1 U-shaped and Y-shaped configuration of third- and fourth-generation agents.

indicated that a "U-shaped" configuration of a pyrimidine core, together with an aniline ring bearing a hydrophilic group and an acrylamide moiety, is propitious to bind with EGFR target (Figs. 6.1 and 6.2) [5]. A great numbers of small molecules were designed and synthesized as selective EGFR inhibitors against the EGFR T790M mutation based on this binding mechanism.

Similarly, they studied the binding interaction of allosteric inhibitors EAI045 using the PDB ID 5D41 [4]. EAI045 was an allosteric, nonATP-competitive inhibitor of mutant T790M EGFR, having a Y-shaped configuration as shown in (Figs. 6.1 and 6.2). Focusing on these molecules and reviewing their design strategies (U- to Y-shaped configuration) along with their typical anticancer activity will provide valuable clues for the further development of more active EGFR T790M inhibitors.

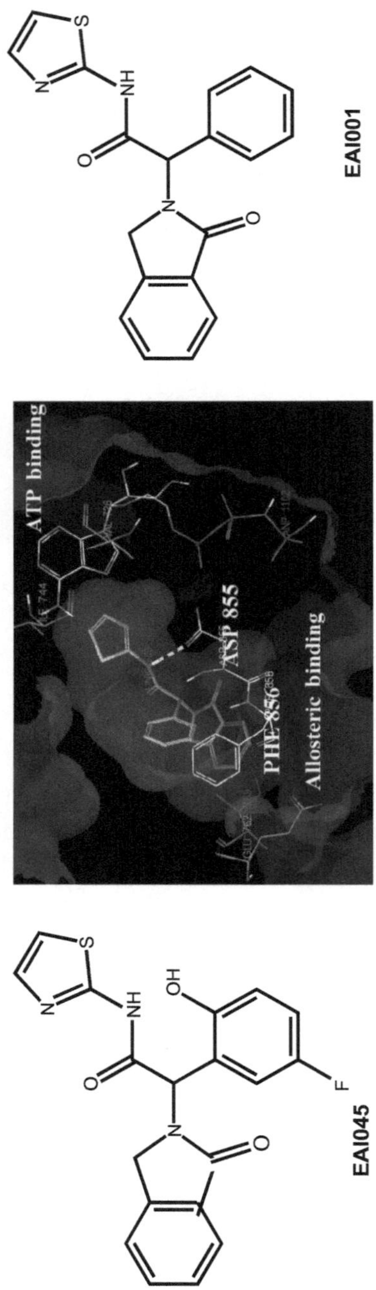

Fig. 6.2 Fourth-generation *C797S* EGFR inhibitors.

REFERENCES

[1] M.J. Niederst, H. Hu, H.E. Mulvey, E.L. Lockerman, A.R. Garcia, Z. Piotrowska, L.V. Sequist, J.A. Engelman, The allelic context of the C797S mutation acquired upon treatment with third-generation EGFR inhibitors impacts sensitivity to subsequent treatment strategies, Clin. Cancer Res. 21 (17) (2015) 3924–3933.
[2] K.S. Thress, C.P. Paweletz, E. Felip, B.C. Cho, D. Stetson, B. Dougherty, Z. Lai, A. Markovets, A. Vivancos, Y. Kuang, D. Ercan, S.E. Matthews, M. Cantarini, J.C. Barrett, P.A. Janne, G.R. Oxnard, Acquired EGFR C797S mutation mediates resistance to AZD9291 in non-small cell lung cancer harboring EGFR T790M, Nat. Med. 21 (6) (2015) 560–562.
[3] H.N. Song, K.S. Jung, K.H. Yoo, J. Cho, J.Y. Lee, S.H. Lim, H.S. Kim, J.M. Sun, S.H. Lee, J.S. Ahn, K. Park, Y.L. Choi, W. Park, M.J. Ahn, Acquired C797S mutation upon treatment with a T790M-specific third-generation EGFR inhibitor (HM61713) in non-small cell lung cancer, J. Thorac. Oncol. 11 (4) (2016) e45–e47.
[4] Y. Jia, C.H. Yunt, E. Park, D. Ercan, M. Manuia, J. Juarez, C. Xu, K. Rhee, T. Chen, I. Zhang, S. Palakurthi, J. Jang, G. Lelais, M. DiDonato, B. Bursulaya, P.Y. Michellys, R. Epple, T.H. Marsilje, M. McNeilP, W. Lu, J. Harris, S. Ilender, Overcoming EGFR(T790M) and EGFR(C797S) resistance with mutant-selective allosteric inhibitors, Nature 534 (7605) (2016) 129–132.
[5] Z. Song, Y. Ge, C. Wang, S. Huang, X. Shu, K. Liu, Y. Zhou, X. Ma, Challenges and perspectives on the development of small-molecule EGFR inhibitors against T790M-mediated resistance in non-small cell lung cancer, J. Med. Chem. 59 (14) (2016) 6580–6594.

CHAPTER 7
Conclusion and Future Perspectives

The tyrosine kinase inhibitors are the first treatment of choice for most advanced NSCLC [1]. The T790M mutation of exon 20 of the EGFR is the most common mechanism of acquired resistance to the first- and second-generation EGFR TKIs [2]. The predominant strategy employed by medicinal chemists to overcome these resistance mechanisms has been the use of covalent inhibitors, which are able to bond with the poorly conserved Cys797 within the EGFR active site, providing both potency and kinase selectivity. Clinical results with these second- and third-generation compounds have been mixed [3, 4]. The clinical efficacy of second-generation EGFR inhibitors containing reactive groups was limited by associated skin rash and gastrointestinal toxicity, possibly because of their potency against wild-type WT EGFR [5, 6]. These results and additional reports of acquired resistance to one such covalent inhibitor via the T790M mutation led to concerns that drug levels sufficient to inhibit T790M mutant forms of EGFR may not be achieved with these agents. Recent results for third-generation covalent inhibitors osimertinib (AZD9291) and rociletinib (CO1686), which have been designed to be selective for T790M containing EGFR mutants over WT EGFR, indicate promising efficacy and tolerability with this approach [7].

The third-generation compounds selectively target EGFR T790M. Their in vitro and in vivo activity in preclinical models is striking; they are 30- to 100-fold more potent against EGFR T790M and up to 100-fold less potent against WT EGFR than quinazoline-based EGFR inhibitors in vitro. On March 30, 2017, the US Food and Drug Administration granted regular approval to osimertinib (TAGRISSO, AstraZeneca Pharmaceuticals, LP) for the treatment of patients with metastatic EGFR-T790M mutation-positive NSCLC [8, 9].

Despite the clinical efficacy of the pyrimidine-based EGFR inhibitors rociletinib and osimertinib, it is fully anticipated that patients will ultimately develop an acquired resistance to these agents. EGFR L718Q, L844V, and

Third-Generation EGFR Inhibitors
https://doi.org/10.1016/B978-0-08-102661-8.00007-X

C797S have been identified as three major drug resistance mutations for these third-generation inhibitors. Unfortunately, when in the presence of Del19 or L858R and T790M, the *C797S* mutation causes resistance to all current EGFR inhibitors [8, 9]. Additional studies with third-generation resistant cell lines have shown that the allelic context of the activating gatekeeper and *C797S* mutations affects the sensitivity to the three generations of inhibitors with no EGFR TKIs alone or in combination able to suppress activity when the mutations are in cis [8, 9]. This data suggests that there is a need for drugs that do not rely on a covalent reaction with Cys797 for potency or selectivity. *C797S* mutation confers resistance to all third-generation irreversible EGFR inhibitors. EA1045 is thus far the first allosteric TKI purposefully engineered to overcome T790M and *C797S* mutations. Mutations in *C797S* are not supposed to affect the efficacy of EA1045 because the residue is away from the allosteric binding pocket. The allosteric inhibitor, however, is ineffective alone due to receptor dimerization; however, in combination with cetuximab, it renders EA1045 fully active against T790M and *C797S* [8, 9]. The clinical efficacy of this compound remains unknown at the moment. *C797S* mutation is not the only mechanism for resistance to third-generation EGFR TKIs. Albeit less common, EGFR L718Q mutation represents another mechanism of resistance to osimertinib (AZD9291) [8, 9]. Hence it is possible to develop TKIs targeting this particular mutant. The combination of EGFR TKI with other agents should be one of the future directions in order to overcome the diverse resistant tumor clones. Besides cetuximab, MET and MEK inhibitors may be of value in combination therapy. Immune checkpoint blockers have been approved for the treatment of a variety of advanced cancers; these may be considered for combination therapy to overcome the resistance to TKIs.

REFERENCES

[1] V. Miller, V. Hirsh, J. Cadranel, Y. Chen, K. Park, S. Kim, C. Zhou, W. Su, M. Wang, Y. Sun, D. Heo, L. Crino, E. Tan, T. Chao, M. Shahidi, X. Cong, R. Lorence, J. Yang, Afatinib versus placebo for patients with advanced, metastatic non-small-cell lung cancer after failure of erlotinib, gefitinib, or both, and one or two lines of chemotherapy (LUX-lung 1): a phase 2B/3 randomised trial, Lancet Oncol. 13 (2012) 528–538.

[2] D. Camidge, W. Pao, L. Sequist, Acquired resistance to TKIs in solid tumours: learning from lung cancer, Nat. Rev. Clin. Oncol. 11 (8) (2014) 473–481.

[3] W. Zhou, D. Ercan, L. Chen, C.H. Yun, D. Li, M. Capelletti, A.B. Cortot, L. Chirieac, R.E. Iacob, R. Padera, J.R. Engen, K.K. Wong, M.J. Eck, N.S. Gray, P.A. Janne, Novel mutant-selective EGFR kinase inhibitors against EGFR T790M, Nature 462 (7276) (2009) 1070–1074.

[4] A.O. Walter, R.T. Sjin, H.J. Haringsma, K. Ohashi, J. Sun, K. Lee, A. Dubrovskiy, M. Labenski, Z. Zhu, Z. Wang, M. Sheets, T.S. Martin, R. Karp, D.V. Kalken, P. Chaturvedi, D. Niu, M. Nacht, R.C. Petter, W. Westlin, K. Lin, S. Jaw-Tsai, M. Raponi, T.V. Dyke, J. Etter, Z. Weaver, W. Pao, J. Singh, A.D. Simmons, T.C. Harding, A. Allen, Discovery of a mutant-selective covalent inhibitor of EGFR that overcomes T790M-mediated resistance in NSCLC, Cancer Discov. 3 (12) (2013) 1404–1415.

[5] H. Shigematsu, L. Lin, T. Takahashi, et al., Clinical and biological features associated with epidermal growth factor receptor gene mutations in lung cancers, J. Natl. Cancer Inst. 97 (2005) 339–346.

[6] L.V. Sequist, D.W. Bell, T.J. Lynch, et al., Molecular predictors of response to epidermal growth factor receptor antagonists in non-small-cell lung cancer, J. Clin. Oncol. 25 (2007) 587–595.

[7] K.S. Thress, C.P. Paweletz, E. Felip, B.C. Cho, D. Stetson, B. Dougherty, Z. Lai, A. Markovets, A. Vivancos, Y. Kuang, D. Ercan, S.E. Matthews, M. Cantarini, J.C. Barrett, P.A. Janne, G.R. Oxnard, Acquired EGFR C797S mutation mediates resistance to AZD9291 in non-small cell lung cancer harboring EGFR T790M, Nat. Med. 21 (6) (2015) 560–562.

[8] H.N. Song, K.S. Jung, K.H. Yoo, J. Cho, J.Y. Lee, S.H. Lim, H.S. Kim, J.M. Sun, S.H. Lee, J.S. Ahn, K. Park, Y.L. Choi, W. Park, M.J. Ahn, Acquired C797S mutation upon treatment with a T790M-specific third-generation EGFR inhibitor (HM61713) in non-small cell lung cancer, J. Thorac. Oncol. 11 (4) (2016) e45–e47.

[9] Y. Jia, C.H. Yunt, E. Park, D. Ercan, M. Manuia, J. Juarez, C. Xu, K. Rhee, T. Chen, I. Zhang, S. Palakurthi, J. Jang, G. Lelais, M. DiDonato, B. Bursulaya, P.Y. Michellys, R. Epple, T.H. Marsilje, M. McNeilP, W. Lu, J. Harris, S. Ilender, Overcoming EGFR(T790M) and EGFR(C797S) resistance with mutant-selective allosteric inhibitors, Nature 534 (7605) (2016) 129–132.

INDEX

Note: Page numbers followed by *f* indicate figures, *t* indicate tables and *s* indicate schemes.

FSC
www.fsc.org
MIX
Papier | Fördert gute Waldnutzung
FSC® C083411

Kolibri 360 GmbH
Ferdinand-Jühlke-Straße 7
99095 Erfurt, Deutschland
produktsicherheit@kolibri360.de